Numerical Analysis 2000, Volume 2

Interpolation and Extrapolation

Numerical Analysis 2000, Volume 2

Interpolation and Extrapolation

Edited by

C. Brezinski
Univ. des Sciences et Techn. de Lille
Lab. d'Analyse Numérique et d'Optimisation
UFR IEEA - M3, B.P. 36
Villeneuve d'Ascq Cedex 59655
France

2000
ELSEVIER
Amsterdam - London - New York - Oxford - Paris - Shannon - Tokyo

ELSEVIER SCIENCE B.V.
Sara Burgerhartstraat 25
P.O. Box 211, 1000 AE Amsterdam, The Netherlands

First edition 2000

Library of Congress Cataloging in Publication Data
A catalog record from the Library of Congress has been applied for.

ISBN: 0 444 50597 0

♾ The paper used in this publication meets the requirements of ANSI/NISO Z39.48-1992 (Permanence of Paper).
Transferred to digital printing 2006

JOURNAL OF COMPUTATIONAL AND APPLIED MATHEMATICS

Volume 122, Numbers 1–2, 1 October 2000

Contents

ELSEVIER
Journal of Computational and Applied Mathematics 122 (2000) ix–xi
JOURNAL OF COMPUTATIONAL AND APPLIED MATHEMATICS
www.elsevier.nl/locate/cam

Preface

Numerical Analysis 2000
Vol. II: Interpolation and extrapolation

C. Brezinski

Laboratoire d'Analyse Numérique et d'Optimisation, Université des Sciences et Technologies de Lille, 59655 Villeneuve d'Ascq Cedex, France

This volume is dedicated to two closely related subjects: interpolation and extrapolation. The papers can be divided into three categories: historical papers, survey papers and papers presenting new developments.

Interpolation is an old subject since, as noticed in the paper by M. Gasca and T. Sauer, the term was coined by John Wallis in 1655. Interpolation was the first technique for obtaining an approximation of a function. Polynomial interpolation was then used in quadrature methods and methods for the numerical solution of ordinary differential equations.

Obviously, some applications need interpolation by functions more complicated than polynomials. The case of rational functions with prescribed poles is treated in the paper by G. Mühlbach. He gives a survey of interpolation procedures using Cauchy–Vandermonde systems. The well-known formulae of Lagrange, Newton and Neville–Aitken are generalized. The construction of rational B-splines is discussed.

Trigonometric polynomials are used in the paper by T. Strohmer for the reconstruction of a signal from non-uniformly spaced measurements. They lead to a well-posed problem that preserves some important structural properties of the original infinite dimensional problem.

More recently, interpolation in several variables was studied. It has applications in finite differences and finite elements for solving partial differential equations. Following the pioneer works of P. de Casteljau and P. Bézier, another very important domain where multivariate interpolation plays a fundamental role is computer-aided geometric design (CAGD) for the approximation of surfaces.

The history of multivariate polynomial interpolation is related in the paper by M. Gasca and T. Sauer.

The paper by R.A. Lorentz is devoted to the historical development of multivariate Hermite interpolation by algebraic polynomials.

In his paper, G. Walz treats the approximation of multivariate functions by multivariate Bernstein polynomials. An asymptotic expansion of these polynomials is given and then used for building, by extrapolation, a new approximation method which converges much faster.

E-mail address: Claude.Brezinski@univ-lille1.fr (C. Brezinski).

PII: S0377-0427(00)00352-6

Extrapolation is based on interpolation. In fact, extrapolation consists of interpolation at a point outside the interval containing the interpolation points. Usually, this point is either zero or infinity. Extrapolation is used in numerical analysis to improve the accuracy of a process depending of a parameter or to accelerate the convergence of a sequence. The most well-known extrapolation processes are certainly Romberg's method for improving the convergence of the trapezoidal rule for the computation of a definite integral and Aitken's Δ^2 process which can be found in any textbook of numerical analysis.

An historical account of the development of the subject during the 20th century is given in the paper by C. Brezinski.

The theory of extrapolation methods lays on the solution of the system of linear equations corresponding to the interpolation conditions. In their paper, M. Gasca and G. Mühlbach show, by using elimination techniques, the connection between extrapolation, linear systems, totally positive matrices and CAGD.

There exist many extrapolation algorithms. From a finite section $S_n, \ldots, S_{n+k}$ of the sequence (S_n), they built an improved approximation of its limit S. This approximation depends on n and k. When at least one of these indexes goes to infinity, a new sequence is obtained with, possibly, a faster convergence.

In his paper, H.H.H. Homeier studies scalar Levin-type acceleration methods. His approach is based on the notion of remainder estimate which allows to use asymptotic information on the sequence to built an efficient extrapolation process.

The most general extrapolation process known so far is the sequence transformation known under the name of E-algorithm. It can be implemented by various recursive algorithms. In his paper, N. Osada proved that the E-algorithm is mathematically equivalent to the Ford–Sidi algorithm. A slightly more economical algorithm is also proposed.

When S depends on a parameter t, some applications need the evaluation of the derivative of S with respect to t. A generalization of Richardson extrapolation process for treating this problem is considered in the paper by A. Sidi.

Instead of being used for estimating the limit S of a sequence from $S_n, \ldots, S_{n+k}$, extrapolation methods can also be used for predicting the next unknown terms $S_{n+k+1}, S_{n+k+2}, \ldots$. The prediction properties of some extrapolation algorithms are analyzed in the paper by E.J. Weniger.

Quite often in numerical analysis, sequences of vectors have to be accelerated. This is, in particular, the case in iterative methods for the solution of systems of linear and nonlinear equations.

Vector acceleration methods are discussed in the paper by K. Jbilou and H. Sadok. Using projectors, they derive a different interpretation of these methods and give some theoretical results. Then, various algorithms are compared when used for the solution of large systems of equations coming out from the discretization of partial differential equations.

Another point of view is taken in the paper by P.R. Graves-Morris, D.E. Roberts and A. Salam. After reminding, in the scalar case, the connection between the ε-algorithm, Padé approximants and continued fractions, these authors show that the vector ε-algorithm is the best all-purpose algorithm for the acceleration of vector sequences.

There is a subject which can be related either to interpolation (more precisely, Hermite interpolation by a rational function at the point zero) and to convergence acceleration: it is Padé approximation. Padé approximation is strongly connected to continued fractions, one of the oldest subject in mathematics since Euclid g.c.d. algorithm is an expansion into a terminating continued fraction.

Although they were implicitly known before, Padé approximants were really introduced by Johan Heinrich Lambert in 1758 and Joseph Louis Lagrange in 1776. Padé approximants have important applications in many branches of applied sciences when the solution of a problem is obtained as a power series expansion and some of its properties have to be guessed from its first Taylor coefficients. In this volume, two papers deal with nonclassical applications of Padé approximation.

M. Prévost shows how Padé approximants can be used to obtain Diophantine approximations of real and complex numbers and then proving irrationality. Padé approximation of the asymptotic expansion of the remainder of a series also provides Diophantine approximations.

The solution of a discrete dynamical system can be related to matrix Hermite–Padé approximants, an approach developed in the paper by V. Sorokin and J. van Iseghem. Spectral properties of the band operator are investigated. The inverse spectral method is used for the solution of dynamical systems defined by a Lax pair.

Obviously, all aspects of interpolation and extrapolation have not been treated in this volume. However, many important topics have been covered.

I would like to thank all authors for their efforts.

ELSEVIER Journal of Computational and Applied Mathematics 122 (2000) 1–21

JOURNAL OF COMPUTATIONAL AND APPLIED MATHEMATICS

www.elsevier.nl/locate/cam

Convergence acceleration during the 20th century

C. Brezinski

Laboratoire d'Analyse Numérique et d'Optimisation, UFR IEEA, Université des Sciences et Technologies de Lille, 59655–Villeneuve d'Ascq cedex, France

Received 8 March 1999; received in revised form 12 October 1999

1. Introduction

In numerical analysis many methods produce sequences, for instance iterative methods for solving systems of equations, methods involving series expansions, discretization methods (that is methods depending on a parameter such that the approximate solution tends to the exact one when the parameter tends to zero), perturbation methods, etc. Sometimes, the convergence of these sequences is slow and their effective use is quite limited. Convergence acceleration methods consist of transforming a slowly converging sequence (S_n) into a new sequence (T_n) converging to the same limit faster than the initial one. Among such *sequence transformations*, the most well known are certainly Richardson's extrapolation algorithm and Aitken's Δ^2 process. All known methods are constructed by extrapolation and they are often called *extrapolation methods*. The idea consists of interpolating the terms $S_n, S_{n+1}, \ldots, S_{n+k}$ of the sequence to be transformed by a sequence satisfying a certain relationship depending on parameters. This set of sequences is called *kernel* of the transformation and every sequence of this set is transformed into a constant sequence by the transformation into consideration. For example, as we will see below, the kernel of Aitken's Δ^2 process is the set of sequences satisfying $\forall n, a_0(S_n - S) + a_1(S_{n+1} - S) = 0$, where a_0 and a_1 are parameters such that $a_0 + a_1 \neq 0$. If Aitken's process is applied to such a sequence, then the constant sequence $(T_n = S)$ is obtained. The parameters involved in the definition of the kernel are uniquely determined by the interpolation conditions and then the limit of the interpolating sequence of the kernel is taken as an approximation of the limit of the sequence to be accelerated. Since this limit depends on the index n, it will be denoted by T_n. Effectively, the sequence (S_n) has been transformed into a new sequence (T_n).

This paper, which is based on [31], but includes new developments obtained since 1995, presents my personal views on the historical development of this subject during the 20th century. I do not pretend to be exhaustive nor even to quote every important contribution (if a reference does not

E-mail address: claude.brezinski@univ-lille1.fr (C. Brezinski)

PII: S0377-0427(00)00360-5

appear below, it does not mean that it is less valuable). I refer the interested reader to the literature and, in particular to the recent books [55,146,33,144]. For an extensive bibliography, see [28].

I will begin with scalar sequences and then treat the case of vector ones. As we will see, a sequence transformation able to accelerate the convergence of *all* scalar sequences cannot exist. Thus, it is necessary to obtain many different convergence acceleration methods, each being suitable for a particular class of sequences. Many authors have studied the properties of these procedures and proved some important classes of sequences to be accelerable by a given algorithm. Scalar sequence transformations have also been extensively studied from the theoretical point of view.

The situation is more complicated and more interesting for vector sequences. In the case of a sequence of vectors, it is always possible to apply a scalar acceleration procedure componentwise. However, such a strategy does not take into account connections which may exist between the various components, as in the important case of sequences arising from iterative methods for solving a system of linear or nonlinear equations.

2. Scalar sequences

Let (S_n) be a scalar sequence converging to a limit S. As explained above, an extrapolation method consists of transforming this sequence into a new one, (T_n), by a sequence transformation $T:(S_n) \to (T_n)$. The transformation T is said to *accelerate the convergence* of the sequence (S_n) if and only if

$$\lim_{n\to\infty} \frac{T_n - S}{S_n - S} = 0.$$

We can then say that (T_n) *converges* (to S) *faster than* (S_n).

The first methods to have been used were linear transformations

$$T_n = \sum_{i=0}^{\infty} a_{ni} S_i, \quad n = 0, 1, \ldots,$$

where the numbers a_{ni} are constants independent of the terms of the sequence (S_n). Such a linear transformation is usually called a *summation process* and its properties are completely determined by the matrix $A = (a_{ni})$. For practical reasons, only a finite number of the coefficients a_{ni} are different from zero for each n. Among such processes are those named after Euler, Cesaro and Hölder. In the case of linear methods, the convergence of the sequence (T_n) to S for any converging sequence (S_n) is governed by the Toeplitz summability theorem; see [115] for a review. Examples of such processes are

$$T_n = \frac{1}{n+1} \sum_{i=0}^{n} S_i$$

or

$$T_n = \frac{1}{k+1} \sum_{i=n}^{n+k} S_i.$$

In the second case, the sequence (T_n) also depends on a second index, k, and the convergence has to be studied either when k is fixed and n tends to infinity, or when n is fixed and k tends to infinity.

With respect to convergence acceleration, summation processes are usually only able to accelerate the convergence of restricted classes of sequences and this is why the numerical analysts of the 20th century turned their efforts to nonlinear transformations. However, there is one exception: Richardson's extrapolation process.

2.1. Richardson's process

It seems that the first appearance of a particular case of what is now called the Richardson extrapolation process is due to Christian Huygens (1629–1695). In 1903, Robert Moir Milne (1873) applied the idea of Huygens for computing π [101]. The same idea was exploited again by Karl Kommerell (1871–1948) in his book of 1936 [78]. As explained in [143], Kommerell can be considered as the real discoverer of Romberg's method although he used this scheme in the context of approximating π.

Let us now come to the procedures used for improving the accuracy of the trapezoidal rule for computing approximations to a definite integral. In the case of a sufficiently smooth function, the error of this method is given by the Euler–Maclaurin expansion. In 1742, Colin Maclaurin (1698–1746) [90] showed that its precision could be improved by forming linear combinations of the results obtained with various stepsizes. His procedure can be interpreted as a preliminary version of Romberg's method; see [49] for a discussion.

In 1900, William Fleetwood Sheppard (1863–1936) used an elimination strategy in the Euler–Maclaurin quadrature formula with $h_n = r_n h$ and $1 = r_0 < r_1 < r_2 < \cdots$ to produce a better approximation to the given integral [132].

In 1910, combining the results obtained with the stepsizes h and $2h$, Lewis Fry Richardson (1881–1953) eliminated the first term in a discretization process using central differences [119]. He called this procedure the *deferred approach to the limit* or *h^2-extrapolation*. The transformed sequence (T_n) is given by

$$T_n = \frac{h_{n+1}^2 S(h_n) - h_n^2 S(h_{n+1})}{h_{n+1}^2 - h_n^2}.$$

In a 1927 paper [120] he used the same technique to solve a 6th order differential eigenvalue problem. His process was called *(h^2, h^4)-extrapolation*. Richardson extrapolation consists of computing the value at 0, denoted by $T_k^{(n)}$, of the interpolation polynomial of the degree at most k, which passes through the points $(x_n, S_n), \ldots, (x_{n+k}, S_{n+k})$. Using the Neville–Aitken scheme for these interpolation polynomials, we immediately obtain

$$T_{k+1}^{(n)} = \frac{x_{n+k+1} T_k^{(n)} - x_n T_k^{(n+1)}}{x_{n+k+1} - x_n}$$

with $T_0^{(n)} = S_n$.

Let us mention that Richardson referred to a 1926 paper by Nikolai Nikolaevich Bogolyubov (born in 1909) and Nikolai Mitrofanovich Krylov (1879–1955) where the procedure (often called the *deferred approach to the limit*) can already be found [11].

In 1955, Werner Romberg (born in 1909) was the first to use repeatedly an elimination approach for improving the accuracy of the trapezoidal rule [121]. He himself refers to the book of Lothar Collatz (1910–1990) of 1951 [50]. The procedure became widely known after the rigorous error analysis given in 1961 by Friedrich L. Bauer [3] and the work of Eduard L. Stiefel (1909–1978) [138]. Romberg's derivation of his process was heuristic. It was proved by Pierre-Jean Laurent in 1963 [81] that the process comes out from the Richardson process by choosing $x_n = h_n^2$ and $h_n = h_0/2^n$. Laurent also gave conditions on the choice of the sequence (x_n) in order that the sequences $(T_k^{(n)})$ tend to S either when k or n tends to infinity. Weaker conditions were given by Michel Crouzeix and Alain L. Mignot in [52, pp. 52–55]. As we shall see below, extensions of Romberg's method to nonsmooth integrands leads to a method called the E-algorithm.

Applications of extrapolation to the numerical solution of ordinary differential equations were studied by H.C. Bolton and H.I. Scoins in 1956 [12], Roland Bulirsch and Josef Stoer in 1964–1966 [47] and William B. Gragg [65] in 1965. The case of difference methods for partial differential equations was treated by Guriĭ Ivanovich Marchuk and V.V. Shaidurov [91]. Sturm–Liouville problems are discussed in [117]. Finally, we mention that Heinz Rutishauser (1918–1970) pointed out in 1963 [122] that Romberg's idea can be applied to any sequence as long as the error has an asymptotic expansion of a form similar to the Euler–Maclaurin's.

For a detailed history of the Richardson method, its developments and applications, see [57,77,143].

2.2. Aitken's process and the ε-algorithm

The most popular nonlinear acceleration method is certainly Aitken's Δ^2 process which is given by

$$T_n = \frac{S_n S_{n+2} - S_{n+1}^2}{S_{n+2} - 2S_{n+1} + S_n} = S_n - \frac{(S_{n+1} - S_n)^2}{S_{n+2} - 2S_{n+1} + S_n}, \qquad n = 0, 1, \ldots$$

The method was stated by Alexander Craig Aitken (1895–1967) in 1926 [1], who used it to accelerate the convergence of Bernoulli's method for computing the dominant zero of a polynomial. Aitken pointed out that the same method was obtained by Hans von Naegelsbach (1838) in 1876 in his study of Furstenau's method for solving nonlinear equations [104]. The process was also given by James Clerk Maxwell (1831–1879) in his *Treatise on Electricity and Magnetism* of 1873 [95]. However, neither Naegelsbach nor Maxwell used it for the purpose of acceleration. Maxwell wanted to find the equilibrium position of a pointer oscillating with an exponentially damped simple harmonic motion from three experimental measurements. It is surprising that Aitken's process was known to Takakazu Seki (1642–1708), often considered the greatest Japanese mathematician. In his book *Katsuyō Sanpō*, Vol. IV, he used this process to compute the value of π, the length of a chord and the volume of a sphere. This book was written around 1680 but only published in 1712 by his disciple Murahide Araki. Parts of it can be found in [73]. Let us mention that the Japanese characters corresponding to Takakazu have another pronounciation which is Kōwa. This is the reason why this mathematician is often called, erroneously as in [29,31] Seki Kōwa.

What makes Aitken's process so popular is that it accelerates the convergence of all linearly converging sequences, that is sequences such that $\exists a \neq 1$

$$\lim_{n\to\infty} \frac{S_{n+1} - S}{S_n - S} = a.$$

It can even accelerate some logarithmic sequences (that is corresponding to $a = 1$) which are those with the slowest convergence and the most difficult to accelerate.

Aitken's Δ^2 process is exact (which means that $\forall n, T_n = S$) for sequences satisfying, $a_0(S_n - S) + a_1(S_{n+1} - S) = 0$, $\forall n$, $a_0 a_1 \neq 0$, $a_0 + a_1 \neq 0$. Such sequences form the kernel of Aitken's process. The idea naturally arose of finding a transformation with the kernel

$$a_0(S_n - S) + \cdots + a_k(S_{n+k} - S) = 0, \quad \forall n,$$

$a_0 a_k \neq 0$, $a_0 + \cdots + a_k \neq 0$. A particular case of $k = 2$ was already treated by Maxwell in his book of 1873 and a particular case of an arbitrary value of k was studied by T.H. O'Beirne in 1947 [107]. This last work remains almost unknown since it was published only as an internal report. The problem was handled in full generality by Daniel Shanks (1917–1996) in 1949 [130] and again in 1955 [131]. He obtained the sequence transformation defined by

$$T_n = e_k(S_n) = \frac{\begin{vmatrix} S_n & S_{n+1} & \cdots & S_{n+k} \\ S_{n+1} & S_{n+2} & \cdots & S_{n+k+1} \\ \vdots & \vdots & & \vdots \\ S_{n+k} & S_{n+k+1} & \cdots & S_{n+2k} \end{vmatrix}}{\begin{vmatrix} \Delta^2 S_n & \cdots & \Delta^2 S_{n+k-1} \\ \vdots & & \vdots \\ \Delta^2 S_{n+k-1} & \cdots & \Delta^2 S_{n+2k-2} \end{vmatrix}}.$$

When $k = 1$, Shanks transformation reduces to the Aitken's Δ^2 process. It can be proved that $e_k(S_n) = S$, $\forall n$ if and only if (S_n) belongs to the kernel of the transformation given above. The same ratios of determinants were obtained by R.J. Schmidt in 1941 [127] in his study of a method for solving systems of linear equations.

The determinants involved in the definition of $e_k(S_n)$ have a very special structure. They are called *Hankel determinants* and were studied by Hermann Hankel (1839–1873) in his thesis in 1861 [72]. Such determinants satisfy a five-term recurrence relationship. This relation was used by O'Beirne and Shanks to implement the transformation by computing separately the numerators and the denominators of the $e_k(S_n)$'s. However, numerical analysts know it is difficult to compute determinants (too many arithmetical operations are needed and rounding errors due to the computer often lead to a completely wrong result). A recursive procedure for computing the $e_k(S_n)$'s without computing the determinants involved in their definition was needed. This algorithm was obtained in 1956 by Peter Wynn. It is called the ε-algorithm [147]. It is as follows. One starts with

$$\varepsilon_{-1}^{(n)} = 0, \quad \varepsilon_0^{(n)} = S_n$$

and then

$$\varepsilon_{k+1}^{(n)} = \varepsilon_{k-1}^{(n+1)} + \frac{1}{\varepsilon_k^{(n+1)} - \varepsilon_k^{(n)}}.$$

Note that the numbers $\varepsilon_k^{(n)}$'s fill out a two-dimensional array. The ε-algorithm is related to Shanks transformation by

$$\varepsilon_{2k}^{(n)} = e_k(S_n) \quad \text{and} \quad \varepsilon_{2k+1}^{(n)} = 1/e_k(\Delta S_n).$$

Thus, the ε's with an odd lower index are only auxiliary quantities. They can be eliminated from the algorithm, thus leading to the so-called *cross rule* due to Wynn [153].

When implementing the ε-algorithm or using Wynn's cross rule, division by zero can occur and the algorithm must be stopped. However, if the singularity is confined, a term that will again be used in Section 1.6, that is if it occurs only for some adjacent values of the indexes k and n, one may jump over it by using *singular rules* and continue the computation. If a division by a number close to zero arises, the algorithm becomes numerically unstable due to the cancellation errors. A similar situation holds for the other convergence acceleration algorithms. The study of such problems was initiated by Wynn in 1963 [151], who proposed particular rules for the ε-algorithm which are more stable than the usual rules. They were extended by Florent Cordellier in 1979 [51,151]. Particular rules for the θ-algorithm were obtained by Redivo Zaglia [155].

The convergence and acceleration properties of the ε-algorithm have only been completely described only for two classes of sequences, namely totally monotonic and totally oscillating sequences [154,15,16].

Shanks' transformation and the ε-algorithm have close connections to Padé approximants, continued fractions and formal orthogonal polynomials; see, for example [18].

2.3. Subsequent developments

The Shanks transformation and the ε-algorithm sparked the rebirth of the study of nonlinear acceleration processes. They now form an independent chapter in numerical analysis with connections to other important topics such as orthogonal and biorthogonal polynomials, continued fractions, and Padé approximants. They also have applications to the solution of systems of linear and nonlinear equations, the computation of the eigenvalues of a matrix, the solution of systems of linear and nonlinear equations, and many other topics, see [40]. Among other acceleration methods which were obtained and studied, are the W-process of Samuel Lubkin [89], the method of Kjell J. Overholt [110], the ρ-algorithm of Wynn [148], the G-transformation of H.L. Gray, T.A. Atchison and G.V. McWilliams [70], the θ-algorithm of Claude Brezinski [14], the transformations of Bernard Germain–Bonne [63] and the various transformations due to David Levin [85]. To my knowledge, the only known acceleration theorem for the ρ-algorithm was obtained by Naoki Osada [108]. Simultaneously, several applications began to appear. For example, the ε-algorithm provides a quadratically convergent method for solving systems of nonlinear equations and its does not require the knowledge of any derivative. This procedure was proposed simultaneously by Brezinski [13] and Eckhart Gekeler [61]. It has important applications to the solution of boundary value problems for ordinary differential equations [44]. Many other algorithms are given in the work of Ernst Joachim Weniger [145], which also contains applications to physics, or in the book of Brezinski and Michela Redivo Zaglia [40] where applications to various domains of numerical analysis can be found. The authors of this book provide FORTRAN subroutines. The book of Annie Cuyt and Luc Wuytack must also be mentioned [53]. The ε-algorithm has been applied to statistics, see the work of Alain Berlinet [9], and to the acceleration of the convergence of sequences of random variables, considered by Hélène Lavastre [82]. Applications to optimization were proposed by Le Ferrand [84] and Bouchta Rhanizar [118].

Instead of using a quite complicated algorithm, such as the ε-algorithm, it can be interesting to use a simpler one (for instance, Aitken's Δ^2 process) iteratively. Such a use consists of applying the algorithm to (S_n) to produce a new sequence (T_n), then to apply the same algorithm to (T_n), and so on. For example, applying the iterated Δ^2 process to the successive convergents of a periodic continued fraction produces a better acceleration than using the ε-algorithm [24]. In particular, the iterated Δ^2 process transforms a logarithmic sequence into a sequence converging linearly and linear convergence into superlinear, to my knowledge the only known cases of such transformations.

The experience gained during these years lead to a deeper understanding of the subject. Research workers began to study more theoretical and general questions related to the theory of convergence acceleration. The first attempt was made by R. Pennacchi in 1968 [114], who studied rational sequence transformations. His work was generalized by Germain–Bonne in 1973 [62], who proposed a very general framework and showed how to construct new algorithms for accelerating some classes of sequences. However, a ground breaking discovery was made by Jean Paul Delahaye and Germain–Bonne in 1980 [56]. They proved that if a set of sequences satisfies a certain property, called *remanence* (too technical to be explained here), then a universal algorithm, i.e. one able to accelerate *all* sequences of this set, cannot exist. This result shows the limitations of acceleration methods. Many sets of sequences were proved to be remanent, for example, the sets of monotonic or logarithmic sequences. Even some subsets of the set of logarithmic sequences are remanent.

Moulay Driss Benchiboun [5] observed that all the sequence transformations found in the literature could be written as

$$T_n = \frac{f(S_n, \ldots, S_{n+k})}{Df(S_n, \ldots, S_{n+k})}$$

with $D^2 f \equiv 0$, where Df denotes the sum of the partial derivatives of the function f. The reason for that fact was explained by Brezinski [26], who showed that it is related to the translativity property of sequence transformations. Hassane Sadok [123] extended these results to the vector case. Abderrahim Benazzouz [7] proved that quasilinear transformations can be written as the composition of two projections.

In many transformations, such as Shanks', the quantities computed are expressed as a ratio of determinants. This property is related to the existence of a triangular recurrence scheme for their computation as explained by Brezinski and Guido Walz [46].

Herbert Homeier [74] studied a systematic procedure for constructing sequences transformations. He considered iterated transformations which are *hierarchically consistent*, which means that the kernel of the basic transformation is the lowest one in the hierarchy. The application of the basic transformation to a sequence which is higher in the hierarchy leads to a new sequence belonging to a kernel lower in the hierarchy. Homeier wrote several papers on this topics.

Thus, the theory of convergence acceleration methods has progressed impressively. The practical side was not forgotten and authors obtained a number of special devices for improving their efficiency. For example, when a certain sequence is to be accelerated, it is not obvious to know in advance which method will give the best result unless some properties of the sequence are already known. Thus, Delahaye [54] proposed using simultaneously several transformations and selecting, at each step of the procedure, one answer among the answers provided by the various algorithms. He proved that, under some assumptions, some tests are able to find automatically the best answer. The work of Delahaye was extended by Abdelhak Fdil [58,59]. The various answers could also be combined leading to *composite* transformations [23]. It is possible, in some cases, to extract a linear subsequence from the original one and then to accelerate it, for example, by Aitken's Δ^2 process [37]. Devices for controlling the error were also constructed [21].

When faced with the problem of accelerating the convergence of a given sequence, two approaches are possible. The first is to use a known extrapolation procedure and to try to prove that it accelerates the convergence of the given sequence. The second possibility is to construct an extrapolation procedure especially for that sequence. Convergence tests for sequences and series can be used for that purpose as explained by Brezinski [25]. This approach was mostly developed by Ana Cristina Matos [92]. Special extrapolation procedures for sequences such that $\forall n, S_n - S = a_n D_n$, where (D_n) is a known sequence and (a_n) an unknown one, can also be constructed from the asymptotic properties of the sequences (a_n) and (D_n). Brezinski and Redivo Zaglia did this in [39].

A.H. Bentbib [10] considered the acceleration of sequences of intervals. Mohammed Senhadji [129] defined and studied the condition number of a sequence transformation.

2.4. The E-algorithm

As we see above, the quantities involved in Shanks transformation are expressed as a ratio of determinants and the ε-algorithm allows one to compute them recursively. It is well known that an interpolation polynomial can be expressed as a ratio of determinants. Thus polynomial extrapolation also leads to such a ratio and the Neville–Aitken scheme can be used to avoid the computation of these determinants which leads to the Richardson extrapolation algorithm. A similar situation arises for many other transformations: in each case, the quantities involved are expressed as a ratio of special determinants and, in each case, one seeks for a special recursive algorithm for the practical implementation of the transformation. Thus, there was a real need for a general theory of such sequence transformations and for a single general recursive algorithm for their implementation. This work was performed independently between 1973 and 1980 by five different people. It is now known as the E-algorithm.

It seems that the first appearance of this algorithm is due to Claus Schneider in a paper received on December 21, 1973 [128]. The quantities $S(h_i)$ being given for $i = 0, 1, \ldots,$ Schneider looked for $S'(h) = S' + a_1 g_1(h) + \cdots + a_k g_k(h)$ satisfying the interpolation conditions $S'(h_i) = S(h_i)$ for $i = n, \ldots, n+k$, where the g_j's are given functions of h. Of course, the value of the unknown S' thus obtained will depend on the indexes k and n. Assuming that $\forall j, g_j(0) = 0$, we have $S' = S'(0)$. Denoting by ϕ_k^n the extrapolation functional on the space of functions f defined at the points $h_0 > h_1 > \cdots > 0$ and at the point 0 and such that $\phi_k^n f = f(0)$, we have

$$\phi_k^n S' = c_0 S(h_n) + \cdots + c_k S(h_{n+k})$$

with $c_0 + \cdots + c_k = 1$. The interpolation conditions become

$$\phi_k^n E = 1, \quad \text{and} \quad \phi_k^n g_j = 0, \quad j = 1, \ldots, k$$

with $E(h) \equiv 1$. Schneider wanted to express the functional ϕ_k^n in the form $\phi_k^n = a\phi_{k-1}^n + b\phi_{k-1}^{n+1}$. He obtained the two conditions

$$\phi_k^n E = a + b = 1$$

and

$$\phi_k^n g_k = a\phi_{k-1}^n g_k + b\phi_{k-1}^{n+1} g_k = 0.$$

The values of a and b follow immediately and we have

$$\phi_k^n = \frac{[\phi_{k-1}^{n+1} g_k]\phi_{k-1}^n - [\phi_{k-1}^n g_k]\phi_{k-1}^{n+1}}{[\phi_{k-1}^{n+1} g_k] - [\phi_{k-1}^n g_k]}.$$

Thus, the quantities $\phi_k^n S'$ can be recursively computed by this scheme. The auxiliary quantities $\phi_k^n g_j$ needed in this formula must be computed separately by the same scheme using a different initialization. As we shall see below, this algorithm is just the E-algorithm. In a footnote, Schneider mentioned that this representation for ϕ_k^n was suggested by Börsch–Supan from Johannes Gutenberg Universität in Mainz.

In 1976, Günter Meinardus and G.D. Taylor wrote a paper [97] on best uniform approximation by functions from $\mathrm{span}(g_1,\ldots,g_N)\subset C[a,b]$. They defined the linear functionals L_n^k on $C[a,b]$ by

$$L_n^k(f)=\sum_{i=n}^{n+k} c_i f(h_i),$$

where $a\leqslant h_1<h_2<\cdots<h_{N+1}\leqslant b$ and where the coefficients c_i, which depend on n and k, are such that $c_n>0, c_i\neq 0$ for $i=n,\ldots,n+k$, sign $c_i=(-1)^{i-n}$ and

$$\sum_{i=n}^{n+k} |c_i|=1,$$

$$\sum_{i=n}^{n+k} c_i g_j(h_i)=0, \quad j=1,\ldots,k.$$

By using Gaussian elimination to solve the system of linear equations

$$\sum_{i=n}^{N} a_i g_i(h_j)+(-1)^j\lambda=f(h_j), \quad j=1,\ldots,k,$$

Meinardus and Taylor obtained a recursive scheme

$$L_i^k(f)=\frac{L_{i+1}^{k-1}(g_k)L_i^{k-1}(f)-L_i^{k-1}(g_k)L_{i+1}^{k-1}(f)}{L_{i+1}^{k-1}(g_k)-L_i^{k-1}(g_k)}$$

with $L_i^0(f)=f(h_i)$, $i=n,\ldots,n+k$. This is the same scheme as above.

Newton's formula for computing the interpolation polynomial is well known. It is based on divided differences. One can try to generalize these formulae to the case of interpolation by a linear combination of functions from a complete Chebyshev system (a technical concept which insures the existence and uniqueness of the solution). We seek

$$P_k^{(n)}(x)=a_0g_0(x)+\cdots+a_kg_k(x),$$

satisfying the interpolation conditions

$$P_k^{(n)}(x_i)=f(x_i), \quad i=n,\ldots,n+k,$$

where the x_i's are distinct points and the g_i's given functions. The $P_k^{(n)}$ can be recursively computed by an algorithm which generalizes the Neville–Aitken scheme for polynomial interpolation. This algorithm was obtained by Günter Mühlbach in 1976 [103] from a generalization of the notion of divided differences and their recurrence relationship. This algorithm was called the Mühlbach–Neville–Aitken algorithm, for short the MNA. It is as follows:

$$P_k^{(n)}(x)=\frac{g_{k-1,k}^{(n+1)}(x)P_{k-1}^{(n)}(x)-g_{k-1,k}^{(n)}(x)P_{k-1}^{(n+1)}(x)}{g_{k-1,k}^{(n+1)}(x)-g_{k-1,k}^{(n)}(x)}$$

with $P_0^{(n)}(x)=f(x_n)g_0(x)/g_0(x_n)$. The $g_{k,i}^{(n)}$'s can be recursively computed by a quite similar relationship

$$g_{k,i}^{(n)}(x)=\frac{g_{k-1,k}^{(n+1)}(x)g_{k-1,i}^{(n)}(x)-g_{k-1,k}^{(n)}(x)g_{k-1,i}^{(n+1)}(x)}{g_{k-1,k}^{(n+1)}(x)-g_{k-1,k}^{(n)}(x)}$$

with $g_{0,i}^{(n)}(x)=g_i(x_n)g_0(x)/g_0(x_n)-g_i(x)$. If $g_0(x)\equiv 1$, if it is assumed that $\forall i>0, g_i(0)=0$, the quantities $P_k^{(n)}(0)$ are the same as those obtained by the E-algorithm and the MNA reduces to it. Let us mention that, in fact, the MNA is closely related to the work of Henri Marie Andoyer (1862–1929) which goes back to 1906 [2]; see [30] for detailed explanations.

We now come to the work of Tore Håvie. We already mentioned Romberg's method for accelerating the convergence of the trapezoidal rule. The success of this procedure is based on the existence of the Euler–Maclaurin expansion for the error. This expansion only holds if the function to be integrated has no singularity in the interval. In the presence of singularities, the expansion of the error is no longer a series in h^2 (the stepsize) but a more complicated one depending on the singularity. Thus, Romberg's scheme has to be modified to incorporate the various terms appearing in the expansion of the error. Several authors worked on this question, treating several types of singularities. In particular, Håvie began to study this question under Romberg (Romberg emigrated to Norway and came to Trondheim in 1949). In 1978, Håvie wrote a report, published one year later [71], where he treated the most general case of an error expansion of the form

$$S(h)-S=a_1g_1(h)+a_2g_2(h)+\cdots,$$

where $S(h)$ denotes the approximation obtained by the trapezoidal rule with step size h to the definite integral S and the g_i are the known functions (forming an asymptotic sequence when h tends to zero) appearing in the expansion of the error. Let $h_0>h_1>\cdots>0$, $S_n=S(h_n)$ and $g_i(n)=g_i(h_n)$. Håvie set

$$E_1^{(n)}=\frac{g_1(n+1)S_n-g_1(n)S_{n+1}}{g_1(n+1)-g_1(n)}.$$

Replacing S_n and S_{n+1} by their expansions, he obtained

$$E_1^{(n)}=S+a_2g_{1,2}^{(n)}+a_3g_{1,3}^{(n)}+\cdots$$

with

$$g_{1,i}^{(n)}=\frac{g_1(n+1)g_i(n)-g_1(n)g_i(n+1)}{g_1(n+1)-g_1(n)}.$$

The same process can be repeated for eliminating $g_{1,2}^{(n)}$ in the the expansion of $E_1^{(n)}$, and so on. Thus, once again we obtain the E-algorithm

$$E_k^{(n)}=\frac{g_{k-1,k}^{(n+1)}E_{k-1}^{(n)}-g_{k-1,k}^{(n)}E_{k-1}^{(n+1)}}{g_{k-1,k}^{(n+1)}-g_{k-1,k}^{(n)}}$$

with $E_0^{(n)}=S_n$ and $g_{0,i}^{(n)}=g_i(n)$. The auxiliary quantities $g_{k,i}^{(n)}$ are recursively computed by the quite similar rule

$$g_{k,i}^{(n)}=\frac{g_{k-1,k}^{(n+1)}g_{k-1,i}^{(n)}-g_{k-1,k}^{(n)}g_{k-1,i}^{(n+1)}}{g_{k-1,k}^{(n+1)}-g_{k-1,k}^{(n)}}$$

with $g_{0,i}^{(n)}=g_i(n)$.

Håvie gave an interpretation of this algorithm in terms of the Gaussian elimination process for solving the system

$$E_k^{(n)} + b_1 g_1(n+i) + \cdots + b_k g_k(n+i) = S_{n+i}, \quad i = 0, \ldots, k$$

for the unknown $E_k^{(n)}$.

In 1980, Brezinski took up the same problem, but from the point of view of extrapolation [19]. Let (S_n) be the sequence to be accelerated. Interpolating it by a sequence of the form $S'_n = S + a_1 g_1(n) + \cdots + a_k g_k(n)$, where the g_i's are known sequences which can depend on the sequence (S_n) itself, leads to

$$S_{n+i} = S'_{n+i}, \quad i = 0, \ldots, k.$$

Solving this system directly for the unknown S (which, since it depends on n and k, will be denoted by $E_k^{(n)}$) gives

$$E_k^{(n)} = \frac{\begin{vmatrix} S_n & \cdots & S_{n+k} \\ g_1(n) & \cdots & g_1(n+k) \\ \vdots & & \vdots \\ g_k(n) & \cdots & g_k(n+k) \end{vmatrix}}{\begin{vmatrix} 1 & \cdots & 1 \\ g_1(n) & \cdots & g_1(n+k) \\ \vdots & & \vdots \\ g_k(n) & \cdots & g_k(n+k) \end{vmatrix}}.$$

Thus $E_k^{(n)}$ is given as a ratio of determinants which is very similar to the ratios previously mentioned. Indeed, for the choice $g_i(n) = \Delta S_{n+i}$, the ratio appearing in Shanks transformation results while, when $g_i(n) = x_n^i$, we obtain the ratio expressing the quantities involved in the Richardson extrapolation process. Other algorithms may be similarly derived.

Now the problem is to find a recursive algorithm for computing the $E_k^{(n)}$'s. Applying Sylvester's determinantal identity, Brezinski obtained the two rules of the above E-algorithm. His derivation of the E-algorithm is closely related to Håvie's since Sylvester's identity can be proved by using Gaussian elimination. Brezinski also gave convergence and acceleration results for this algorithm when the $(g_i(n))$ satisfy certain conditions [19]. These results show that, for accelerating the convergence of a sequence, it is necessary to know the expansion of the error $S_n - S$ with respect to some asymptotic sequence $(g_1(n)), (g_2(n)), \ldots$. The $g_i(n)$ are those to be used in the E-algorithm. It can be proved that, $\forall k$

$$\lim_{n\to\infty} \frac{E_{k+1}^{(n)} - S}{E_k^{(n)} - S} = 0.$$

These results were refined by Avram Sidi [134–136]. Thus the study of the asymptotic expansion of the error of the sequences to be accelerated is of primary importance, see Walz [144]. For example, Mohammed Kzaz [79,80] and Pierre Verlinden [142] applied this idea to the problem of accelerating the convergence of Gaussian quadrature formulae [79] and Pedro Lima and Mario Graça to boundary value problems with singularities [88,87] (see also the works of Lima and Diogo [87], and Lima and Carpentier [86]). Other acceleration results were obtained by Matos and Marc Prévost [94], Prévost

[116] and Pascal Mortreux and Prévost [102]. An algorithm, more economical than the E-algorithm, was given by William F. Ford and Avram Sidi [60]. The connection between the E-algorithm and the ε-algorithm was studied by Bernhard Beckermann [4]. A general ε-algorithm connected to the E-algorithm was given by Carsten Carstensen [48]. See [27] for a more detailed review on the E-algorithm.

Convergence acceleration algorithms can also be used for predicting the unknowns terms of a series or a sequence. This idea, introduced by Jacek Gilewicz [64], was studied by Sidi and Levin [137], Brezinski [22] and Denis Vekemans [141].

2.5. A new approach

Over the years, a quite general framework was constructed for the theory of extrapolation algorithms. The situation was quite different for the practical construction of extrapolation algorithms and there was little systematic research in their derivation. However, thanks to a formalism due to Weniger [145], such a construction is now possible, see Brezinski and Matos [38]. It is as follows. Let us assume that the sequence (S_n) to be accelerated satisfies, $\forall n, S_n - S = a_n D_n$ where (D_n) is a known sequence, called a *remainder* (or error) *estimate* for the sequence (S_n), and (a_n) an unknown sequence. It is possible to construct a sequence transformation such that its kernel is precisely this set of sequences. For that purpose, we have to assume that a difference operator L (that is a linear mapping of the set of sequences into itself) exists such that $\forall n, L(a_n) = 0$. This means that the sequence obtained by applying L to the sequence (a_n) is identically zero. Such a difference operator is called an *annihilation* operator for the sequence (a_n). We have

$$\frac{S_n}{D_n} - \frac{S}{D_n} = a_n.$$

Applying L and using linearity leads to

$$L\left(\frac{S_n}{D_n}\right) - SL\left(\frac{1}{D_n}\right) = L(a_n) = 0.$$

We solve for S and designate it by the sequence transformation

$$T_n = \frac{L(S_n/D_n)}{L(1/D_n)}.$$

The sequence (T_n) is be such that $\forall n, T_n = S$ if and only if $\forall n, S_n - S = a_n D_n$. This approach is highly versatile.

All the algorithms described above and the related devices such as error control, composite sequence transformations, least squares extrapolation, etc., can be put into this framework. Moreover, many new algorithms can be obtained using this approach. The E-algorithm can also be put into this framework which provides a deeper insight and leads to new properties [41]. Matos [93], using results from the theory of difference equations, obtained new and general convergence and acceleration results when (a_n) has an asymptotic expansion of a certain form.

2.6. Integrable systems

The connection between convergence acceleration algorithms and discrete integrable systems is a subject whose interest is rapidly growing among physicists. When a numerical scheme is used for integrating a partial differential evolution equation, it is important that it preserves the quantities that are conserved by the partial differential equation itself. An important character is the *integrability* of the equation. Although this term has not yet received a completely satisfactory definition (see [66]), it can be understood as the ability to write the solution explicitly in terms of a finite number of functions or as the confinement of singularities in finite domains. The construction of integrable discrete forms of integrable partial differential equations is highly nontrivial. A major discovery in the field of integrability was the occurrence of a solitary wave (called a *soliton*) in the Korteweg–de Vries (KdV) equation. Integrability is a rare phenomenon and the typical dynamical system is nonintegrable. A test of integrability, called *singularity confinement*, was given by B. Grammaticos, A. Ramani and V. Papageorgiou [67]. It turns out that this test is related to the existence of singular rules for avoiding a division by zero in convergence acceleration algorithms (see Section 1.2).

The literature on this topic is vast and we cannot enter into the details of it. We only want to give an indication of the connection between these two subjects since both domains could benefit from it.

In the rule for the ε-algorithm, V. Papageorgiou, B. Grammaticos and A. Ramani set $m=k+n$ and replaced $\varepsilon_k^{(n)}$ by $u(n,m)+mp+nq$, where p and q satisfy $p^2-q^2=1$. They obtained [111]

$$[p-q+u(n,m+1)-u(n+1,m)][p+q+u(n+1,m+1)-u(n,m)]=p^2-q^2.$$

This is the discrete lattice KdV equation. Since this equation is integrable, one can expect integrability to hold also for the ε-algorithm, and, thanks to the singular rules of Wynn and Cordellier mentioned at the end of Subsection 1.2, this is indeed the case.

In the rule of the ε-algorithm, making the change of variable $k=t/\varepsilon^3$ and $n-1/2=x/\varepsilon-ct/\varepsilon^3$ and replacing $\varepsilon_k^{(n)}$ by $p+\varepsilon^2 u(x-\varepsilon/2,t)$ where c and p are related by $1-2c=1/p^2$, A. Nagai and J. Satsuma obtained [105]

$$\varepsilon^2 u(x-\varepsilon/2+c\varepsilon,t+\varepsilon^3)-\varepsilon^2 u(x+\varepsilon/2-c\varepsilon,t-\varepsilon^3)=\frac{1}{p+\varepsilon^2 u(x+\varepsilon/2,t)}-\frac{1}{p+\varepsilon^2 u(x-\varepsilon/2,t)}.$$

We have, to terms of order ε^5, the KdV equation

$$u_t-\frac{1}{p^3}uu_x+\frac{1}{48p^2}(1-p^{-4})u_{xxx}=0.$$

Other discrete numerical algorithms, such as the *qd*, *LR*, and ρ-algorithms are connected to other discrete or continuous integrable equations (see, for example [112]). Formal orthogonal polynomials, continued fractions, Padé approximation also play a rôle in this topic [113].

By replacing the integer n in the ε-algorithm by a continuous variable, Wynn derived the *confluent form* of the ε-algorithm [149]

$$\varepsilon_{k+1}(t)=\varepsilon_{k-1}(t)+\frac{1}{\varepsilon_k'(t)}$$

with $\varepsilon_{-1}(t)\equiv 0$ and $\varepsilon_0(t)=f(t)$. This algorithm is the continuous counterpart of the ε-algorithm and its aim is to compute $\lim_{t\to\infty}f(t)$. Setting $N_k(t)=\varepsilon_k'(t)\varepsilon_{k+1}'(t)$, A. Nagai, T. Tokihiro and J. Satsuma [106] obtained

$$N_k'(t)=N_k(t)[N_{k-1}(t)-N_{k+1}(t)].$$

The above equation is the Bäcklund transformation of the discrete Toda molecule equation [139].

So, we see that some properties of integrable systems are related to properties of convergence acceleration algorithms. On the other hand, discretizing integrable partial differential equations leads to new sequence transformations which have to be studied from the point of view of their algebraic and acceleration properties. Replacing the second integer k in the confluent form of the ε-algorithm by a continuous variable, Wynn obtained a partial differential equation [152]. Its relation with integrable systems is an open question.

The connection between integrable systems and convergence acceleration algorithms needs to be investigated in more details to fully understand its meaning which is not clear yet.

3. The vector case

In numerical analysis, many iterative methods lead to vector sequences. To accelerate the convergence of such sequences, it is always possible to apply a scalar algorithm componentwise. However, vector sequence transformations, specially built for that purpose, are usually more powerful. The first vector algorithm to be studied was the vector ε-algorithm. It was obtained by Wynn [150] by replacing, in the rule of the scalar ε-algorithm, $1/\Delta\varepsilon_k^{(n)}$ by $(\Delta\varepsilon_k^{(n)})^{-1}$ where the inverse y^{-1} of a vector y is defined by $y^{-1} = y/(y,y)$. Thus, with this definition, the rule of the ε-algorithm can be applied to vector sequences. Using Clifford algebra, J.B. McLeod proved in 1971 [96] that $\forall n, \varepsilon_{2k}^{(n)} = S$ if the sequence (S_n) satisfies $a_0(S_n - S) + \cdots + a_k(S_{n+k} - S) = 0$, $\forall n$ with $a_0 a_k \neq 0$, $a_0 + \cdots + a_k \neq 0$. This result is only valid for real sequences (S_n) and real a_i's. Moreover, contrary to the scalar case, this condition is only sufficient. In 1983, Peter R. Graves–Morris [68] extended this result to the complex case using a quite different approach.

A drawback to the development of the theory of the vector ε-algorithm was that it was not known whether a corresponding generalization of Shanks transformation was underlying the algorithm, that is whether the vectors $\varepsilon_k^{(n)}$ obtained by the algorithm could be expressed as ratios of determinants (or some kind of generalization of determinants). This is why Brezinski [17], following the same path as Shanks, tried to construct a vector sequence transformation with the kernel $a_0(S_n - S) + \cdots + a_k(S_{n+k} - S) = 0$. He obtained a transformation expressed as a ratio of determinants. He then had to develop a recursive algorithm for avoiding their computation. This was the so-called topological ε-algorithm. This algorithm has many applications, in particular, to the solution of systems of linear equations (it is related to the biconjugate gradient algorithm [18, pp. 185ff]). In the case of a system of nonlinear equations, it gave rise to a generalization of Steffensen's method [13]. That algorithm has a quadratic convergence under some assumptions as established by Hervé Le Ferrand [83] following the ideas presented by Khalide Jbilou and Sadok [75]. The denominator of the vector $\varepsilon_{2k}^{(n)}$ obtained by the vector ε-algorithm was first written as a determinant of dimension $2k+1$ by Graves–Morris and Chris Jenkins in [69]. The numerator follows immediately by modifying the first row of the denominator, a formula given by Ahmed Salam and Graves–Morris [126]. However, the dimension of the corresponding determinants in the scalar case is only $k+1$. It was proved by Salam [124] that the vectors $\varepsilon_{2k}^{(n)}$ computed by the vector ε-algorithm can be expressed as a ratio of two *designants* of dimension $k+1$. A designant is a generalization of a determinant when solving a system of linear equations in a noncommutative algebra. An algebraic approach to this algorithm was given in [125]. This approach, which involves the use of a Clifford algebra, was used in [45] for extending the mechanism given in [41] to the vector and matrix cases. The vector generalization of the E-algorithm [19] can be explained similarly. This algorithm makes use of a fixed vector y. Jet Wimp [146, pp. 176–177] generalized it using a sequence (y_n) instead of y. Jeannette van Iseghem [140] gave an algorithm for accelerating vector sequences based on the vector orthogonal polynomials she introduced for generalizing Padé approximants to the vector case. Other vector sequence transformations are due to Osada [109] and Jbilou and Sadok [76]. Benchiboun [6] and Abderrahim Messaoudi [100] studied matrix extrapolation algorithms.

We have seen that, in the scalar case, the kernels of sequence transformations may be expressed as relationships with constant coefficients. This is also the case for the vector and the topological ε-algorithms and the vector E-algorithm. The first (and, to my knowledge, only) transformation treating a relationship with varying coefficients was introduced in [42]. The theory developed there also explains why the case of a relationship with non-constant coefficients is a difficult problem in the scalar case and why it could be solved, on the contrary, in the vector case. The reason is that the number of unknown coefficients appearing in the expression of the kernel must be strictly less than the dimension of the vectors. Brezinski in [34] proposed a general methodology for constructing vector sequence transformations. It leads to a unified presentation of several approaches to the subject and to new results. He also discussed applications to linear systems. In fact, as showed by Sidi [133], and Jbilou and Sadok [75], vector sequence transformations are closely related to projection methods for the solution of systems of equations. In particular, the RPA, a vector sequence transformation defined by Brezinski [20] was extensively studied by Messaoudi who showed its connections to direct and iterative methods for solving systems of linear equations [98,99].

Vector sequence transformations lead to new methods for the solution of systems of nonlinear equations. They also have other applications. First of all, it is quite important to accelerate the convergence of iterative methods for the solution of systems of linear equations, see [32,33,36]. Special vector extrapolation techniques were designed for the regularization of ill-posed linear systems in [43] and the idea of extrapolation was used in [35] to obtain estimates of the norm of the error when solving a system of linear equations by an arbitrary method, direct or iterative.

General theoretical results similar to those obtained in the scalar case are still lacking in the vector case although some partial results have been obtained. Relevant results on quasilinear transformations are in the papers by Sadok [123] and Benazzouz [8]. The present author proposed a mechanism for vector sequence transformations in [45,34].

4. Conclusions and perspectives

In this paper, I have tried to give a survey of the development of convergence acceleration methods for scalar and vector sequences in the 20th century. These methods are based on the idea of extrapolation. Since a universal algorithm for accelerating the convergence of all sequences cannot exist (and this is even true for some restricted classes of sequences), it was necessary to define and study a large variety of algorithms, each of them being appropriate for some special subsets of sequences.

It is, of course, always possible to construct other convergence acceleration methods for scalar sequences. However, to be of interest, such new processes must provide a major improvement over existing ones. For scalar sequence transformations, the emphasis must be placed on the theory rather than on special devices (unless a quite powerful one is found) and on the application of new methods to particular algorithms in numerical analysis and to various domains of applied sciences. In particular, the connection between convergence acceleration algorithms and continuous and discrete integrable systems brings a different and fresh look to both domains and could be of benefit to them.

An important problem in numerical analysis is the solution of large, sparse systems of linear equations. Most of the methods used nowadays are projection methods. Often the iterates obtained in such problems must be subject to acceleration techniques. However, many of the known vector convergence acceleration algorithms require the storage of too many vectors to be useful. New and cheaper acceleration algorithms are required. This difficult project, in my opinion, offers many opportunities for future research.

In this paper, I only briefly mentioned the confluent algorithms whose aim is the computation of the limit of a function when the variable tends to infinity (the continuous analog of the problem of convergence acceleration for a sequence). This subject and its applications will provide fertile ground for new discoveries.

Acknowledgements

I would like to thank Jet Wimp for his careful reading of the paper. He corrected my English in many places, he asked me to provide more explanations when needed, and suggested many improvements in the presentation. I am also indebted to Naoki Osada for his informations about Takakazu Seki.

References

[1] A.C. Aitken, On Bernoulli's numerical solution of algebraic equations, Proc. Roy. Soc. Edinburgh 46 (1926) 289–305.
[2] H. Andoyer, Interpolation, in: J. Molk (Ed.), Encyclopédie des Sciences Mathématiques Pures et Appliquées, Tome I, Vol. 4, Fasc. 1, I–21, Gauthier–Villars, Paris, 1904–1912, pp.127–160; (reprint by Editions Gabay, Paris, 1993).
[3] F.L. Bauer, La méthode d'intégration numérique de Romberg, in: Colloque sur l'Analyse Numérique, Librairie Universitaire, Louvain, 1961, pp. 119–129.
[4] B. Beckermann, A connection between the E-algorithm and the epsilon-algorithm, in: C. Brezinski (Ed.), Numerical and Applied Mathematics, Baltzer, Basel, 1989, pp. 443–446.
[5] M.D. Benchiboun, Etude de Certaines Généralisations du Δ^2 d'Aitken et Comparaison de Procédés d'Accélération de la Convergence, Thèse 3ème cycle, Université de Lille I, 1987.
[6] M.D. Benchiboun, Extension of Henrici's method to matrix sequences, J. Comput. Appl. Math. 75 (1996) 1–21.
[7] A. Benazzouz, Quasilinear sequence transformations, Numer. Algorithms 15 (1997) 275–285.
[8] A. Benazzouz, GL(E)-quasilinear transformations and acceleration, Appl. Numer. Math. 27 (1998) 109–122.
[9] A. Berlinet, Sequence transformations as statistical tools, Appl. Numer. Math. 1 (1985) 531–544.
[10] A.H. Bentbib, Acceleration of convergence of interval sequences, J. Comput. Appl. Math. 51 (1994) 395–409.
[11] N. Bogolyubov, N. Krylov, On Rayleigh's principle in the theory of differential equations of mathematical physics and upon Euler's method in the calculus of variation, Acad. Sci. Ukraine (Phys. Math.) 3 (1926) 3–22 (in Russian).
[12] H.C. Bolton, H.I. Scoins, Eigenvalues of differential equations by finite-difference methods, Proc. Cambridge Philos. Soc. 52 (1956) 215–229.
[13] C. Brezinski, Application de l'ε-algorithme à la résolution des systèmes non linéaires, C.R. Acad. Sci. Paris 271 A (1970) 1174–1177.
[14] C. Brezinski, Accélération de suites à convergence logarithmique, C. R. Acad. Sci. Paris 273 A (1971) 727–730.
[15] C. Brezinski, Etude sur les ε et ρ-algorithmes, Numer. Math. 17 (1971) 153–162.

[16] C. Brezinski, L'ε-algorithme et les suites totalement monotones et oscillantes, C.R. Acad. Sci. Paris 276 A (1973) 305–308.
[17] C. Brezinski, Généralisation de la transformation de Shanks, de la table de Padé et de l'ε-algorithme, Calcolo 12 (1975) 317–360.
[18] C. Brezinski, Padé-Type Approximation and General Orthogonal Polynomials, Birkhäuser, Basel, 1980.
[19] C. Brezinski, A general extrapolation algorithm, Numer. Math. 35 (1980) 175–187.
[20] C. Brezinski, Recursive interpolation, extrapolation and projection, J. Comput. Appl. Math. 9 (1983) 369–376.
[21] C. Brezinski, Error control in convergence acceleration processes, IMA J. Numer. Anal. 3 (1983) 65–80.
[22] C. Brezinski, Prediction properties of some extrapolation methods, Appl. Numer. Math. 1 (1985) 457–462.
[23] C. Brezinski, Composite sequence transformations, Numer. Math. 46 (1985) 311–321.
[24] C. Brezinski, A. Lembarki, Acceleration of extended Fibonacci sequences, Appl. Numer. Math. 2 (1986) 1–8.
[25] C. Brezinski, A new approach to convergence acceleration methods, in: A. Cuyt (Ed.), Nonlinear Numerical Methods and Rational Approximation, Reidel, Dordrecht, 1988, pp. 373–405.
[26] C. Brezinski, Quasi-linear extrapolation processes, in: R.P. Agarwal et al. (Eds.), Numerical Mathematics, Singapore 1988, International Series of Numerical Mathematics, Vol. 86, Birkhäuser, Basel, 1988, pp. 61–78.
[27] C. Brezinski, A survey of iterative extrapolation by the E-algorithm, Det Kong. Norske Vid. Selsk. Skr. 2 (1989) 1–26.
[28] C. Brezinski, A Bibliography on Continued Fractions, Padé Approximation, Extrapolation and Related Subjects, Prensas Universitarias de Zaragoza, Zaragoza, 1991.
[29] C. Brezinski, History of Continued Fractions and Padé Approximants, Springer, Berlin, 1991.
[30] C. Brezinski, The generalizations of Newton's interpolation formula due to Mühlbach and Andoyer, Electron Trans. Numer. Anal. 2 (1994) 130–137.
[31] C. Brezinski, Extrapolation algorithms and Padé approximations: a historical survey, Appl. Numer. Math. 20 (1996) 299–318.
[32] C. Brezinski, Variations on Richardson's method and acceleration, in: Numerical Analysis, A Numerical Analysis Conference in Honour of Jean Meinguet, Bull. Soc. Math. Belgium 1996, pp. 33–44.
[33] C. Brezinski, Projection Methods for Systems of Equations, North-Holland, Amsterdam, 1997.
[34] C. Brezinski, Vector sequence transformations: methodology and applications to linear systems, J. Comput. Appl. Math. 98 (1998) 149–175.
[35] C. Brezinski, Error estimates for the solution of linear systems, SIAM J. Sci. Comput. 21 (1999) 764–781.
[36] C. Brezinski, Acceleration procedures for matrix iterative methods, Numer. Algorithms, to appear.
[37] C. Brezinski, J.P. Delahaye, B. Germain-Bonne, Convergence acceleration by extraction of linear subsequences, SIAM J. Numer. Anal. 20 (1983) 1099–1105.
[38] C. Brezinski, A.C. Matos, A derivation of extrapolation algorithms based on error estimates, J. Comput. Appl. Math. 66 (1996) 5–26.
[39] C. Brezinski, M. Redivo Zaglia, Construction of extrapolation processes, Appl. Numer. Math. 8 (1991) 11–23.
[40] C. Brezinski, M. Redivo Zaglia, Extrapolation Methods, Theory and Practice, North-Holland, Amsterdam, 1991.
[41] C. Brezinski, M. Redivo Zaglia, A general extrapolation procedure revisited, Adv. Comput. Math. 2 (1994) 461–477.
[42] C. Brezinski, M. Redivo Zaglia, Vector and matrix sequence transformations based on biorthogonality, Appl. Numer. Math. 21 (1996) 353–373.
[43] C. Brezinski, M. Redivo Zaglia, G. Rodriguez, S. Seatzu, Extrapolation techniques for ill-conditioned linear systems, Numer. Math. 81 (1998) 1–29.
[44] C. Brezinski, A.C. Rieu, The solution of systems of equations using the vector ε-algorithm, and an application to boundary value problems, Math. Comp. 28 (1974) 731–741.
[45] C. Brezinski, A. Salam, Matrix and vector sequence transformation revisited, Proc. Edinburgh Math. Soc. 38 (1995) 495–510.
[46] C. Brezinski, G. Walz, Sequences of transformations and triangular recursion schemes with applications in numerical analysis, J. Comput. Appl. Math. 34 (1991) 361–383.
[47] R. Bulirsch, J. Stoer, Numerical treatment of ordinary differential equations by extrapolation methods, Numer. Math. 8 (1966) 1–13.
[48] C. Carstensen, On a general epsilon algorithm, in: C. Brezinski (Ed.), Numerical and Applied Mathematics, Baltzer, Basel, 1989, pp. 437–441.

[49] J.L. Chabert et al., Histoire d'Algorithmes, Belin, Paris, 1994.
[50] L. Collatz, Numerische Behandlung von Differentialgleichungen, Springer, Berlin, 1951.
[51] F. Cordellier, Démonstration algébrique de l'extension de l'identité de Wynn aux tables de Padé non normales, in: L. Wuytack (Ed.), Padé Approximation and its Applications, Lecture Notes in Mathematics, Vol. 765, Springer, Berlin, 1979, pp. 36–60.
[52] M. Crouzeix, A.L. Mignot, Analyse Numérique des Equations Différentielles, 2nd Edition, Masson, Paris, 1989.
[53] A. Cuyt, L. Wuytack, Nonlinear Methods in Numerical Analysis, North-Holland, Amsterdam, 1987.
[54] J.P. Delahaye, Automatic selection of sequence transformations, Math. Comp. 37 (1981) 197–204.
[55] J.P. Delahaye, Sequence Transformations, Springer, Berlin, 1988.
[56] J.P. Delahaye, B. Germain-Bonne, Résultats négatifs en accélération de la convergence, Numer. Math. 35 (1980) 443–457.
[57] J. Dutka, Richardson-extrapolation and Romberg-integration, Historia Math. 11 (1984) 3–21.
[58] A. Fdil, Sélection entre procédés d'accélération de la convergence, M2AN 30 (1996) 83–101.
[59] A. Fdil, A new technique of selection between sequence transformations, Appl. Numer. Math. 25 (1997) 21–40.
[60] W.F. Ford, A. Sidi, An algorithm for a generalization of the Richardson extrapolation process, SIAM J. Numer. Anal. 24 (1987) 1212–1232.
[61] E. Gekeler, On the solution of systems of equations by the epsilon algorithm of Wynn, Math. Comp. 26 (1972) 427–436.
[62] B. Germain-Bonne, Transformations de suites, RAIRO R1 (1973) 84–90.
[63] B. Germain-Bonne, Estimation de la Limite de Suites et Formalisation de Procédés d'Accélération de la Convergence, Thèse d'Etat, Université de Lille I, 1978.
[64] J. Gilewicz, Numerical detection of the best Padé approximant and determination of the Fourier coefficients of insufficiently sampled function, in: P.R. Graves-Morris (Ed.), Padé Approximants and their Applications, Academic Press, New York, 1973, pp. 99–103.
[65] W.B. Gragg, On extrapolation algorithms for initial-value problems, SIAM J. Numer. Anal. 2 (1965) 384–403.
[66] B. Grammaticos, A. Ramani, Integrability – and how to detect it, in: Y. Kosmann-Schwarzbach et al. (Eds.), Integrability of Nonlinear Systems, Springer, Berlin, 1997, pp. 30–94.
[67] B. Grammaticos, A. Ramani, V.G. Papageorgiou, Do integrable mappings have the Painlevé property? Phys. Rev. Lett. 67 (1991) 1825–1828.
[68] P.R. Graves-Morris, Vector valued rational interpolants I, Numer. Math. 42 (1983) 331–348.
[69] P.R. Graves-Morris, C.D. Jenkins, Vector valued rational interpolants III, Constr. Approx. 2 (1986) 263–289.
[70] H.L. Gray, T.A. Atchison, G.V. McWilliams, Higher order G – transformations, SIAM J. Numer. Anal. 8 (1971) 365–381.
[71] T. Håvie, Generalized Neville type extrapolation schemes, BIT 19 (1979) 204–213.
[72] H. Hankel, Ueber eine besondere Classe der symmetrischen Determinanten, Inaugural Dissertation, Universität Göttingen, 1861.
[73] A. Hirayama, K. Shimodaira, H. Hirose, Takakazu Seki's Collected Works Edited with Explanations, Osaka Kyoiku Tosho, Osaka, 1974.
[74] H.H.H. Homeier, A hierarchically consistent, iterative sequence transformation, Numer. Algorithms 8 (1994) 47–81.
[75] K. Jbilou, H. Sadok, Some results about vector extrapolation methods and related fixed-point iterations, J. Comput. Appl. Math. 36 (1991) 385–398.
[76] K. Jbilou, H. Sadok, Hybrid vector transformations, J. Comput. Appl. Math. 81 (1997) 257–267.
[77] D.C. Joyce, Survey of extrapolation processes in numerical analysis, SIAM Rev. 13 (1971) 435–490.
[78] K. Kommerell, Das Grenzgebiet der Elementaren und Höheren Mathematik, Verlag Köhler, Leipzig, 1936.
[79] M. Kzaz, Gaussian quadrature and acceleration of convergence, Numer. Algorithms 15 (1997) 75–89.
[80] M. Kzaz, Convergence acceleration of the Gauss–Laguerre quadrature formula, Appl. Numer. Math. 29 (1999) 201–220.
[81] P.J. Laurent, Un théorème de convergence pour le procédé d'extrapolation de Richardson, C.R. Acad. Sci. Paris 256 (1963) 1435–1437.
[82] H. Lavastre, On the stochastic acceleration of sequences of random variables, Appl. Numer. Math. 15 (1994) 77–98.
[83] H. Le Ferrand, Convergence of the topological ε-algorithm for solving systems of nonlinear equations, Numer. Algorithms 3 (1992) 273–284.

[84] H. Le Ferrand, Recherches d'extrema par des méthodes d'extrapolation, C.R. Acad. Sci. Paris, Sér. I 318 (1994) 1043–1046.
[85] D. Levin, Development of non-linear transformations for improving convergence of sequences, Int. J. Comput. Math. B3 (1973) 371–388.
[86] P.M. Lima, M.P. Carpentier, Asymptotic expansions and numerical approximation of nonlinear degenerate boundary-value problems, Appl. Numer. Math. 30 (1999) 93–111.
[87] P. Lima, T. Diogo, An extrapolation method for a Volterra integral equation with weakly singular kernel, Appl. Numer. Math. 24 (1997) 131–148.
[88] P.M. Lima, M.M Graça, Convergence acceleration for boundary value problems with singularities using the E-algorithm, J. Comput. Appl. Math. 61 (1995) 139–164.
[89] S. Lubkin, A method of summing infinite series, J. Res. Natl. Bur. Standards 48 (1952) 228–254.
[90] C. Maclaurin, Treatise of Fluxions, Edinburgh, 1742.
[91] G.I. Marchuk, V.V. Shaidurov, Difference Methods and their Extrapolations, Springer, Berlin, 1983.
[92] A.C. Matos, Acceleration methods based on convergence tests, Numer. Math. 58 (1990) 329–340.
[93] A.C. Matos, Linear difference operators and acceleration methods, IMA J. Numer. Anal., to appear.
[94] A.C. Matos, M. Prévost, Acceleration property for the columns of the E-algorithm, Numer. Algorithms 2 (1992) 393–408.
[95] J.C. Maxwell, A Treatise on Electricity and Magnetism, Oxford University Press, Oxford, 1873.
[96] J.B. McLeod, A note on the ε-algorithm, Computing 7 (1971) 17–24.
[97] G. Meinardus, G.D. Taylor, Lower estimates for the error of the best uniform approximation, J. Approx. Theory 16 (1976) 150–161.
[98] A. Messaoudi, Recursive interpolation algorithm: a formalism for solving systems of linear equations – I, Direct methods, J. Comput. Appl. Math. 76 (1996) 13–30.
[99] A. Messaoudi, Recursive interpolation algorithm: a formalism for solving systems of linear equations – II, Iterative methods, J. Comput. Appl. Math. 76 (1996) 31–53.
[100] A. Messaoudi, Matrix extrapolation algorithms, Linear Algebra Appl. 256 (1997) 49–73.
[101] R.M. Milne, Extension of Huygens' approximation to a circular arc, Math. Gaz. 2 (1903) 309–311.
[102] P. Mortreux, M. Prévost, An acceleration property for the E-algorithm for alternate sequences, Adv. Comput. Math. 5 (1996) 443–482.
[103] G. Mühlbach, Neville-Aitken algorithms for interpolating by functions of Čebyšev-systems in the sense of Newton and in a generalized sense of Hermite, in: A.G. Law, B.N. Sahney (Eds.), Theory of Approximation with Applications, Academic Press, New York, 1976, pp. 200–212.
[104] H. Naegelsbach, Studien zu Fürstenau's neuer Methode der Darstellung und Berechnung der Wurzeln algebraischer Gleichungen durch Determinanten der Coefficienten, Arch. Math. Phys. 59 (1876) 147–192; 61 (1877) 19–85.
[105] A. Nagai, J. Satsuma, Discrete soliton equations and convergence acceleration algorithms, Phys. Lett. A 209 (1995) 305–312.
[106] A. Nagai, T. Tokihiro, J. Satsuma, The Toda molecule equation and the ε-algorithm, Math. Comp. 67 (1998) 1565–1575.
[107] T.H. O'Beirne, On linear iterative processes and on methods of improving the convergence of certain types of iterated sequences, Technical Report, Torpedo Experimental Establishment, Greenock, May 1947.
[108] N. Osada, An acceleration theorem for the ρ-algorithm, Numer. Math. 73 (1996) 521–531.
[109] N. Osada, Vector sequence transformations for the acceleration of logarithmic convergence, J. Comput. Appl. Math. 66 (1996) 391–400.
[110] K.J. Overholt, Extended Aitken acceleration, BIT 5 (1965) 122–132.
[111] V. Papageorgiou, B. Grammaticos, A. Ramani, Integrable difference equations and numerical analysis algorithms, in: D. Levi et al. (Eds.), Symmetries and Integrability of Difference Equations, CRM Proceedings and Lecture Notes, Vol. 9, AMS, Providence, RI, 1996, pp. 269–279.
[112] V. Papageorgiou, B. Grammaticos, A. Ramani, Integrable lattices and convergence acceleration algorithms, Phys. Lett. A 179 (1993) 111–115.
[113] V. Papageorgiou, B. Grammaticos, A. Ramani, Orthogonal polynomial approach to discrete Lax pairs for initial-boundary value problems of the QD algorithm, Lett. Math. Phys. 34 (1995) 91–101.
[114] R. Pennacchi, Le trasformazioni razionali di una successione, Calcolo 5 (1968) 37–50.

[115] R. Powell, S.M. Shah, Summability Theory and its Applications, Van Nostrand Reinhold, London, 1972.
[116] M. Prévost, Acceleration property for the E-algorithm and an application to the summation of series, Adv. Comput. Math. 2 (1994) 319–341.
[117] J.D. Pryce, Numerical Solution of Sturm–Liouville Problems, Clarendon Press, Oxford, 1993.
[118] B. Rhanizar, On extrapolation methods in optimization, Appl. Numer. Math. 25 (1997) 485–498.
[119] L.F. Richardson, The approximate arithmetical solution by finite difference of physical problems involving differential equations, with an application to the stress in a masonry dam, Philos. Trans. Roy. Soc. London, Ser. A 210 (1910) 307–357.
[120] L.F. Richardson, The deferred approach to the limit. I: Single lattice, Philos. Trans. Roy. Soc. London, Ser. A 226 (1927) 299–349.
[121] W. Romberg, Vereinfachte numerische Integration, Kgl. Norske Vid. Selsk. Forsk. 28 (1955) 30–36.
[122] H. Rutishauser, Ausdehnung des Rombergschen Prinzips, Numer. Math. 5 (1963) 48–54.
[123] H. Sadok, Quasilinear vector extrapolation methods, Linear Algebra Appl. 190 (1993) 71–85.
[124] A. Salam, Non-commutative extrapolation algorithms, Numer. Algorithms 7 (1994) 225–251.
[125] A. Salam, An algebraic approach to the vector ε-algorithm, Numer. Algorithms 11 (1996) 327–337.
[126] P.R. Graves-Morris, D.E. Roberts, A. Salam, The epsilon algorithm and related topics, J. Comput. Appl. Math. 122 (2000).
[127] J.R. Schmidt, On the numerical solution of linear simultaneous equations by an iterative method, Philos. Mag. 7 (1941) 369–383.
[128] C. Schneider, Vereinfachte Rekursionen zur Richardson-Extrapolation in Spezialfällen, Numer. Math. 24 (1975) 177–184.
[129] M.N. Senhadji, On condition numbers of some quasi-linear transformations, J. Comput. Appl. Math. 104 (1999) 1–19.
[130] D. Shanks, An analogy between transient and mathematical sequences and some nonlinear sequence-to-sequence transforms suggested by it, Part I, Memorandum 9994, Naval Ordnance Laboratory, White Oak, July 1949.
[131] D. Shanks, Non linear transformations of divergent and slowly convergent sequences, J. Math. Phys. 34 (1955) 1–42.
[132] W.F. Sheppard, Some quadrature formulas, Proc. London Math. Soc. 32 (1900) 258–277.
[133] A. Sidi, Extrapolation vs. projection methods for linear systems of equations, J. Comput. Appl. Math. 22 (1988) 71–88.
[134] A. Sidi, On a generalization of the Richardson extrapolation process, Numer. Math. 57 (1990) 365–377.
[135] A. Sidi, Further results on convergence and stability of a generalization of the Richardson extrapolation process, BIT 36 (1996) 143–157.
[136] A. Sidi, A complete convergence and stability theory for a generalized Richardson extrapolation process, SIAM J. Numer. Anal. 34 (1997) 1761–1778.
[137] A. Sidi, D. Levin, Prediction properties of the t-transformation, SIAM J. Numer. Anal. 20 (1983) 589–598.
[138] E. Stiefel, Altes und neues über numerische Quadratur, Z. Angew. Math. Mech. 41 (1961) 408–413.
[139] M. Toda, Waves in nonlinear lattice, Prog. Theor. Phys. Suppl. 45 (1970) 174–200.
[140] J. van Iseghem, Convergence of vectorial sequences, applications, Numer. Math. 68 (1994) 549–562.
[141] D. Vekemans, Algorithm for the E-prediction, J. Comput. Appl. Math. 85 (1997) 181–202.
[142] P. Verlinden, Acceleration of Gauss–Legendre quadrature for an integral with an endpoint singularity, J. Comput. Appl. Math. 77 (1997) 277–287.
[143] G. Walz, The history of extrapolation methods in numerical analysis, Report No. 130, Universität Mannheim, Fakultät für Mathematik und Informatik, 1991.
[144] G. Walz, Asymptotics and Extrapolation, Akademie, Berlin, 1996.
[145] E.J. Weniger, Nonlinear sequence transformations for the acceleration of convergence and the summation of divergent series, Comput. Phys. Rep. 10 (1989) 189–371.
[146] J. Wimp, Sequence Transformations and their Applications, Academic Press, New York, 1981.
[147] P. Wynn, On a device for computing the $e_m(S_n)$ transformation, MTAC 10 (1956) 91–96.
[148] P. Wynn, On a procrustean technique for the numerical transformation of slowly convergent sequences and series, Proc. Cambridge Philos. Soc. 52 (1956) 663–671.
[149] P. Wynn, Confluent forms of certain non-linear algorithms, Arch. Math. 11 (1960) 223–234.

[150] P. Wynn, Acceleration techniques for iterated vector and matrix problems, Math. Comput. 16 (1962) 301–322.
[151] P. Wynn, Singular rules for certain non-linear algorithms, BIT 3 (1963) 175–195.
[152] P. Wynn, Partial differential equations associated with certain non-linear algorithms, ZAMP 15 (1964) 273–289.
[153] P. Wynn, Upon systems of recursions which obtain among the quotients of the Padé table, Numer. Math. 8 (1966) 264–269.
[154] P. Wynn, On the convergence and stability of the epsilon algorithm, SIAM J. Numer. Anal. 3 (1966) 91–122.
[155] M. Redivo Zaglia, Particular rules for the θ-algorithm, Numer. Algorithms 3 (1992) 353–370.

ELSEVIER

Journal of Computational and Applied Mathematics 122 (2000) 23–35

JOURNAL OF COMPUTATIONAL AND APPLIED MATHEMATICS

www.elsevier.nl/locate/cam

On the history of multivariate polynomial interpolation

Mariano Gasca[a,*], Thomas Sauer[b]

[a] *Department of Applied Mathematics, University of Zaragoza, 50009 Zaragoza, Spain*

[b] *Institute of Mathematics, University Erlangen-Nürnberg, Bismarckstr. $1\frac{1}{2}$, D-91054 Erlangen, Germany*

Received 7 June 1999; received in revised form 8 October 1999

Abstract

Multivariate polynomial interpolation is a basic and fundamental subject in Approximation Theory and Numerical Analysis, which has received and continues receiving not deep but constant attention. In this short survey, we review its development in the first 75 years of this century, including a pioneering paper by Kronecker in the 19th century. © 2000 Elsevier Science B.V. All rights reserved.

1. Introduction

Interpolation, by polynomials or other functions, is a rather old method in applied mathematics. This is already indicated by the fact that, apparently, the word "interpolation" itself has been introduced by J. Wallis as early as 1655 as it is claimed in [13]. Compared to this, polynomial interpolation in *several variables* is a relatively new topic and probably only started in the second-half of the last century with work in [6,22]. If one considers, for example, the *Encyklopädie der Mathematischen Wissenschaften* [13] (Encyclopedia of Math. Sciences), originated by the *Preußische Akademie der Wissenschaften* (Prussian Academy of Sciences) to sum up the "state of art" of mathematics at its time, then the part on interpolation, written by J. Bauschinger (Bd. I, Teil 2), mentions only one type of multivariate interpolation, namely (tensor) products of sine and cosine functions in two variables, however, without being very specific. The French counterpart, the *Encyclopédie de Sciences Mathematiques* [14], also contains a section on interpolation (Tome I, vol. 4), where Andoyer translated and extended Bauschinger's exposition. Andoyer is even more

* Corressponding author.

E-mail addresses: gasca@posta.unizar.es (M. Gasca), sauer@mi.uni-erlangen.de (T. Sauer).

0377-0427/00/$ - see front matter
PII: S0377-0427(00)00353-8

explicit with his opinion on multivariate polynomial interpolation, by making the following statement which we think that time has contradicted:

> Il est manifeste que l'interpolation des fonctions de plusiers variables ne demande aucun principe nouveau, car dans tout ce qui précède le fait que la variable indépendante était unique n'a souvent joué aucun rôle. [1]

Nevertheless, despite of Andoyer's negative assessment, multivariate polynomial interpolation has received not deep but constant attention from one part of the mathematical community and is today a basic subject in Approximation Theory and Numerical Analysis with applications to many mathematical problems. Of course, this field has definitely been influenced by the availability of computational facilities, and this is one of the reasons that more papers have been published about this subject in the last 25 years than in the preceding 75 ones.

To our knowledge, there is not any paper before the present one surveying the early papers and books on multivariate polynomial interpolation. Our aim is a first, modest attempt to cover this gap. We do not claim to be exhaustive and, in particular, recognize our limitations with respect to the Russian literature. Moreover, it has to be mentioned that the early results on multivariate interpolation usually appear in the context of many different subjects. For example, papers on cubature formulas frequently have some part devoted to it. Another connection is Algebraic Geometry, since the solvability of a multivariate interpolation problem relies on the fact that the interpolation points do not lie on an algebraic surface of a certain type. So it is difficult to verify precisely if and when a result appeared somewhere for the first time or if it had already appeared, probably even in an implicit way, in a different context. We remark that another paper in this volume [25] deals, complementarily, with recent results in the subject, see also [16].

Along the present paper we denote by Π_k^d the space of d-variate polynomials of total degree not greater than k.

2. Kronecker, Jacobi and multivariate interpolation

Bivariate interpolation by the tensor product of univariate interpolation functions, that is when the variables are treated separately, is the classical approach to multivariate interpolation. However, when the set of interpolation points is not a Cartesian product grid, it is impossible to use that idea. Today, given any set of interpolation points, there exist many methods [2] to construct an adequate polynomial space which guarantees unisolvence of the interpolation problem. Surprisingly, this idea of constructing an appropriate interpolation space was already pursued by Kronecker [22] in a widely unknown paper from 1865, which seems to be the first treatment of multivariate polynomial interpolation with respect to fairly arbitrary point configurations. Besides the mathematical elegance of this approach, we think it is worthwhile to devote some detailed attention to this paper and to resolve its main ideas in today's terminology, in particular, as it uses the "modern" approach of connecting polynomial interpolation to the theory of polynomial ideals.

---

[1] It is clear that the interpolation of functions of several variables does not demand any new principles because in the above exposition the fact that the variable was unique has not played frequently any role.

[2] See [16,25] for exposition and references.

Kronecker's method to construct an interpolating polynomial assumes that the disjoint nodes $z_1,\ldots,z_N \in \mathbb{C}^d$ are given in *implicit* form, i.e., they are (all) the common *simple* zeros of d polynomials $f_1,\ldots,f_d \in \mathbb{C}[z] = \mathbb{C}[\zeta_1,\ldots,\zeta_d]$. Note that the nonlinear system of equations

$$f_j(\zeta_1,\ldots,\zeta_d) = 0, \quad j = 1,\ldots,d, \tag{1}$$

is a *square* one, that is, the number of equations and the number of variables coincide. We are interested in the *finite* variety V of solutions of (1) which is given as

$$V := \{z_1,\ldots,z_N\} = \{z \in \mathbb{C}^d : f_1(z) = \cdots = f_d(z) = 0\}. \tag{2}$$

The *primary decomposition* according to the variety V allows us to write the ideal $\mathscr{I}(V) = \{p : p(z) = 0,\ z \in V\}$ as

$$\mathscr{I}(V) = \bigcap_{k=1}^{N} \langle \zeta_1 - \zeta_{k,1},\ldots,\zeta_d - \zeta_{k,d}\rangle,$$

where $z_k = (\zeta_{k,1},\ldots,\zeta_{k,d})$. In other words, since $f_k \in \mathscr{I}(V)$, $k = 1,\ldots,d$, any of the polynomials $f_1,\ldots,f_d$ can be written, for $k = 1,\ldots,N$, as

$$f_j = \sum_{i=1}^{d} g_{i,j}^{k}(\cdot)(\zeta_i - \zeta_{k,i}), \tag{3}$$

where $g_{i,j}^k$ are appropriate polynomials. Now consider the $d \times d$ square matrices of polynomials

$$G_k = [g_{i,j}^k : i,j = 1,\ldots,d], \quad k = 1,\ldots,N$$

and note that, due to (3), and the assumption that $f_j(z_k) = 0$, $j = 1,\ldots,d$, $k = 1,\ldots,N$, we have

$$0 = \begin{bmatrix} f_1(Z_j) \\ \vdots \\ f_d(z_j) \end{bmatrix} = G_k(z_j) \begin{bmatrix} (\zeta_{j,1} - \zeta_{k,1}) \\ \vdots \\ (\zeta_{j,d} - \zeta_{k,d}) \end{bmatrix}, \quad k = 1,\ldots,N. \tag{4}$$

Since the interpolation nodes are assumed to be *disjoint*, this means that for all $j \neq k$ the matrix $G_k(z_j)$ is *singular*, hence the determinant of $G_k(z_j)$ has to be zero. Moreover, the assumption that $z_1,\ldots,z_N$ are *simple* zeros guarantees that $\det G_k(z_k) \neq 0$. Then, Kronecker's interpolant takes, for any $f : \mathbb{C}^d \to \mathbb{C}$, the form

$$Kf = \sum_{j=1}^{N} f(z_j) \frac{\det G_k(\cdot)}{\det G_k(z_k)}. \tag{5}$$

Hence,

$$\mathscr{P} = \operatorname{span}\left\{ \frac{\det G_k(\cdot)}{\det G_k(z_k)} : k = 1,\ldots,N \right\}$$

is an interpolation space for the interpolation nodes $z_1,\ldots,z_N$. Note that this method does not give only *one* interpolation polynomial but in general *several different* interpolation spaces, depending on how the representation in (3) is chosen. In any way, note that for each polynomial $f \in \mathbb{C}[z]$ the difference

$$f - \sum_{j=1}^{N} f(z_j) \frac{\det G_k(z)}{\det G_k(z_k)}$$

belongs to the ideal $\langle f_1,\dots,f_d\rangle$, hence there exist polynomials $q_1,\dots,q_d$ such that

$$f-\sum_{j=1}^{N} f(z_j)\frac{\det G_k(z)}{\det G_k(z_k)}=\sum_{j=1}^{d} q_j f_j. \tag{6}$$

Moreover, as Kronecker points out, the "magic" polynomials $g_{i,j}^k$ can be chosen such that their leading homogeneous terms, say $G_{i,j}^k$, coincide with the leading homogeneous terms of $(1/\deg f_j)\partial f_j/\partial\zeta_i$. If we denote by F_j the leading homogeneous term of f_j, $j=1,\dots,d$, then this means that

$$G_{i,j}^k=\frac{1}{\deg F_j}\frac{\partial F_j}{\partial\zeta_i},\quad i,j=1,\dots,d,\quad k=1,\dots,N. \tag{7}$$

But this implies that the homogeneous leading term of the "fundamental" polynomials $\det G_k$ coincides, after this particular choice of $g_{i,j}^k$, with

$$g=\frac{1}{\deg f_1\cdots\deg f_d}\det\left[\frac{\partial F_j}{\partial\zeta_i}:i,j=1,\dots,d\right],$$

which is *independent of* k now; in other words, there exist polynomials $\hat g_k$, $k=1,\dots,N$, such that $\deg\hat g_k<\deg g$ and $\det G_k=g+\hat g_k$. Moreover, g is a homogeneous polynomial of degree at most $\deg f_1+\cdots+\deg f_d-d$. Now, let p be any polynomial, then

$$Kp=\sum_{j=1}^{N} p(z_j)\frac{\det G_j(\cdot)}{\det G_j(z_j)}=g\sum_{j=1}^{N}\frac{p(z_j)}{\det G_j(z_j)}+\sum_{j=1}^{N}\frac{p(z_j)}{\det G_j(z_j)}\hat g_j. \tag{8}$$

Combining (8) with (6) then yields the existence of polynomials $q_1,\dots,q_d$ such that

$$p=g\sum_{j=1}^{N}\frac{p(z_j)}{\det G_j(z_j)}+\sum_{j=1}^{N}\frac{p(z_j)}{\det G_j(z_j)}\hat g_j+\sum_{j=1}^{d} q_j f_j$$

and comparing homogeneous terms of degree $\deg g$ Kronecker realized that either, for any p such that $\deg p<\deg g$,

$$\sum_{j=1}^{N}\frac{p(z_j)}{\det G_j(z_j)}=0 \tag{9}$$

or there exist homogeneous polynomials $h_1,\dots,h_d$ such that

$$g=\sum_{j=1}^{d} h_j\det F_j. \tag{10}$$

The latter case, Eq. (10), says (in algebraic terminology) that there is a *syzygy* among the leading terms of the polynomials F_j, $j=1,\dots,d$, and is equivalent to the fact that $N<\deg f_1\cdots\deg f_d$, while (9) describes and even characterizes the *complete intersection case* that $N=\deg f_1\cdots\deg f_d$. In his paper, Kronecker also mentions that the condition (10) has been overlooked in [21]. Jacobi dealt there with the common zeros of two bivariate polynomials and derived *explicit* representations for the functional

$$[z_1,\dots,z_N]f:=\sum_{j=1}^{N}\frac{f(z_j)}{\det G_j(z_j)}, \tag{11}$$

which behaves very much like a divided difference, since it is a combination of point evaluations which, provided that (9) hold true, annihilates $\Pi^d_{\deg g-1}$.

In addition, Kronecker refers to a paper [6] which he says treats the case of symmetric functions, probably elementary symmetric polynomials. Unfortunately, this paper is unavailable to us so far.

3. Bivariate tables, the natural approach

Only very few research papers on multivariate polynomial interpolation were published during the first part of this century. In the classical book *Interpolation* [45], where one section (Section 19) is devoted to this topic, the author only refers to two related papers, recent at that time (1927), namely [27,28]. The latter one [28], turned out to be inaccessible to us, unfortunately, but it is not difficult to guess that it might have pursued a tensor product approach, because this is the unique point of view of [45] (see also [31]).

The formulas given in [27] are Newton formulas for tensor product interpolation in two variables, and the author, Narumi, claims (correctly) that they can be extended to "many variables". Since it is a tensor product approach, the interpolation points are of the form (x_i, y_j), $0 \leqslant i \leqslant m$, $0 \leqslant j \leqslant n$, with x_i, y_j arbitrarily distributed on the axes OX and OY, respectively. Bivariate divided differences for these sets of points are obtained in [27], by recurrence, separately for each variable. With the usual notations, the interpolation formula from [27] reads as

$$p(x,y) = \sum_{i=0}^{m} \sum_{j=0}^{n} f[x_0,\ldots,x_i; y_0,\ldots,y_j] \prod_{h=0}^{i-1}(x-x_h) \prod_{k=0}^{j-1}(y-x_k), \tag{12}$$

where empty products have the value 1. Remainder formulas based on the mean value theorem are also derived recursively from the corresponding univariate error formulas in [27]. For f sufficiently smooth there exist values ξ, ξ', η, η' such that

$$R(x,y) = \frac{\partial^{m+1} f(\xi,y)}{\partial x^{m+1}} \frac{\prod_{h=0}^{m}(x-x_h)}{(m+1)!} + \frac{\partial^{n+1} f(x,\eta)}{\partial y^{n+1}} \frac{\prod_{k=0}^{n}(y-y_k)}{(n+1)!}$$
$$- \frac{\partial^{m+n+2} f(\xi',\eta')}{\partial x^{m+1}\partial y^{n+1}} \frac{\prod_{h=0}^{m}(x-x_h)}{(m+1)!} \frac{\prod_{k=0}^{n}(y-y_k)}{(n+1)!}. \tag{13}$$

The special case of equidistant points on both axes is particularly considered in [27], and since the most popular formulas at that time were based on finite differences with equally spaced arguments, Narumi shows how to extend Gauss, Bessel and Stirling univariate interpolation formulas for equidistant points to the bivariate case by tensor product. He also applies the formulas he obtained to approximate the values of bivariate functions, but he also mentions that some of his formulas had been already used in [49].

In [45], the Newton formula (12) is obtained in the same way, with the corresponding remainder formula (13). Moreover, Steffensen considers a more general case, namely when for each i, $0 \leqslant i \leqslant m$, the interpolation points are of the form $y_0,\ldots,y_{n_i}$, with $0 \leqslant n_i \leqslant n$. Now with a similar argument the interpolating polynomial becomes

$$p(x,y) = \sum_{i=0}^{m} \sum_{j=0}^{n_i} f[x_0,\ldots,x_i; y_0,\ldots,y_j] \prod_{h=0}^{i-1}(x-x_h) \prod_{k=0}^{j-1}(y-x_k) \tag{14}$$

with a slightly more complicated remainder formula. The most interesting particular cases occur when $n_i = n$, which is the Cartesian product considered above, and when $n_i = m - i$. This *triangular* case (triangular not because of the geometrical distribution of the interpolation points, but of the indices (i, j)), gives rise to the interpolating polynomial

$$p(x, y) = \sum_{i=0}^{m} \sum_{j=0}^{m-i} f[x_0, \ldots, x_i; y_0, \ldots, y_j] \prod_{h=0}^{i-1}(x - x_h) \prod_{k=0}^{j-1}(y - x_k), \tag{15}$$

that is

$$p(x, y) = \sum_{0 \leqslant i+j \leqslant m} f[x_0, \ldots, x_i; y_0, \ldots, y_j] \prod_{h=0}^{i-1}(x - x_h) \prod_{k=0}^{j-1}(y - x_k). \tag{16}$$

Steffensen refers for this formula to Biermann's lecture notes [4] from 1905, and actually it seems that Biermann has been the first who considered polynomial interpolation on the triangular grid in a paper [3] from 1903 (cf. [44]) in the context of cubature.

Since the triangular case corresponds to looking at the "lower triangle" of the tensor product situation only, this case can be resolved by tensor product methods. In particular, the respective error formula can be written as

$$R(x, y) = \sum_{i=0}^{m+1} \frac{\partial^{m+1} f(\xi_i, \eta_i)}{\partial x^i \partial y^{m+1-i}} \frac{\prod_{h=0}^{i-1}(x - x_h)}{i!} \frac{\prod_{k=0}^{m-i}(y - y_k)}{(m - i + 1)!}. \tag{17}$$

In the case of Cartesian product Steffensen also provides the Lagrange formula for (12), which can be obviously obtained by tensor product of univariate formulas.

Remainder formulas based on intermediate points (ξ_i, η_i) can be written in many different forms. For them we refer to Stancu's paper [44] which also contains a brief historical introduction where the author refers, among others, to [3,15,27,40,41]. Multivariate remainder formulas with Peano (spline) kernel representation, however, have not been derived until very recently in [42] and, in particular, in [43] which treats the triangular situation.

4. Salzer's papers: from bivariate tables to general sets

In 1944, Salzer [33] considered the interpolation problem at points of the form $(x_1 + s_1 h_1, \ldots, x_n + s_n h_n)$ where

(i) $(x_1, \ldots, x_n)$ is a given point in $\mathbb{R}^n$,
(ii) $h_1, \ldots, h_n$ are given real numbers,
(iii) $s_1, \ldots, s_n$ are nonnegative integers summing up to m.

This is the multivariate extension of the triangular case (16) for equally spaced arguments, where finite differences can be used. Often, different names are used for the classical Newton interpolation formula in the case of equally spaced arguments using forward differences: Newton–Gregory, Harriot–Briggs, also known by Mercator and Leibnitz, etc. See [18] for a nice discussion of this issue. In [33], Salzer takes the natural multivariate extension of this formula considering the polynomial $q(t_1, \ldots, t_n) := p(x_1 + t_1 h_1, \ldots, x_n + t_n h_n)$ of total degree not greater than m in the variables $t_1, \ldots, t_n$, which interpolates a function $f(x_1 + t_1 h_1, \ldots, x_n + t_n h_n)$ at the points corresponding to $t_i = s_i$, $i = 1, \ldots, n$, where the s_i are all nonnegative integers such that $0 \leqslant s_1 + \cdots + s_n \leqslant m$. The formula, which is called in [33] a *multiple Gregory–Newton formula*, is rewritten there in terms of the values of the function f at the interpolation points, i.e., in the form

$$q(t_1,\ldots,t_n) = \sum_{s_1+\cdots+s_n \leqslant m} \binom{t_1}{s_1} \cdots \binom{t_n}{s_n} \binom{m - t_1 - \cdots - t_n}{m - s_1 - \cdots - s_n} f(x_1 + s_1 h_1, \ldots, x_n + s_n h_n). \quad (18)$$

Note that (18) is the Lagrange formula for this interpolation problem. Indeed, each function

$$\binom{t_1}{s_1} \cdots \binom{t_n}{s_n} \binom{m - t_1 - \cdots - t_n}{m - s_1 - \cdots - s_n} \quad (19)$$

is a polynomial in $t_1,\ldots,t_n$ of total degree m which vanishes at all points $(t_1,\ldots,t_n)$ with t_i nonnegative integers $0 \leqslant t_1 + \cdots + t_n \leqslant m$, except at the point $(s_1,\ldots,s_n)$, where it takes the value 1. In particular, for $n = 1$ we get the well-known univariate Lagrange polynomials

$$\ell_s(t) = \binom{t}{s} \binom{m-t}{m-s} = \prod_{\substack{0 \leqslant i \leqslant m,\\ i \neq s}} \frac{t-i}{s-i}$$

for $s = 0,\ldots,m$.

Salzer used these results in [34] to compute tables for the polynomials (18) and, some years later in [35], he studied in a similar form how to get the Lagrange formula for the more general case of formula (16), even starting with this formula. He obtained the multivariate Lagrange polynomials by a rather complicated expression involving the univariate ones.

It should be noted that several books related to computations and numerical methods published around this time include parts on multivariate interpolation to some extent, surprisingly, more than most of the recent textbooks in Numerical Analysis. We have already mentioned Steffensen's book [45], but we should also mention Whittaker and Robinson [51, pp. 371–374], Mikeladze [26, Chapter XVII] and especially Kunz [23, pp. 248–274], but also Isaacson and Keller [20, pp. 294–299] and Berezin and Zhidkov [2, pp. 156–194], although in any of them not really much more than in [45] is told.

In [36,37], Salzer introduced a concept of bivariate divided differences abandoning the idea of iteration for each variable x and y taken separately. Apparently, this was the first time (in spite of the similarity with (11)), that bivariate divided differences were explicitly defined for irregularly distributed sets of points. Divided differences with repeated arguments are also considered in [37] by coalescence of the ones with different arguments. Since [36] was just a first attempt of [37], we only explain the latter one. Salzer considers the set of monomials $\{x^i y^j\}$, with i,j nonnegative integers, ordered in a graded lexical term order, that is,

$$(i,j) < (h,k) \Leftrightarrow i+j < h+k \quad \text{or} \quad i+j = h+k,\ i > h. \quad (20)$$

Hence, the monomials are listed as

$$\{1, x, y, x^2, xy, y^2, x^3, \ldots\}. \quad (21)$$

For any set of $n+1$ points (x_i, y_i), Salzer defines the associated divided difference

$$[01\ldots n]f := \sum_{k=0}^{n} A_k f(x_k, y_k), \quad (22)$$

choosing the coefficients A_k in such a form that (22) vanishes when f is any of the first n monomials of list (21) and takes the value 1 when f is the $(n+1)$st monomial of that list. In other words, the coefficients A_k are the solution of the linear system

$$\sum_{k=0}^{n} A_k x_k^i y_k^j = 0, \quad x^i y^j \text{ any of the first } n \text{ monomials of (21)},$$

$$\sum_{k=0}^{n} A_k x_k^i y_k^j = 1, \quad x^i y^j \text{ the } (n+1)\text{th monomial of (21)}. \tag{23}$$

These generalized divided differences share some of the properties of the univariate ones but not all. Moreover, they have some limitations, for example, they exist only if the determinant of the coefficients in (23) is different from zero, and one has no control of that property in advance. On the other hand, observe that for example the simple divided difference with two arguments (x_0, y_0) and (x, y), which is

$$\frac{f(x,y) - f(x_0, y_0)}{x - x_0},$$

gives, when applied to $f(x, y) = xy$, the rational function

$$\frac{xy - x_0 y_0}{x - x_0}$$

and not a polynomial of lower degree. In fact, Salzer's divided differences did not have great success. Several other definitions of multivariate divided differences had appeared since then, trying to keep as many as possible of the good properties of univariate divided differences, cf. [16].

5. Reduction of a problem to other simpler ones

Around the 1950s an important change of paradigm happened in multivariate polynomial interpolation, as several people began to investigate more general distributions of points, and not only (special) subsets of Cartesian products. So, when studying cubature formulae [32], Radon observed the following in 1948: if a bivariate interpolation problem with respect to a set $T \subset \mathbb{R}^2$ of $\binom{k+2}{2}$ interpolation points is unisolvent in Π_k^2, and U is a set of $k+2$ points on an arbitrary straight line $\ell \subset \mathbb{R}^2$ such that $\ell \cap T = \emptyset$, then the interpolation problem with respect to $T \cup U$ is unisolvent in Π_{k+1}^2. Radon made use of this observation to build up point sets which give rise to unisolvent interpolation problems for Π_m recursively by degree. Clearly, these interpolation points immediately yield interpolatory cubature formulae.

The well-known Bézout theorem, cf. [50], states that two planar algebraic curves of degree m and n, with no common component, intersect each other at exactly mn points in an algebraic closure of the underlying field, counting multiplicities. This theorem has many interesting consequences for bivariate interpolation problems, extensible to higher dimensions. For example, no unisolvent interpolation problem in Π_n^2 can have more than $n+1$ collinear points. Radon's method in [32] is a consequence of this type of observations, and some other more recent results of different authors can also be deduced in a similar form, as we shall see later.

Another example of a result which shows the more general point of view taken in multivariate interpolation at that time is due to Thacher Jr. and Milne [47] (see also [48]). Consider two univariate interpolation problems in Π^1_{n-1}, with T_1, T_2 as respective sets of interpolation points, both of cardinality n. Assume that $T_1 \cap T_2$ has cardinality $n-1$, hence $T = T_1 \cup T_2$ has cardinality $n+1$. The univariate Aitken–Neville interpolation formula combines the solutions of the two smaller problems based on T_1 and T_2 to obtain the solution in Π^1_n of the interpolation problem with T as the underlying set of interpolation points. The main idea is to find a *partition of unity*, in this case affine polynomials ℓ_1, ℓ_2, i.e., $\ell_1 + \ell_2 = 1$, such that

$$\ell_1(T_2 \setminus T_1) = \ell_2(T_1 \setminus T_2) = 0$$

and then combine the solutions p_1, p_2 with respect to T_1, T_2, into the solution $\ell_1 p_1 + \ell_2 p_2$ with respect to T. This method was developed in the 1930s independently by Aitken and Neville with the goal to avoid the explicit use of divided differences in the computation of univariate Lagrange polynomial interpolants.

It was exactly this idea which Thatcher and Milne extended to the multivariate case in [47]. Let us sketch their approach in the bivariate case. For example, consider an interpolation problem with 10 interpolation points, namely, the set $T = \{(i,j) : 0 \leqslant i+j \leqslant 3\}$, where i,j are nonnegative integers, and the interpolation space Π^2_3. The solution p_T of this problem is obtained in [47] from the solutions $p_{T_k} \in \Pi^2_2$, $k = 1,2,3$, of the 3 interpolation problems with respect to the six-point sets $T_k \subset T$, $k = 1,2,3$, where

$$T_1 = \{(i,j) : 0 \leqslant i+j \leqslant 2\},$$

$$T_2 = \{(i,j) : (i,j) \in T,\ i > 0\},$$

$$T_3 = \{(i,j) : (i,j) \in T,\ j > 0\}.$$

Then,

$$p_T = \ell_1 p_{T_1} + \ell_2 p_{T_2} + \ell_3 p_{T_3},$$

where ℓ_k, $k = 1,2,3$ are appropriate polynomials of degree 1. In fact, in this case these polynomials are the barycentric coordinates relative to the simplex (0, 0), (3, 0), (0, 3) and thus a partition of unity. In [47] the problem is studied in d variables and in that case $d+1$ "small" problems, with respective interpolation sets T_k, $k = 1,\ldots,d$, with a simplicial structure (the analogue of the triangular grid), are used to obtain the solution of the full problem with $T = T_1 \cap \cdots \cap T_{d+1}$ as interpolation points.

In 1970, Guenter and Roetman [19], among other observations, made a very interesting remark, which connects to the Radon/Bézout context and deserves to be explained here. Let us consider a set T of $\binom{m+d}{d}$ points in $\mathbb{R}^d$, where exactly $\binom{m+d-1}{d-1}$ of these points lie on a hyperplane H. Then $T \setminus H$ consists of $\binom{m-1+d}{d}$ points. Let us denote by $\Pi^m_{d,H}$ the space of polynomials of Π^m_d with the variables restricted to H, which is isomorphic to Π^m_{d-1}. If the interpolation problems defined by the sets $T \setminus H$ and $T \cap H$ are unisolvent in the spaces Π^{m-1}_d and $\Pi^m_{d,H}$, respectively, then the interpolation problem defined by T is unisolvent in Π^m_d. In other words, the idea is to decompose, whenever possible, a problem of degree m and d variables into two simpler problems, one of degree m and $d-1$ variables and the other one with degree $m-1$ and d variables.

6. The finite element approach

In 1943, Courant [11] suggested a finite difference method applicable to boundary value problems arising from variational problems. It is considered one of the motivations of the finite element method, which emerged from the engineering literature along the 1950s. It is a variational method of approximation which makes use of the Rayleigh–Ritz–Galerkin technique. The method became very successful, with hundreds of technical papers published (see, e.g., the monograph [52]), even before its mathematical basis was completely understood at the end of the 1960s.

Involved in the process of the finite element method there are local polynomial interpolation problems, generally for polynomials of low degree, thus, with only few interpolation data. The global solution obtained by solving all the local interpolation problems is a piecewise polynomial of a certain regularity, depending on the amount and type of interpolation data in the common boundary between pieces. Some of the interest in multivariate polynomial interpolation along the 1960/1970s was due to this method. Among the most interesting mathematical papers of that time in Finite Elements, we can mention [53,5], see also the book [46] by Strang and Fix, but, in our opinion, the most relevant papers and book from the point of view of multivariate polynomial interpolation are due to Ciarlet et al., for example [7–9].

In 1972, Nicolaides [29,30] put the classical problem of interpolation on a simplicial grid of $\binom{m+d}{d}$ points of $\mathbb{R}^d$, regularly distributed, forming what he called a *principal lattice*, into the finite element context. He actually used barycentric coordinates for the Lagrange formula, and moreover gave the corresponding error representations, see also [7]. However, much of this material can already be found in [3]. In general, taking into account that these results appeared under different titles, in a different context and in journals not accessible everywhere, it is not so surprising any more, how often the basic facts on the interpolation problem with respect to the simplicial grid had been rediscovered.

7. Hermite problems

The use of partial or directional derivatives as interpolation data in the multivariate case had not received much attention prior to the finite element method, where they were frequently used. It seems natural to approach partial derivatives by coalescence, as in univariate Hermite interpolation problems. However, things are unfortunately much more complicated in several variables. As it was already pointed out by Salzer and Kimbro [39] in 1958, the Hermite interpolation problem based on the values of a bivariate function $f(x,y)$ at two distinct points $(x_1,y_1),(x_2,y_2)$ and on the values of the partial derivatives $\partial f/\partial x$, $\partial f/\partial x$ at each of these two points is not solvable in the space Π_2^2 for any choice of points, although the number of interpolation conditions coincides with the dimension of the desired interpolation space. Some years later, Ahlin [1] circumvented some of these problems by using a tensor product approach: k^2 derivatives $\partial^{p+q}f/\partial x^p \partial y^q$ with $0 \leqslant p,q \leqslant k-1$ are prescribed at the n^2 points of a Cartesian product. The interpolation space is the one spanned by $x^\alpha y^\beta$ with $0 \leqslant \alpha, \beta \leqslant nk-1$ and a formula for the solution is easily obtained.

We must mention that Salzer came back to bivariate interpolation problems with derivatives in [38] studying *hyperosculatory interpolation* over Cartesian grids, that is, interpolation problems where all partial derivatives of first and second order and the value of the function are known at the interpolation points. Salzer gave some special configurations of points which yield solvability of this type of interpolation problem in an appropriate polynomial space and also provided the corresponding remainder formulae.

Nowadays, Hermite and Hermite–Birkhoff interpolation problems have been studied much more systematically, see [16,25] for references.

8. Other approaches

In 1966, Coatmelec [10] studied the approximation of functions of several variables by linear operators, including interpolation operators. At the beginning of the paper, he only considered interpolation operators based on values of point evaluations of the function, but later he also used values of derivatives. In this framework he obtained some qualitative and quantitative results on the approximation order of polynomial interpolation. At the end of [10], Coatmelec also includes some examples in $\mathbb{R}^2$ of points which are distributed irregularly along lines: $n+1$ of the points on a line r_0, n of them on another line r_1, but not on r_0, and so on until 1 point is chosen on a line r_n but not on $r_0 \cup \cdots \cup r_{n-1}$. He then points out the unisolvence of the corresponding interpolation problem in Π_n^2 which is, in fact, again a consequence of Bézout's theorem as in [32].

In 1971, Glaeser [17] considers Lagrange interpolation in several variables from an abstract algebraic/analytic point of view and acknowledges the inconvenience of working with particular systems of interpolation points due to the possibility of the nonexistence of a solution, in contrast to the univariate case. This is due to the nonexistence of polynomial spaces of dimension $k > 1$ in more than one variable such that the Lagrange interpolation problem has a unique solution for any system of k interpolation points. In other words, there are no nontrivial Haar (or Chebychev) spaces any more for two and more variables, cf. [12] or [24]. In [17], polynomial spaces with dimension greater than the number of interpolation conditions are considered in order to overcome this problem. Glaeser investigated these *underdetermined* systems which he introduced as *interpolation schemes* in [17] and also studied the problem of how to particularize the affine space of all solutions of a given interpolation problem in order to obtain a unique solution. This selection process is done in such a way that it controls the variation of the solution when two systems of interpolation points are very "close" to each other, with the goal to obtain a continuous selection process.

Acknowledgements

1. We thank Carl de Boor for several references, in particular, for pointing out to us the paper [32] with the result mentioned at the beginning of Section 4, related to Bezout's theorem. We are also grateful the help of Elena Ausejo, from the group of History of Sciences of the University of Zaragoza, in the search of references.
2. M. Gasca has been partially supported by the Spanish Research Grant PB96 0730.
3. T. Sauer was supported by a DFG Heisenberg fellowship, Grant SA 627/6.

References

[1] A.C. Ahlin, A bivariate generalization of Hermite's interpolation formula, Math. Comput. 18 (1964) 264–273.
[2] I.S. Berezin, N.P. Zhidkov, Computing Methods, Addison-Wesley, Reading, MA, 1965 (Russian version in 1959).
[3] O. Biermann, Über näherungsweise Kubaturen, Monatshefte Math. Phys. 14 (1903) 211–225.
[4] O. Biermann, Vorlesungen über Mathematische Näherungsmethoden, Vieweg, Braunschweig, 1905.
[5] G. Birkhoff, M.H. Schultz, R.S. Varga, Piecewise Hermite interpolation in one and two variables with applications to partial differential equations, Numer. Math. 11 (1968) 232–256.
[6] W. Borchardt, Über eine Interpolationsformel für eine Art symmetrischer Funktionen und deren Anwendung, Abh. d. Preuß. Akad. d. Wiss. (1860) 1–20.
[7] P.G. Ciarlet, The Finite Element Method for Elliptic Problems, North-Holland, Amsterdam, 1978.
[8] P.G. Ciarlet, P.A. Raviart, General Lagrange and Hermite interpolation in R^n with applications to finite element methods, Arch. Rational Mech. Anal. 46 (1972) 178–199.
[9] P.G. Ciarlet, C. Wagschal, Multipoint Taylor formulas and applications to the finite element method, Numer. Math. 17 (1971) 84–100.
[10] C. Coatmelec, Approximation et interpolation des fonctions differentiables de plusieurs variables, Ann. Sci. Ecole Norm. Sup. 83 (1966) 271–341.
[11] R. Courant, Variational methods for the solution of problems of equilibrium and vibrations, Bull. Amer. Math. Soc. 49 (1943) 1–23.
[12] P.J. Davis, Interpolation and Approximation, Blaisdell, Walthan, MA, 1963 (2nd Edition, Dover, New York, 1975).
[13] Encyklopädie der mathematischen Wissenschaften, Teubner, Leipzig, pp. 1900–1904.
[14] Encyclopédie des Sciences Mathematiques, Gauthier-Villars, Paris, 1906.
[15] I.A. Ezrohi, General forms of the remainder terms of linear formulas in multidimensional approximate analysis I, II Mat. Sb. 38 (1956) 389–416 and 43 (1957) 9–28 (in Russian).
[16] M. Gasca, T. Sauer, Multivariate polynomial interpolation, Adv. Comput. Math., 12 (2000) 377–410.
[17] G. Glaeser, L'interpolation des fonctions differentiables de plusieurs variables, in: C.T.C. Wall (Ed.), Proceedings of Liverpool Singularities Symposium II, Lectures Notes in Mathematics, Vol. 209, Springer, Berlin, 1971, pp. 1–29.
[18] H.H. Goldstine, A History of Numerical Analysis from the 16th Through the 19th Century, Springer, Berlin, 1977.
[19] R.B. Guenter, E.L. Roetman, Some observations on interpolation in higher dimensions, Math. Comput. 24 (1970) 517–521.
[20] E. Isaacson, H.B. Keller, Analysis of Numerical Methods, Wiley, New York, 1966.
[21] C.G.J. Jacobi, Theoremata nova algebraica circa systema duarum aequationum inter duas variabiles propositarum, Crelle J. Reine Angew. Math. 14 (1835) 281–288.
[22] L. Kronecker, Über einige Interpolationsformeln für ganze Funktionen mehrerer Variabeln. Lecture at the academy of sciences, December 21, 1865, in: H. Hensel (Ed.), L. Kroneckers Werke, Vol. I, Teubner, Stuttgart, 1895, pp. 133–141. (reprinted by Chelsea, New York, 1968).
[23] K.S. Kunz, Numerical Analysis, McGraw-Hill, New York, 1957.
[24] G.G. Lorentz, Approximation of Funtions, Chelsea, New York, 1966.
[25] R. Lorentz, Multivariate Hermite interpolation by algebaic polynomials: a survey, J. Comput. Appl. Math. (2000) this volume.
[26] S.E. Mikeladze, Numerical Methods of Mathematical Analysis, Translated from Russian, Office of Tech. Services, Department of Commerce, Washington DC, pp. 521–531.
[27] S. Narumi, Some formulas in the theory of interpolation of many independent variables, Tohoku Math. J. 18 (1920) 309–321.
[28] L. Neder, Interpolationsformeln für Funktionene mehrerer Argumente, Skandinavisk Aktuarietidskrift (1926) 59.
[29] R.A. Nicolaides, On a class of finite elements generated by Lagrange interpolation, SIAM J. Numer. Anal. 9 (1972) 435–445.
[30] R.A. Nicolaides, On a class of finite elements generated by Lagrange interpolation II, SIAM J. Numer. Anal. 10 (1973) 182–189.
[31] K. Pearson, On the construction of tables and on interpolation, Vol. 2, Cambridge University Press, Cambridge, 1920.
[32] J. Radon, Zur mechanischen Kubatur, Monatshefte Math. Physik 52 (1948) 286–300.

[33] H.E. Salzer, Note on interpolation for a function of several variables, Bull. AMS 51 (1945) 279–280.
[34] H.E. Salzer, Table of coefficients for interpolting in functions of two variables, J. Math. Phys. 26 (1948) 294–305.
[35] H.E. Salzer, Note on multivariate interpolation for unequally spaced arguments with an application to double summation, J. SIAM 5 (1957) 254–262.
[36] H.E. Salzer, Some new divided difference algorithms for two variables, in: R.E. Langer (Ed.), On Numerical Approximation, University of Wisconsin Press, Madison, 1959, pp. 61–98.
[37] H.E. Salzer, Divided differences for functions of two variables for irregularly spaced arguments, Numer. Math. 6 (1964) 68–77.
[38] H.E. Salzer, Formulas for bivariate hyperosculatory interpolation, Math. Comput. 25 (1971) 119–133.
[39] H.E. Salzer, G.M. Kimbro, Tables for Bivariate Osculatory Interpolation over a Cartesian Grid, Convair Astronautics, 1958.
[40] A. Sard, Remainders: functions of several variables, Acta Math. 84 (1951) 319–346.
[41] A. Sard, Remainders as integrals of partial derivatives, Proc. Amer. Math. Soc. 3 (1952) 732–741.
[42] T. Sauer, Yuan Xu, On multivariate Lagrange interpolation, Math. Comput. 64 (1995) 1147–1170.
[43] T. Sauer, Yuan Xu, A case study in multivariate Lagrange interpolation, in: S.P. Singh (Ed.), Approximation Theory, Wavelets and Applications, Kluwer Academic Publishers, Dordrecht, 1995, pp. 443–452.
[44] D.D. Stancu, The remainder of certain linear approximation formulas in two variables, J. SIAM Numer. Anal. 1 (1964) 137–163.
[45] I.F. Steffensen, Interpolation, Chelsea, New York, 1927 (2nd Edition, 1950).
[46] G. Strang, G.J. Fix, An Analysis of the Finite Element Method, Prentice-Hall, Englewood Cliffs, NJ, 1973.
[47] H.C. Thacher Jr., W.E. Milne, Interpolation in several variables, J. SIAM 8 (1960) 33–42.
[48] H.C. Thacher Jr., Derivation of interpolation formulas in several independent variables, Ann. N.Y. Acad. Sci. 86 (1960) 758–775.
[49] T.N. Thiele, Interpolationsrechnung, Teubner, Leipzig, 1909.
[50] R.S. Walker, Algebraic Curves, Springer, Berlin, 1978.
[51] E.T. Whittaker, G. Robinson, Calculus of Observations, 4th Edition, Blackie and Sons, London, 1944.
[52] O.C. Zienkiewicz, The Finite Element Method in Structural and Continuum Mechanics, McGraw-Hill, London, 1967.
[53] M. Zlamal, On the finite element method, Numer. Math. 12 (1968) 394–409.

ELSEVIER

Journal of Computational and Applied Mathematics 122 (2000) 37–50

JOURNAL OF COMPUTATIONAL AND APPLIED MATHEMATICS

www.elsevier.nl/locate/cam

Elimination techniques: from extrapolation to totally positive matrices and CAGD

M. Gasca *, G. Mühlbach

Department of Applied Mathematics, University of Zaragoza, 5009 Zaragoza, Spain

Received 20 May 1999; received in revised form 22 September 1999

Abstract

In this survey, we will show some connections between several mathematical problems such as extrapolation, linear systems, totally positive matrices and computer-aided geometric design, with elimination techniques as the common tool to deal with all of them. © 2000 Elsevier Science B.V. All rights reserved.

1. Introduction

Matrix elimination techniques are basic tools in many mathematical problems. In this paper we will show their crucial role in some results that various authors with us have obtained in two problems apparently distant: extrapolation and computer-aided geometric design (CAGD). A brief overview of how things were developed over time will show that, once again, two results which are apparently far from each other, even obtained by different groups in different countries, are the natural consequence of a sequence of intermediate results.

Newton's interpolation formula is a classical tool for constructing an interpolating polynomial by recurrence, by using divided differences. In the 1930s, Aitken [1] and Neville [52] derived independently of each other algorithms to compute the interpolating polynomial from the solutions of two simpler interpolation problems, avoiding the explicit use of divided differences. Some papers, [38,46] among others, extended both approaches at the beginning of the 1970s, to the more general setting of Chebyshev systems. Almost simultaneously, extrapolation methods were being studied and extended by several authors, as Schneider [54], Brezinski [4,5,7], Håvie [31–33], Mühlbach [39–42,48] and Gasca and López-Carmona [19]. For a historical overview of extrapolation methods confer Brezinski's contribution [6] to this volume and the book [8]. It must be remarked that the

* Corresponding author.

E-mail address: gasca@posta.unizar.es (M. Gasca).

PII: S0377-0427(00)00356-3

techniques used by these authors were different, and that frequently the results obtained using one of these techniques induced some progress in the other ones, in a very cooperative form.

However, it is clear that the basic role in all these papers was played by elimination techniques. In [21] we studied general elimination strategies, where one strategy which we called Neville elimination proved to be well suited to work with some special classes of matrices, in particular *totally positive matrices* (that are matrices with all subdeterminants nonnegative).

This was the origin of a series of papers [24–27] where the properties of Neville elimination were carefully studied and its application to totally positive matrices allowed a much better knowledge of these matrices. Since one of the applications of totally positive matrices is CAGD, the results obtained for them have given rise in the last years to several other papers as [28,11,12]. In [11,12] Carnicer and Peña proved the optimality in their respective spaces of some well-known function bases as Bernstein polynomials and B-splines in the context of shape preserving representations. Neville elimination has appeared, once again, as a way to construct other bases with similar properties.

2. Extrapolation and Schur complement

A k-tuple $L=(\ell_1,\dots,\ell_k)$ of natural numbers, with $\ell_1<\cdots<\ell_k$, will be called an *index list* of *length* k over $\mathbb{N}$. For $I=(i_1,\dots,i_m)$ and $J=(j_1,\dots,j_n)$ two index lists over $\mathbb{N}$, we write $I\subset J$ iff every element of I is an element of J. Generally, we shall use for index lists the same notations as for sets emphasizing that $I\setminus J, I\cap J, I\cup J\dots$ always have to be ordered as above.

Let $A=(a_i^j)$ be a real matrix and $I=(i_1,\dots,i_m)$ and $J=(j_1,\dots,j_n)$ index lists contained, repectively, in the index lists of rows and columns of A. By

$$A\begin{pmatrix}J\\I\end{pmatrix}=A\begin{pmatrix}j_1,\dots,j_n\\i_1,\dots,i_m\end{pmatrix}=(a_{i_\mu}^{j_\nu})_{\mu=1,\dots,m}^{\nu=1,\dots,n}\in\mathbb{R}^{m\times n},$$

we denote the *submatrix* of A with list of rows I and list of columns J.

If $I^\circ, I^{\circ\prime}$ and J°, $J^{\circ\prime}$ are partitions of I and J, respectively, i.e., $I^\circ\cup I^{\circ\prime}=I$, $I^\circ\cap I^{\circ\prime}=\emptyset$, $J^\circ\cup J^{\circ\prime}=J$, $J^\circ\cap J^{\circ\prime}=\emptyset$, we represent $A(^J_I)$ in a corresponding partition

$$A\begin{pmatrix}J\\I\end{pmatrix}=\begin{pmatrix}A\begin{pmatrix}J^\circ\\I^\circ\end{pmatrix} & A\begin{pmatrix}J^{\circ\prime}\\I^\circ\end{pmatrix}\\ A\begin{pmatrix}J^\circ\\I^{\circ\prime}\end{pmatrix} & A\begin{pmatrix}J^{\circ\prime}\\I^{\circ\prime}\end{pmatrix}\end{pmatrix}. \tag{1}$$

If $m=n$, then by

$$A\begin{vmatrix}J\\I\end{vmatrix}:=\det A\begin{pmatrix}J\\I\end{pmatrix}=A\begin{vmatrix}j_1,\dots,j_m\\i_1,\dots,i_m\end{vmatrix},$$

we denote the determinant of $A(^J_I)$ which is called a *subdeterminant* of A. Throughout we set $A|^{\emptyset}_{\emptyset}|:=1$.

Let $N\in\mathbb{N}$, $I:=(1,2,\dots,N+1)$ and $I^\circ:=(1,2,\dots,N)$. By a prime we denote ordered complements with respect to I. Given elements $f_1,\dots,f_N$ and $f=:f_{N+1}$ of a linear space E over $\mathbb{R}$, elements $L_1,\dots,L_N$ and $L=:L_{N+1}$ of its dual E^*, consider the problem of finding

$$\langle L, p_1^N(f)\rangle, \tag{2}$$

where $p = p_1^N(f) = c_1 \cdot f_1 + \cdots + c_N \cdot f_N$ satisfies the *interpolation conditions*

$$\langle L_i, p\rangle = \langle L_i, f\rangle \quad i \in I^\circ. \tag{3}$$

Here $\langle \cdot, \cdot \rangle$ means duality between E^* and E. If we write

$$A\begin{pmatrix} j \\ i \end{pmatrix} := \langle L_i, f_j\rangle \quad \text{for } i, j \in I, \quad (i,j) \neq (N+1, N+1),$$

and $\boldsymbol{c}$ is the vector of components c_i, this problem is equivalent to solving the bordered system (cf. [16])

$$B \cdot \boldsymbol{x} = \boldsymbol{y} \quad \text{where} \quad B = \begin{pmatrix} A\begin{pmatrix} I^\circ \\ I^\circ \end{pmatrix} & \mathbf{0} \\ A\begin{pmatrix} I^\circ \\ N+1 \end{pmatrix} & 1 \end{pmatrix}, \quad \boldsymbol{x} = \begin{pmatrix} \boldsymbol{c} \\ \xi \end{pmatrix}, \quad \boldsymbol{y} = \begin{pmatrix} A\begin{pmatrix} N+1 \\ I^\circ \end{pmatrix} \\ A\begin{pmatrix} N+1 \\ N+1 \end{pmatrix} \end{pmatrix}. \tag{4}$$

Assuming $A\binom{I^\circ}{I^\circ}$ nonsingular this can be solved by eliminating the components of $\boldsymbol{c}$ in the last equation by adding a suitable linear combination of the first N equations of (4) to the last one, yielding one equation for one unknown, namely ξ:

$$\xi = A\begin{pmatrix} N+1 \\ N+1 \end{pmatrix} - A\begin{pmatrix} I^\circ \\ N+1 \end{pmatrix} \cdot A\begin{pmatrix} I^\circ \\ I^\circ \end{pmatrix}^{-1} A\begin{pmatrix} N+1 \\ I^\circ \end{pmatrix}. \tag{5}$$

Considering the effect of this block elimination step on the matrix

$$A = \begin{pmatrix} A\begin{pmatrix} I^\circ \\ I^\circ \end{pmatrix} & A\begin{pmatrix} N+1 \\ I^\circ \end{pmatrix} \\ A\begin{pmatrix} I^\circ \\ N+1 \end{pmatrix} & A\begin{pmatrix} N+1 \\ N+1 \end{pmatrix} \end{pmatrix}, \tag{6}$$

we find it transformed to

$$\tilde{A} = \begin{pmatrix} A\begin{pmatrix} I^\circ \\ I^\circ \end{pmatrix} & A\begin{pmatrix} N+1 \\ I^\circ \end{pmatrix} \\ & \xi \end{pmatrix}.$$

If we take

$$A\begin{pmatrix} N+1 \\ N+1 \end{pmatrix} := 0, \tag{7}$$

then we have

$$\xi = -\langle L, p_1^N(f)\rangle. \tag{8}$$

On the other hand, if instead of (7) we take

$$A\begin{pmatrix} N+1 \\ N+1 \end{pmatrix} := \langle L_{N+1}, f_{N+1}\rangle, \tag{9}$$

then, in this frame, we get

$$\xi = \langle L, r_1^N(f)\rangle, \tag{10}$$

where

$$r_1^N(f) := f - p_1^N(f)$$

is the interpolation remainder.

If the systems $(f_1,\ldots,f_N)$ and $(L_1,\ldots,L_N)$ are independent of f and L then these problems are called *general linear extrapolation* problems, and if one or both do depend on $f=f_{N+1}$ or $L=L_{N+1}$ they are called problems of *quasilinear extrapolation.*

Observe, that with regard to determinants the block elimination step above is an elementary operation leaving the value of $\det A$ unchanged. Hence

$$\xi = \frac{\det A\begin{pmatrix} I \\ I \end{pmatrix}}{\det A\begin{pmatrix} I^\circ \\ I^\circ \end{pmatrix}},$$

which is known as the Schur complement of $A\binom{I^\circ}{I^\circ}$ in $A\binom{I}{I}$. This concept, introduced in [34,35] has found many applications in Linear Algebra and Statistics [13,53]. It may be generalized in different ways, see, for example, [21,22,44] where we used the concept of general elimination strategy which is explained in the next section.

3. Elimination strategies

In this section and the next two let $k,m,n\in\mathbb{N}$ such that $k+m=n$ and $I=(1,\ldots,n)$. Given a square matrix $A=A\binom{I}{I}$ over $\mathbb{R}$, how can we simplify $\det A$ by elementary operations, not altering the value of $\det A$, producing zeros in prescribed columns, e.g. in columns 1 to k?. Take a permutation of all rows, $M=(m_1,\ldots,m_n)$ say, then look for a linear combination of k rows from $(m_1,\ldots,m_{n-1})$ which, when added to row m_n, will produce zeros in columns 1 to k. Then add to row m_{n-1} a linear combination of k of its predecessors in M, to produce zeros in columns 1 to k, etc. Finally, add to row m_{k+1} a suitable linear combination of rows $m_1,\ldots,m_k$ to produce zeros in columns 1 to k. Necessarily,

$$A\left|\begin{matrix} 1,\ldots,k \\ j_1^r,\ldots,j_k^r \end{matrix}\right| \neq 0$$

is assumed when a linear combination of rows $j_1^r,\ldots,j_k^r$ is added to row m_r $(r=n,n-1,\ldots,k+1)$ to generate zeros in columns 1 to k, and $j_q^r < m_r$ $(q=1,\ldots,k;\ r=n,n-1,\ldots,k+1)$ in order that in each step an elementary operation will be performed.

Let us give a formal description of this general procedure. Suppose that (I_s,I_s°) $(s=1,\ldots,m)$ are pairs of ordered index lists of length $k+1$ and k, respectively, over a basic index list M with $I_s^\circ\subset I_s$. Then the family

$$\Sigma := ((I_s,I_s^\circ))_{s=1,\ldots,m}$$

will be called a *(k,m)-elimination strategy* over $I := I_1\cup\cdots\cup I_m$ provided that for $s=2,\ldots,m$

(i) $\mathrm{card}(I_1\cup\cdots\cup I_s)=k+s$,

(ii) $I_s^\circ\subset I_s\cap(I_1\cup\cdots\cup I_{s-1})$.

By $E(k,m,I)$ we denote the set of all (k,m)-*elimination strategies* over I. $I^\circ := I_1^\circ$ is called the *basic index list* of the strategy Σ. For each s, the zeros in the row $\alpha_\Sigma(s) := I_s \setminus I_s^\circ$ are produced with the rows of I_s°. For shortness, we shall abbreviate the phrase "elimination strategy" by e.s. Notice that, when elimination is actually performed, it is done in the reverse ordering: first in row $\alpha_\Sigma(m)$, then in row $\alpha_\Sigma(m-1)$, etc.

The simplest example of e.s. over $I = (1,\dots,m+k)$, is *Gauss elimination*:

$$\Gamma = ((G_s, G_s^\circ))_{s=1,\dots,m}, \quad G^\circ = G_s^\circ = \{1,\dots,k\}, \quad G_s = G^\circ \cup \{k+s\}. \tag{11}$$

For this strategy it is irrelevant in which order elimination is performed. This does not hold for another useful strategy over I:

$$\mathcal{N} = ((N_s, N_s^\circ))_{s=1,\dots,m} \tag{12}$$

with $N_s^\circ = (s,\dots,s+k-1), N_s = (s,\dots,s+k)$, $s = 1,\dots,m$, which we called [21,43,44] the *Neville* (k,m)–e.s. Using this strategy elimination must be performed from bottom to top. The reason for the name Neville is their relationship with Neville interpolation algorithm, based on consecutivity, see [43,23].

4. Generalized Schur complements

Suppose that $\Sigma = ((I_s, I_s^\circ))_{s=1,\dots,m} \in E(k,m,I)$ and that $\mathcal{K}^\circ \subset I$ is a fixed index list of length k. We assume that the submatrices $A(\begin{smallmatrix}\mathcal{K}^\circ \\ I_s^\circ\end{smallmatrix})$ of a given matrix $A = A(\begin{smallmatrix}I \\ I\end{smallmatrix}) \in \mathbb{R}^{n\times n}$ are nonsingular for $s = 1,\dots,m$. Then the elimination strategy transforms A into the matrix $\tilde{A}$ which, partitioned with respect to $I^\circ \cup I^{\circ\prime} = I, \mathcal{K}^\circ \cup \mathcal{K}^{\circ\prime} = I$, can be written as

$$\tilde{A} = \begin{pmatrix} \tilde{A}\begin{pmatrix}\mathcal{K}^\circ \\ I^\circ\end{pmatrix} & \tilde{A}\begin{pmatrix}\mathcal{K}^{\circ\prime} \\ I^\circ\end{pmatrix} \\ 0 & \tilde{A}\begin{pmatrix}\mathcal{K}^{\circ\prime} \\ I^{\circ\prime}\end{pmatrix} \end{pmatrix}$$

with

$$\tilde{A}\begin{pmatrix}\mathcal{K}^\circ \\ I^\circ\end{pmatrix} = A\begin{pmatrix}\mathcal{K}^\circ \\ I^\circ\end{pmatrix}, \quad \tilde{A}\begin{pmatrix}\mathcal{K}^{\circ\prime} \\ I^\circ\end{pmatrix} = A\begin{pmatrix}\mathcal{K}^{\circ\prime} \\ I^\circ\end{pmatrix}.$$

The submatrix $\tilde{S} := \tilde{A}(\begin{smallmatrix}\mathcal{K}^{\circ\prime} \\ I^{\circ\prime}\end{smallmatrix})$ of $\tilde{A}$ is called the *Schur complement* of $A(\begin{smallmatrix}\mathcal{K}^\circ \\ I^\circ\end{smallmatrix})$ in A with respect to the e.s. Σ and the column list $\mathcal{K}^\circ$, and is also denoted by

$$\tilde{S} = \left[A\begin{pmatrix}I \\ I\end{pmatrix} \Big/ A\begin{pmatrix}\mathcal{K}^\circ \\ I^\circ\end{pmatrix}\right]_\Sigma.$$

When $\Sigma = \Gamma$ as in (11) and $\mathcal{K}^\circ = \{1,\dots,k\}$, then $\tilde{S}$ is the classical Schur complement, which can also be written as

$$\tilde{A}\begin{pmatrix}\mathcal{K}^{\circ\prime} \\ I^{\circ\prime}\end{pmatrix} = A\begin{pmatrix}\mathcal{K}^{\circ\prime} \\ I^{\circ\prime}\end{pmatrix} - A\begin{pmatrix}\mathcal{K}^\circ \\ I^{\circ\prime}\end{pmatrix} A\begin{pmatrix}\mathcal{K}^\circ \\ I^\circ\end{pmatrix}^{-1} A\begin{pmatrix}\mathcal{K}^{\circ\prime} \\ I^\circ\end{pmatrix}.$$

When $\Sigma = \mathcal{N}$ is the Neville (k,m)–e.s. (12) and $\mathcal{K}^\circ = \{1,\dots,k\}$, then the rows of the Schur complement $\tilde{S} = \tilde{A}\binom{\mathcal{K}^{\circ\prime}}{I^{\circ\prime}}$ are

$$\tilde{A}\binom{\mathcal{K}^{\circ\prime}}{k+s} = A\binom{\mathcal{K}^{\circ\prime}}{k+s} - A\binom{\mathcal{K}^{\circ}}{k+s} A\binom{\mathcal{K}^{\circ}}{s,\dots,s+k-1}^{-1} A\binom{\mathcal{K}^{\circ\prime}}{s,\dots,s+k-1} \quad s=1,\dots,m.$$

Whereas, the Schur complement of a submatrix depends essentially on the elimination strategies used, its determinant does not! There holds the following generalization of Schur's classical determinantal identity [21,22,44]:

$$\det A\binom{I}{I} = (-1)^\beta \det A\binom{\mathcal{K}^\circ}{I^\circ} \det\left[A\binom{I}{I} \Big/ A\binom{\mathcal{K}^\circ}{I^\circ}\right]_\Sigma$$

for all e.s. $\Sigma \in E(k,m,I)$, where β is an integer depending only on Σ and $\mathcal{K}^\circ$.

Also, Sylvester's classical determinantal identity [55,56] has a corresponding generalization, see [18,21,22,43,44] for details. In the case of Gauss elimination we get Sylvester's classical identity [9,10,55,56]

$$\det\left(A\begin{vmatrix}1,\dots,k,\ k+t\\ 1,\dots,k,\ k+s\end{vmatrix}\right)_{s=1,\dots,m}^{t=1,\dots,m} = \det A\left(A\begin{vmatrix}1,\dots,k\\ 1,\dots,k\end{vmatrix}\right)^{m-1}.$$

In the case of Neville elimination one has

$$\det\left(A\begin{vmatrix}1,\dots,k, & k+t\\ s,\dots,s+k-1, & s+k\end{vmatrix}\right)_{s=1,\dots,m}^{t=1,\dots,m} = \det A \prod_{s=2}^{m} A\begin{vmatrix}1,\dots,k\\ s,\dots,s+k-1\end{vmatrix}.$$

Another identity of Sylvester's type has been derived in [3]. Also some applications to the E-algorithm [5] are given there.

As we have seen, the technique of e.s. has led us in particular to general determinantal identities of Sylvester's type. It can also be used to extend determinantal identities in the sense of Muir [51], see [47].

5. Application to quasilinear extrapolation problems

Suppose we are given elements $f_1,\dots,f_N$ of a linear space E and elements $L_1,\dots,L_N$ of its dual E^*. Consider furthermore elements $f =: f_{N+1}$ of E and $L =: L_{N+1}$ of E^*. Setting $I = (1,\dots,N+1)$, by A we denote the *generalized Vandermonde matrix*

$$A = A\binom{I}{I} = V\binom{f_1,\dots,f_N,f_{N+1}}{L_1,\dots,L_N,L_{N+1}} := (\langle L_i, f_j\rangle)_{i=1,\dots,N+1}^{j=1,\dots,N+1}. \tag{13}$$

Assume now that $k,m \in \mathbb{N}, m \leqslant N+1-k$ and that

$$\Sigma = ((I_s, I_s^\circ))_{s=1,\dots,m} \tag{14}$$

is a $(k-1,m)$–e.s. over $\bigcup_{s=1}^m I_s \subset (1,\dots,N)$. Let $G := (1,\dots,k)$. If the submatrices

$$A\binom{G}{I_s} \quad \text{are nonsingular for } s=1,\dots,m, \tag{15}$$

then for $s = 1,\dots,m$ the interpolants

$$p_s^k(f) := \sum_{j=1}^{k} c_{s,j}^k(f) \cdot f_j, \tag{16}$$

satisfying the interpolation conditions

$$\langle L_i, p_s^k(f)\rangle = \langle L_i, f\rangle \quad \text{for } i \in I_s$$

are well defined as well as

$$\tau_s^k(f) := \langle L, p_s^k(f)\rangle.$$

Clearly, in case of general linear extrapolation the mapping

$$E \ni f \xrightarrow{p_s^k} p_s^k(f)$$

is a linear projection onto $\text{span}\{f_1,\dots,f_N\}$ and

$$E \ni f \xrightarrow{c_{s,j}^k} c_{s,j}^k(f)$$

is a linear functional. In case of quasilinear extrapolation we assume that, as a function of $f \in E$, p_s^k remains idempotent. Then, as a function of $f \in E$, in general the coefficients $c_{s,j}^k(f)$ are not linear. We assume that, as functions of $f \in \text{span}\{f_1,\dots,f_N\}$, $c_{s,j}^k(f)$ remain linear.

The task is

(i) to find conditions, such that $p_1^N(f), \tau_1^N(f)$ are well defined, and
(ii) to find methods to compute these quantities from $p_s^k(f), \tau_s^k(f) (s = 1,\dots,m)$, respectively.

When translated into pure terms of Linear Algebra these questions mean: Consider matrix (13) and assume (15),

(i) under which conditions can we ensure that $A\binom{1,\dots,N}{1,\dots,N}$ is nonsingular?

The coefficient problem reads:

(ii′) Suppose that we do know the solutions

$$\boldsymbol{c}_s^k(f) = (c_{s,j}^k(f))_{j=1,\dots,k}$$

of the linear systems

$$A\begin{pmatrix} G \\ I_s \end{pmatrix} \cdot \boldsymbol{c}_s^k(f) = A\begin{pmatrix} N+1 \\ I_s \end{pmatrix}, \quad s = 1,\dots,m.$$

How to get from these the solution $\boldsymbol{c}_1^N(f) = (c_{1,j}^N(f))_{j=1,\dots,N}$ of

$$A\begin{pmatrix} 1,\dots,N \\ 1,\dots,N \end{pmatrix} \cdot \boldsymbol{c}_1^N(f) = A\begin{pmatrix} N+1 \\ 1,\dots,N \end{pmatrix}?$$

The value problem reads:

(iii) Suppose that we do know the values

$$\tau_s^k(f) = \langle L, p_s^k(f)\rangle, \quad s = 1,\dots,m.$$

How to get from these the value $\tau_1^N(f) = \langle L, p_1^N(f)\rangle$?

A *dual coefficient problem* can be also considered interchanging the roles of the spaces E and E^*. These problems were considered and solved in [20,7,19,31,40–42,45,48,50].

6. Applications to special classes of matrices

General elimination strategies, in particular the Neville e.s. and generalized Schur complements have found other applications in matrix theory and related problems.

In [21,22,44] we have considered some classes $\mathscr{L}_n$ of real $n \times n$-matrices A including the classes

(i) $\mathscr{C}_n$ of matrices satisfying $\det A(\begin{smallmatrix} J \\ J \end{smallmatrix}) > 0$ for all $J \subset (1, \dots, n)$, $\det A(\begin{smallmatrix} K \\ J \end{smallmatrix}) \cdot \det A(\begin{smallmatrix} J \\ K \end{smallmatrix}) > 0$ for all $J, K \subset (1, \dots, n)$ of the same cardinality, which was considered in [36];
(ii) of symmetric positive-definite matrices;
(iii) of *strictly totally positive matrices* (STP), which are defined by the property that all square submatrices have positive determinants [36];
(iv) of Minkowski matrices, defined by

$$A\begin{pmatrix} j \\ i \end{pmatrix} < 0 \quad \text{for all } i \neq j, \quad \det A\begin{pmatrix} 1, \dots, k \\ 1, \dots, k \end{pmatrix} > 0 \quad \text{for all } 1 \leqslant k \leqslant n.$$

In [21] we have proved that

$$A \in \mathscr{L}_n \Rightarrow \tilde{S} \in \mathscr{L}_m,$$

where $m = n - k$ and $\tilde{S}$ denotes the classical Schur complement of $A(\begin{smallmatrix} 1,\dots,k \\ 1,\dots,k \end{smallmatrix})$ in A. For STP matrices also generalized Schur complements with respect to the Neville e.s. are STP. Using the Neville e.s. in [21,49] tests of algorithmic complexity $O(N^4)$ for matrices being STP were derived for the first time. Neville elimination, based on consecutivity, proved to be especially well suited for STP matrices, because these matrices were characterized in [36] by the property of having all subdeterminants with *consecutive* rows and columns positive.

Elimination by consecutive rows is not at all new in matrix theory. It has been used to prove some properties of special classes of matrices, for example, totally positive (TP) matrices, which, as it has already been said, are matrices with all subdeterminants nonnegative. However, motivated by the above mentioned algorithm for testing STP matrices, Gasca and Peña [24] initiated an exhaustive study of Neville elimination in an algorithmic way, of the pivots and multipliers used in the proccess to obtain new properties of totally positive matrices and to improve and simplify the known characterizations of these matrices.

Totally positive matrices have interesting applications in many fields, as, for example, vibrations of mechanical systems, combinatorics, probability, spline functions, computer-aided geometric design, etc., see [36,37]. For this reason, remarkable papers on total positivity due to specialists on these fields have appeared, see for example the ones collected in [29].

The important survey [2] presents a complete list of references on totally positive matrices before 1987. One of the main points in the recent study of this class of matrices has been that of characterizing them in practical terms, by factorizations or by the nonnegativity of some minors (instead of all of them, as claimed in the definition).

In [24] for example, it was proved that a matrix is STP if and only if all subdeterminants with lists of consecutive rows and consecutive columns, starting at least one of these lists by 1, are positive. Necessarily, one of the lists must start with 1. Observe, that the new characterization considerably decreases the number of subdeterminants to be checked, compared with the classical characterization, due to Fekete and Pólya [17], which used all subdeterminants with consecutive rows and columns.

This result means that the set of all subdeterminants of a matrix A with consecutive rows and columns, of the form

$$A\begin{vmatrix} 1,\dots,j \\ i,\dots,i+j-1 \end{vmatrix}, \quad A\begin{vmatrix} i,\dots,i+j-1 \\ 1,\dots,j \end{vmatrix},$$

called in [24] column- and row-initial minors, play in total positivity a similar role to that of the leading principal minors

$$A\begin{vmatrix} 1,\dots,j \\ 1,\dots,j \end{vmatrix}$$

in positive definiteness of symmetric real matrices. An algorithm based on Neville elimination was given in [24] with a complexity $O(N^3)$ for a matrix of order N, instead of the one with $O(N^4)$ previously obtained in [21,49]. Other similar simplifications were obtained in [24] for the characterization of totally positive matrices (not strictly).

Concerning factorizations, in [26] Neville elimination was described in terms of a product by bidiagonal unit-diagonal matrices. Some of the most well-known characterizations of TP and STP matrices are related to their LU factorization. Cryer [14,15], in the 1970s, extended to TP matrices what was previously known for STP matrices, thus obtaining the following result.

A square matrix A is TP (resp. STP) iff it has an LU factorization such that L and U are TP (ΔSTP).

Here, as usual, L (resp. U) denotes a lower (upper) triangular matrix and ΔSTP means triangular nonnegative matrices with all the nontrivial subdeterminants of any order strictly positive.

Also Cryer pointed out that the matrix A is STP iff it can be written in the form

$$A = \prod_{r=1}^{N} L_r \prod_{s=1}^{M} U_s$$

where each L_r (resp. U_s) is a lower (upper) ΔSTP matrix. Observe that this result does not mention the relation of N or M with the order n of the matrix A.

The matricial description of Neville elimination obtained in [26] produced in the same paper the following result.

Let A be a nonsingular matrix of order n. Then A is STP iff it can be expressed in the form:

$$A = F_{n-1}\cdots F_1 D G_1 \cdots G_{n-1},$$

where, for each $i=1,2,\dots,n-1$, F_i is a bidiagonal, lower triangular, unit diagonal matrix, with zeros in positions $(2,1),\dots,(i,i-1)$ and positive entries in $(i+1,i),\dots,(n,n-1)$, G_i has the transposed form of F_i and D is a diagonal matrix with positive diagonal.

Similar results were obtained in [26] for TP matrices. In that paper all these new characterizations were collected in three classes: characterizations in terms of determinants, in terms of algorithms and in terms of factorizations.

7. Variation diminution and computer-aided geometric design

An $n \times n$ matrix A is *sign-regular* (SR) if for each $1 \leqslant k \leqslant n$ all its minors of order k have the same (non strict) sign (in the sense that the product of any two of them is greater than or equal to zero). The matrix is *strictly sign-regular* (SSR) if for each $1 \leqslant k \leqslant n$ all its minors of order k are different from zero and have the same sign. In [27] a test for strict sign regularity is given.

The importance of these types of matrices comes from their *variation diminishing properties*. By a *sign sequence* of a vector $x=(x_1,\ldots,x_n)^{\mathrm{T}} \in \mathbb{R}^n$ we understand any signature sequence ε for which $\varepsilon_i x_i = |x_i|$, $i=1,2,\ldots,n$. The number of sign changes of x associated to ε, denoted by $\mathscr{C}(\varepsilon)$, is the number of indices i such that $\varepsilon_i\varepsilon_{i+1} < 0$, $1 \leqslant i \leqslant n-1$. The *maximum* (resp. *minimum*) *variation of signs*, $V_+(x)$ (resp. $V_-(x)$), is by definition the maximum (resp. minimum) of $\mathscr{C}(\varepsilon)$ when ε runs over all sign sequences of x. Let us observe that if $x_i \neq 0$ for all i, then $V_+(x) = V_-(x)$ and this value is usually called the exact variation of signs. The next result (see [2, Theorems 5.3 and 5.6]) characterizes sign-regular and strictly sign-regular matrices in terms of their variation diminishing properties.

Let A be an $n \times n$ nonsingular matrix. Then:

(i) *A is SR $\Leftrightarrow V_-(Ax) \leqslant V_-(x)\ \forall x \in \mathbb{R}^n$.*
(ii) *A is SR $\Leftrightarrow V_+(Ax) \leqslant V_+(x)\ \forall x \in \mathbb{R}^n$.*
(iii) *A is SSR $\Leftrightarrow V_+(Ax) \leqslant V_-(x)\ \forall x \in \mathbb{R}^n \setminus \{0\}$.*

The above matricial definitions lead to the corresponding definitions for systems of functions. A system of functions $(u_0,\ldots,u_n)$ is *sign-regular* if all its collocation matrices are sign-regular of the same kind. The system is *strictly sign-regular* if all its collocation matrices are strictly sign-regular of the same kind. Here a *collocation matrix* is defined to be a matrix whose (i,j)-entry is of the form $u_i(x_j)$ with any system of strictly increasing points x_j.

Sign-regular systems have important applications in CAGD. Given $u_0,\ldots,u_n$, functions defined on $[a,b]$, and $P_0,\ldots,P_n \in \mathbb{R}^k$, we may define a curve $\gamma(t)$ by

$$\gamma(t) = \sum_{i=0}^{n} u_i(t) P_i.$$

The points $P_0,\ldots,P_n$ are called *control points*, because we expect to modify the shape of the curve by changing these points adequately. The polygon with vertices $P_0,\ldots,P_n$ is called *control polygon* of γ.

In CAGD the functions $u_0,\ldots,u_n$ are usually nonnegative and normalized ($\sum_{i=0}^{n} u_i(t)=1\ \forall t \in [a,b]$). In this case they are called *blending functions*. These requirements imply that the curve lies in the convex hull of the control polygon (*convex hull property*). Clearly, $(u_0,\ldots,u_n)$ is a system of blending functions if and only if all the collocation matrices are stochastic (that is, they are nonnegative matrices such that the elements of each row sum up to 1). For design purposes, it is desirable that the curve imitates the control polygon and that the control polygon even “exaggerates” the shape of the curve, and this holds when the system satisfies variation diminishing properties. If $(u_0,\ldots,u_n)$ is a sign-regular system of blending functions then the curve γ preserves many shape properties of the control polygon, due to the variation diminishing properties of $(u_0,\ldots,u_n)$. For instance, any line intersects the curve no more often than it intersects the control polygon.

A characterization of SSR matrices A by the Neville elimination of A and of some submatrices of A is obtained in [26, Theorem 4.1].

A system of functions $(u_0,\ldots,u_n)$ is said to be *totally positive* if all its collocation matrices are totally positive. The system is *normalized totally positive* (NTP) if it is totally positive and $\sum_{i=0}^{n} u_i = 1$.

Normalized totally positive systems satisfy an interesting shape-preserving property, which is very convenient for design purposes and which we call *endpoint interpolation property*: the initial and final endpoints of the curve and the initial and final endpoints (respectively) of the control polygon coincide. In summary, these systems are characterized by the fact that they always generate curves γ satisfying simultaneously the convex hull, variation diminishing and endpoint interpolation properties.

Now the following question arises. Given a system of functions used in CAGD to generate curves, does there exist a basis of the space generated by that system with optimal shape preserving properties? Or equivalently, is there a basis such that the generated curves γ imitate better the form of the corresponding control polygon than the form of the corresponding control polygon for any other basis?

In the space of polynomials of degree less than or equal to n on a compact interval, the Bernstein basis is optimal. This was conjectured by Goodman and Said in [30], and it was proved in [11]. In [12], there is also an affirmative answer to the above questions for any space with TP basis. Moreover, Neville elimination provides a constructive way to obtain optimal bases. In the space of polynomial splines, B-splines form the optimal basis.

Since the product of TP matrices is a TP matrix, if $(u_0,\dots,u_n)$ is a TP system of functions and A is a TP matrix of order $n+1$, then the new system $(u_0,\dots,u_n)A$ is again a TP system (which satisfies a "stronger" variation diminishing property than $(u_0,\dots,u_n)$). If we obtain from a basis $(u_0,\dots,u_n)$, in this way, all the totally positive bases of the space, then $(u_0,\dots,u_n)$ will be the "least variation diminishing" basis of the space. In consequence, the control polygons with respect to $(u_0,\dots,u_n)$ will imitate the form of the curve better than the control polygons with respect to other bases of the space. Therefore, we may reformulate the problem of finding an optimal basis $(b_0,\dots,b_n)$ in the following way:

Given a vector space $\mathscr{U}$ with a TP basis, is there a TP basis $(b_0,\dots,b_n)$ of $\mathscr{U}$ such that, for any TP basis $(v_0,\dots,v_n)$ of $\mathscr{U}$ there exists a TP matrix K satisfying $(v_0,\dots,v_n)=(b_0,\dots,b_n)K$?.

The existence of such *optimal basis* $(b_0,\dots,b_n)$ was proved in [12], where it was called *B-basis*. In the same paper, a method of construction, inspired by the Neville elimination process, was given. As mentioned above, Bernstein polynomials and B-splines are examples of B-bases.

Another point of view for B-bases is closely related to corner cutting algorithms, which play an important role in CAGD.

Given two NTP bases, $(p_0,\dots,p_n)$, $(b_0,\dots,b_n)$, let K be the nonsingular matrix such that

$$(p_0,\dots,p_n)=(b_0,\dots,b_n)K.$$

Since both bases are normalized, if K is a nonnegative matrix, it is clearly stochastic.

A curve γ can be expressed in terms of both bases

$$\gamma(t)=\sum_{i=0}^{n}B_ib_i(t)=\sum_{i=0}^{n}P_ip_i(t),\quad t\in[a,b],$$

and the matrix K gives the relationship between both control polygons

$$(B_0,\dots,B_n)^{\mathrm{T}}=K(P_0,\dots,P_n)^{\mathrm{T}}.$$

An *elementary corner cutting* is a transformation which maps any polygon $P_0 \cdots P_n$ into another polygon $B_0 \cdots B_n$ defined by:

$$B_j = P_j, \quad j \neq i,$$
$$B_i = (1-\lambda)P_i + \lambda P_{i+1}, \quad \text{for one} \quad i \in \{0,\ldots,n-1\} \tag{17}$$

or

$$B_j = P_j, \quad j \neq i,$$
$$B_i = (1-\lambda)P_i + \lambda P_{i-1}, \quad \text{for one} \quad i \in \{1,\ldots,n\}. \tag{18}$$

Here $\lambda \in (0,1)$.

A *corner-cutting algorithm* is the algorithmic description of a corner cutting transformation, which is any composition of elementary corner cutting transformations.

Let us assume now that the matrix K above is TP. Since it is stochastic, nonsingular and TP, it can be factorized as a product of bidiagonal nonnegative matrices, (as we have mentioned in Section 6), which can be interpreted as a corner cutting transformation. Such factorizations are closely related to the Neville elimination of the matrix [28]. From the variation diminution produced by the totally positive matrices of the process, it can be deduced that the curve γ imitates better the form of the control polygon $B_0 \cdots B_n$ than that of the control polygon $P_0 \cdots P_n$. Therefore, we see again that an NTP basis $(b_0,\ldots,b_n)$ of a space $\mathcal{U}$ has *optimal shape-preserving properties* if for any other NTP basis $(p_0,\ldots,p_n)$ of $\mathcal{U}$ there exists a (stochastic) TP matrix K such that

$$(p_0,\ldots,p_n) = (b_0,\ldots,b_n)K. \tag{19}$$

Hence, a basis has optimal shape preserving properties if and only if it is a normalized B-basis. Neville elimination has also inspired the construction of B-bases in [11,12]. Many of these results and other important properties and applications of totally positive matrices have been collected, as we have already said in [28, Section 6].

References

[1] A.G. Aitken, On interpolation by iteration of proportional parts without the use of differences, Proc. Edinburgh Math. Soc. 3 (1932) 56–76.
[2] T. Ando, Totally positive matrices, Linear Algebra Appl. 90 (1987) 165–219.
[3] B. Beckermann, G. Mühlbach, A general determinantal identity of Sylvester type and some applications, Linear Algebra Appl. 197,198 (1994) 93–112.
[4] Cl. Brezinski, The Mühlbach–Neville–Aitken-algorithm and some extensions, BIT 20 (1980) 444–451.
[5] Cl. Brezinski, A general extrapolation algorithm, Numer. Math. 35 (1980) 175–187.
[6] Cl. Brezinski, Convergence acceleration during the 20th century, this volume, J. Comput. Appl. Math. 122 (2000) 1–21.
[7] Cl. Brezinski, Recursive interpolation, extrapolation and projection, J. Comput. Appl. Math. 9 (1983) 369–376.
[8] Cl. Brezinski, M. Redivo Zaglia, Extrapolation methods, theory and practice, North-Holland, Amsterdam, 1991.
[9] R.A. Brualdi, H. Schneider, Determinantal identities: Gauss, Schur, Cauchy, Sylvester, Kronecker, Jacobi, Binet, Laplace, Muir and Cayley, Linear Algebra Appl. 52/53 (1983) 769–791.
[10] R.A. Brualdi, H. Schneider, Determinantal identities revisited, Linear Algebra Appl. 59 (1984) 183–211.
[11] J.M. Carnicer, J.M. Peña, Shape preserving representations and optimality of the Bernstein basis, Adv. Comput. Math. 1 (1993) 173–196.

[12] J.M. Carnicer, J.M. Peña, Totally positive bases for shape preserving curve design and optimality of B-splines, Comput. Aided Geom. Design 11 (1994) 633–654.
[13] R.W. Cottle, Manifestations of the Schur complement, Linear Algebra Appl. 8 (1974) 189–211.
[14] C. Cryer, The LU-factorization of totally positive matrices, Linear Algebra Appl. 7 (1973) 83–92.
[15] C. Cryer, Some poperties of totally positive matrices, Linear algebra Appl. 15 (1976) 1–25.
[16] D.R. Faddeev, U.N. Faddeva, Computational Methods of Linear Algebra, Freeman, San Francisco, 1963.
[17] M. Fekete, G. Pólya, Über ein Problem von Laguerre, Rend. C.M. Palermo 34 (1912) 89–120.
[18] M. Gasca, A. López-Carmona, V. Ramírez, A generalized Sylvester's identity on determinants and 1st application to interpolation problems, in: W. Schempp, K. Zeller (Eds.), Multivariate Approximation Theory II, ISNM, Vol. 61, Biskhäuser, Basel, 1982, pp. 171–184.
[19] M. Gasca, A. López-Carmona, A general interpolation formula and its application to multivariate interpolation, J. Approx. Theory 34 (1982) 361–374.
[20] M. Gasca, E. Lebrón, Elimination techniques and interpolation, J. Comput. Appl. Math. 19 (1987) 125–132.
[21] M. Gasca, G. Mühlbach, Generalized Schur-complements and a test for total positivity, Appl. Numer. Math. 3 (1987) 215–232.
[22] M. Gasca, G. Mühlbach, Generalized Schur-complements, Publicacciones del Seminario Matematico Garcia de Galdeano, Serie II, Seccion 1, No. 17, Universidad de Zaragoza, 1984.
[23] M. Gasca, J.M. Peña, Neville elimination and approximation theory, in: S.P. Singh (Ed.), Approximation Theory, Wavelets and Applications, Kluwer Academic Publishers, Dordrecht, 1995, pp. 131–151.
[24] M. Gasca, J.M. Peña, Total positivity and Neville elimination, Linear Algebra Appl. 165 (1992) 25–44.
[25] M. Gasca, J.M. Peña, On the characterization of TP and STP matrices, in: S.P. Singh (Ed.), Aproximation Theory, Spline Functions and Applications, Kluwer Academic Publishers, Dordrecht, 1992, pp. 357–364.
[26] M. Gasca, J.M. Peña, A matricial description of Neville elimination with applications to total positivity, Linear Algebra Appl. 202 (1994) 33–54.
[27] M. Gasca, J.M. Peña, A test for strict sign-regularity, Linear Algebra Appl. 197–198 (1994) 133–142.
[28] M. Gasca, J.M. Peña, Corner cutting algorithms and totally positive matrices, in: P.J. Laurent, A. Le Méhauté, L.L. Schumaker (Eds.), Curves and Surfaces II, 177–184, A.K. Peters, Wellesley, MA, 1994.
[29] M. Gasca, C.A. Micchelli (Eds.), Total Positivity and its Applications, Kluwer Academic Publishers, Dordrecht, 1996.
[30] T.N.T. Goodman, H.B. Said, Shape preserving properties of the generalized ball basis, Comput. Aided Geom. Design 8 (115–121) 1991.
[31] T. Håvie, Generalized Neville type extrapolation schemes, BIT 19 (1979) 204–213.
[32] T. Håvie, Remarks on a unified theory of classical and generalized interpolation and extrapolation, BIT 21 (1981) 465–474.
[33] T. Håvie, Remarks on the Mühlbach–Neville–Aitken-algorithm, Math. a. Comp. Nr. 2/80, Department of Numerical Mathematics, The University of Trondheim, 1980.
[34] E. Haynsworth, Determination of the inertia of a partitioned Hermitian matrix, Linear Algebra Appl. 1 (1968) 73–81.
[35] E. Haynsworth, On the Schur Complement, Basel Mathemathics Notes No. 20, June 1968.
[36] S. Karlin, Total Positivity, Stanford University Press, Standford, 1968.
[37] S. Karlin, W.J. Studden, Tchebycheff Systems: with Applications in Analysis and Statistics, Interscience, New York, 1966.
[38] G. Mühlbach, Neville–Aitken Algorithms for interpolation by functions of Čebyšev-systems in the sense of Newton and in a generalized sense of hermite, in: A.G. Law, B.N. Sahney (Eds.), Theory of Approximation, with Applications, Proceedings of the International Congress on Approximation Theory in Calgary, 1975, Academic Press, New York, 1976, pp. 200–212.
[39] G. Mühlbach, The general Neville–Aitken-algorithm and some applications, Numer. Math. 31 (1978) 97–110.
[40] G. Mühlbach, On two general algorithms for extrapolation with applications to numerical differentiation and integration, in: M.G. de Bruin, H. van Rossum (Eds.), Padé Approximation and its Applications, Lecture Notes in Mathematics, Vol. 888, Springer, Berlin, 1981, pp. 326–340.
[41] G. Mühlbach, Extrapolation algorithms as elimination techniques with applications to systems of linear equations, Report 152, Institut für Mathematik der Universität Hannover, 1982, pp. 1–47.
[42] G. Mühlbach, Algorithmes d'extrapolation, Publication ANO 118, Université de Lille 1, January 1984.

[43] G. Mühlbach, Sur une identité généralisée de Sylvester, Publication ANO 119, Université de Lille 1, January 1984.

[44] G. Mühlbach, M. Gasca, A generalization of Sylvester's identity on determinants and some applications, Linear Algebra Appl. 66 (1985) 221–234.

[45] G. Mühlbach, Two composition methods for solving certain systems of linear equations, Numer. Math. 46 (1985) 339–349.

[46] G. Mühlbach, A recurrence formula for generalized divided differences and some applications, J. Approx. Theory. 9 (1973) 165–172.

[47] G. Mühlbach, On extending determinantal identities, Publicaciones del Seminario Matematico Garcia de Galdeano, Serie II, Seccion 1, No. 139, Universidad de Zaragoza, 1987.

[48] G. Mühlbach, Linear and quasilinear extrapolation algorithms, in: R. Vichnevetsky, J. Vignes (Eds.), Numerical Mathematics and Applications, Elsevier, North-Holland, Amsterdam, IMACS, 1986, pp. 65–71.

[49] G. Mühlbach, M. Gasca, A test for strict total positivity via Neville elimination, in: F. Uhlig, R. Grone (Eds.), Current Trends in Matrix Theory, North-Holland, Amsterdam, 1987, pp. 225–232.

[50] G. Mühlbach, Recursive triangles, in: D. Beinov, V. Covachev (Eds.), Proceedings of the third International Colloquium on Numerical Analysis, Utrecht, VSP, 1995, pp. 123–134.

[51] T. Muir, The law of extensible minors in determinants, Trans. Roy. Soc. Edinburgh 30 (1883) 1–4.

[52] E.H. Neville, Iterative interpolation, J. Indian Math. Soc. 20 (1934) 87–120.

[53] D.V. Ouellette, Schur complements and statistics, Linear Algebra Appl. 36 (1981) 186–295.

[54] C. Schneider, Vereinfachte rekursionen zur Richardson-extrapolation in spezialfällen, Numer. Math. 24 (1975) 177–184.

[55] J.J. Sylvester, On the relation between the minor determinants of linearly equivalent quadratic functions, Philos. Mag. (4) (1851) 295–305.

[56] J.J. Sylvester, Collected Mathematical Papers, Vol. 1, Cambridge University Press, Cambridge, 1904, pp. 241–250.

ELSEVIER

Journal of Computational and Applied Mathematics 122 (2000) 51–80

JOURNAL OF COMPUTATIONAL AND APPLIED MATHEMATICS

www.elsevier.nl/locate/cam

The epsilon algorithm and related topics

P.R. Graves-Morris[a,]*, D.E. Roberts[b], A. Salam[c]

[a] *School of Computing and Mathematics, University of Bradford, Bradford, West Yorkshire BD7 1DP, UK*
[b] *Department of Mathematics, Napier University, Colinton Road, Edinburgh, EH14 1DJ Scotland, UK*
[c] *Laboratoire de Mathématiques Pures et Appliquées, Université du Littoral, BP 699, 62228 Calais, France*

Received 7 May 1999; received in revised form 27 December 1999

Abstract

The epsilon algorithm is recommended as the best *all-purpose* acceleration method for slowly converging sequences. It exploits the numerical precision of the data to extrapolate the sequence to its limit. We explain its connections with Padé approximation and continued fractions which underpin its theoretical base. Then we review the most recent extensions of these principles to treat application of the epsilon algorithm to vector-valued sequences, and some related topics. In this paper, we consider the class of methods based on using generalised inverses of vectors, and the formulation specifically includes the complex case wherever possible. © 2000 Elsevier Science B.V. All rights reserved.

Keywords: Epsilon algorithm; qd algorithm; Padé; Vector-valued approximant; Wynn; Cross rule; Star identity; Compass identity; Designant

1. Introduction

A sequence with a limit is as basic a topic in mathematics as it is a useful concept in science and engineering. In the applications, it is usually the limit of a sequence, or a fixed point of its generator, that is required; the existence of the limit is rarely an issue, and rapidly convergent sequences are welcomed. However, if one has to work with a sequence that converges too slowly, the epsilon algorithm is arguably the best all-purpose method for accelerating its convergence. The algorithm was discovered by Wynn [54] and his review article [59] is highly recommended. The epsilon algorithm can also be used for weakly diverging sequences, and for these the desired limit is usually defined as being a fixed point of the operator that generates the sequence. There are interesting exceptional cases, such as quantum well oscillators [51], where the epsilon algorithm is not powerful enough and we refer to the companion paper by Homeier [33] in which the more

* Corresponding author.
E-mail address: p.r.graves-morris@bradford.ac.uk (P.R. Graves-Morris).

PII: S0377-0427(00)00355-1

powerful Levin-type algorithms, etc., are reviewed. The connections between the epsilon algorithm and similar algorithms are reviewed by Weniger [50,52].

This paper is basically a review of the application of the epsilon algorithm, with an emphasis on the case of complex-valued, vector-valued sequences. There are already many reviews and books which include sections on the scalar epsilon algorithm, for example [1,2,9,17,53]. In the recent past, there has been progress with the problem of numerical breakdown of the epsilon algorithm. Most notably, Cordellier's algorithm deals with both scalar and vector cases [13–16]. This work and its theoretical basis has been extensively reviewed [26,27]. In this paper, we focus attention on how the epsilon algorithm is used for sequences $(\boldsymbol{s}_i)$ in which $\boldsymbol{s}_i \in \mathbb{C}^d$. The case $d = 1$ is the scalar case, and the formulation for $\boldsymbol{s}_i \in \mathbb{C}$ is essentially the same as that for $\boldsymbol{s}_i \in \mathbb{R}$. Not so for the vector case, and we give full details of how the vector epsilon and vector qd algorithms are implemented when $\boldsymbol{s}_i \in \mathbb{C}^d$, and of the connections with vector Padé approximation. Understanding these connections is essential for specifying the range of validity of the methods. Frequently, the word "normally" appears in this paper to indicate that the results may not apply in degenerate cases. The adaptations for the treatment of degeneracy are almost the same for both real and complex cases, and so we refer to [25–27] for details.

In Section 2, we formulate the epsilon algorithm, and we explain its connection with Padé approximation and the continued fractions called C-fractions. We give an example of how the epsilon algorithm works in ideal circumstances, without any significant loss of numerical precision (which is an unusual outcome).

In Section 3, we formulate the vector epsilon algorithm, and we review its connection with vector-valued Padé approximants and with vector-valued C-fractions. There are two major generalisations of the scalar epsilon algorithm to the vector case. One of them is Brezinski's topological epsilon algorithm [5,6,35,48,49]. This algorithm has two principal forms, which might be called the forward and backward versions; and the backward version has the orthogonality properties associated with Lanczos methods [8]. The denominator polynomials associated with all forms of the topological epsilon algorithm have degrees which are the same as those for the scalar case [2,5,8]. By contrast, the other generalisation of the scalar epsilon algorithm to the vector case can be based on using generalised inverses of vectors, and it is this generalisation which is the main topic of this paper. We illustrate how the vector epsilon algorithm works in a two-dimensional real space, and we give a realistic example of how it works in a high-dimensional complex space. The denominator polynomials used in the scalar case are generalised both to operator polynomials of the same degree and to scalar polynomials of double the degree in the vector case, and we explain the connections between these twin generalisations. Most of the topics reviewed in Section 3 have a direct generalisation to the rational interpolation problem [25]. We also note that the method of GIPAs described in Section 3 generalises directly to deal with sequences of functions in $L_2(a,b)$ rather than vectors $\mathbb{C}^d$; in this sense, the vectors are regarded as discretised functions [2].

In Section 4 we review the use of the vector qd algorithm for the construction of vector-valued C-fractions, and we note the connections between vector orthogonal polynomials and the vector epsilon algorithm. We prove the cross-rule (4.18), (4.22) using a Clifford algebra. For real-valued vectors, we observe that it is really an overlooked identity amongst Hankel designants. Here, the Cross Rule is proved as an identity amongst complex-valued vectors using Moore–Penrose inverses.

The importance of studying the vector epsilon algorithm lies partly in its potential [20] for application to the acceleration of convergence of iterative solution of discretised PDEs. For example, Gauss–Seidel iteration generates sequences of vectors which often converge too slowly to be useful. SOR, multigrid and Lanczos methods are alternative approaches to the problem which are currently popular, but the success of the techniques like CGS and LTPMs (see [31] for an explanation of the techniques and the acronyms) indicates the need for continuing research into numerical methods for the acceleration of convergence of vector-valued sequences.

To conclude this introductory section, we recall that all algorithms have their domains of validity. The epsilon algorithm fails for logarithmically convergent sequences (which converge too slowly) and it fails to find the fixed point of the generator of sequences which diverge too fast. For example, if

$$s_n - s = \frac{C}{n} + O(n^{-2}), \quad C \neq 0,$$

the sequence (s_n) is logarithmically convergent to s. More precisely, a sequence is defined to converge logarithmically to s if it converges to s at a rate governed by

$$\lim_{n\to\infty} \frac{s_{n+1} - s}{s_n - s} = 1.$$

Not only does the epsilon algorithm usually fail for such sequences, but Delahaye and Germain-Bonne [18,19] have proved that there is no universal accelerator for logarithmically convergent sequences.

Reviews of series transformations, such as those of the energy levels of the quantum-mechanical harmonic oscillator [21,50,51], and of the Riemann zeta function [34], instructively show the inadequacy of the epsilon algorithm when the series coefficients diverge too fast. Information about the asymptotic form of the coefficients and scaling properties of the solution is exploited to create purpose-built acceleration methods. Exotic applications of the ε-algorithm appear in [55].

2. The epsilon algorithm

The epsilon algorithm was discovered by Wynn [54] as an efficient implementation of Shanks' method [47]. It is an algorithm for acceleration of convergence of a sequence

$$S = (s_0, s_1, s_2, \ldots, s_i \in \mathbb{C}) \tag{2.1}$$

and it comprises the following initialisation and iterative phases:

Initialisation: For $j = 0, 1, 2, \ldots$

$$\varepsilon_{-1}^{(j)} = 0 \quad \text{(artificially)}, \tag{2.2}$$

$$\varepsilon_0^{(j)} = s_j. \tag{2.3}$$

Iteration: For $j, k = 0, 1, 2, \ldots$

$$\varepsilon_{k+1}^{(j)} = \varepsilon_{k-1}^{(j+1)} + [\varepsilon_k^{(j+1)} - \varepsilon_k^{(j)}]^{-1}. \tag{2.4}$$

The entries $\varepsilon_k^{(j)}$ are displayed in the epsilon table on the left-hand side of Fig. 1, and the initialisation has been built in.

	s_0			
0		$\varepsilon_1^{(0)}$		
	s_1		$\varepsilon_2^{(0)}$	
0		$\varepsilon_1^{(1)}$		$\ddots$
	s_2		$\varepsilon_2^{(1)}$	
0		$\varepsilon_1^{(2)}$		
	s_3		$\varepsilon_2^{(2)}$	
0		$\vdots$		$\ddots$
	$\vdots$			

	4·000				
0		−0·75			
	2·667		3·167		
0		1·25		−28·75	
	3·467		3·133		3·142
0		−1·75		82·25	
	2·895		3·145		
0		2·25			

Fig. 1. The epsilon table, and a numerical example of it.

Example 2.1. Gregory's series for $\tan^{-1} z$ is

$$\tan^{-1} z = z - \frac{z^3}{3} + \frac{z^5}{5} - \frac{z^7}{7} + \cdots . \tag{2.5}$$

This series can be used to determine the value of π by evaluating its MacLaurin sections at $z = 1$:

$$s_j := \left. [4\tan^{-1}(z)]_0^{2j+1} \right|_{z=1}, \quad j = 0, 1, 2, \ldots . \tag{2.6}$$

Nuttall's notation is used here and later on. For a function whose MacLaurin series is

$$\phi(z) = \phi_0 + \phi_1 z + \phi_2 z^2 + \cdots,$$

its sections are defined by

$$[\phi(z)]_j^k = \sum_{i=j}^{k} \phi_i z^i \quad \text{for } 0 \leqslant j \leqslant k. \tag{2.7}$$

In fact, $s_j \to \pi$ as $j \to \infty$ [2] but sequence (2.6) converges slowly, as is evidenced in the column $k = 0$ of entries $s_j = \varepsilon_0^{(j)}$ in Fig. 1. The columns of odd index have little significance, whereas the columns of even index can be seen to converge to π, which is the correct limit [2], increasingly fast, as far as the table goes. Some values of $\varepsilon_{2k}^{(j)}$ are also shown on the bar chart (Fig. 2). Notice that $\varepsilon_2^{(2)} = 3.145$ and $\varepsilon_4^{(0)} = 3.142$ cannot be distinguished visually on this scale.

In Example 2.1, convergence can be proved and the rate of convergence is also known [2]. From the theoretical viewpoint, Example 2.1 is ideal for showing the epsilon algorithm at its best. It is noticeable that the entries in the columns of odd index are large, and this effect warns us to beware of possible loss of numerical accuracy. Like all algorithms of its kind (which use reciprocal differences of convergent sequences) the epsilon algorithm uses (and usually uses up) numerical precision of the data to do its extrapolation. In this case, there is little loss of numerical precision using 16 decimal place (MATLAB) arithmetic, and $\varepsilon_{22}^{(0)} = \pi$ almost to machine precision. In this case, the epsilon algorithm converges with great numerical accuracy because series (2.5) is a totally oscillating series [4,7,17,59].

To understand in general how and why the epsilon algorithm converges, whether we are referring to its even columns ($\varepsilon_{2k}^{(j)}$, $j = 0, 1, 2, \ldots, k$ fixed) or its diagonals ($\varepsilon_{2k}^{(j)}$, $k = 0, 1, 2, \ldots, j$ fixed) or any

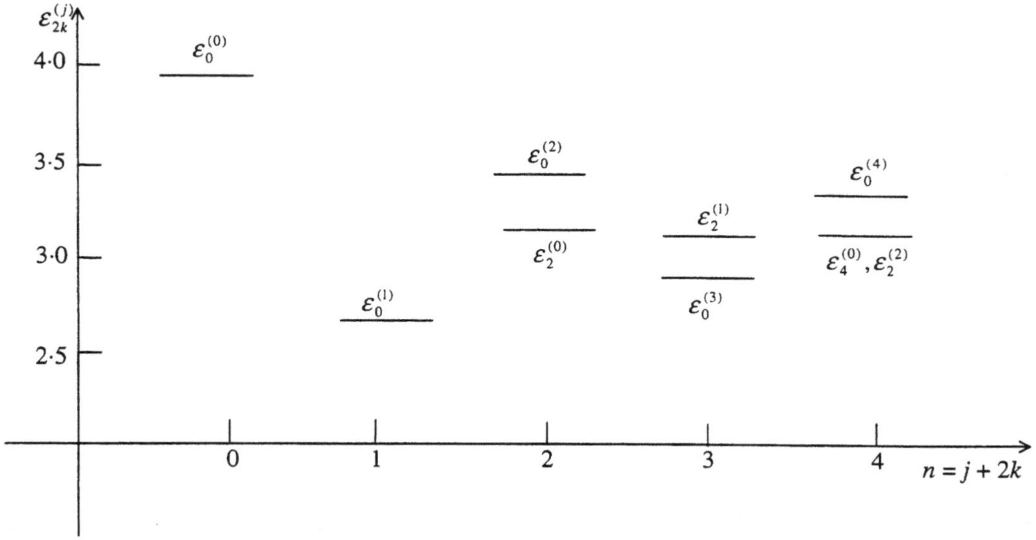

Fig. 2. Values of $\varepsilon_{2k}^{(j)}$ for Example 2.1, showing the convergence rate of the epsilon algorithm using $n+1 = 1,2,3,4,5$ terms of the given sequence.

other sequence, the connection with Padé approximation is essential [1,2,56]. Given a (possibly formal) power series

$$f(z) = c_0 + c_1 z + c_2 z^2 + \cdots, \tag{2.8}$$

the rational function

$$A(z)B(z)^{-1} \equiv [\ell/m](z) \tag{2.9}$$

is defined as a Padé approximant for $f(z)$ of type $[\ell/m]$ if

$$\text{(i) } \deg\{A(z)\} \leqslant \ell, \quad \deg\{B(z)\} \leqslant m, \tag{2.10}$$

$$\text{(ii) } f(z)B(z) - A(z) = \mathrm{O}(z^{\ell+m+1}), \tag{2.11}$$

$$\text{(iii) } B(0) \neq 0. \tag{2.12}$$

The Baker condition

$$B(0) = 1 \tag{2.13}$$

is often imposed for reliability in the sense of (2.14) below and for a definite specification of $A(z)$ and $B(z)$. The definition above contrasts with the classical (Frobenius) definition in which axiom (iii) is waived, and in this case the existence of $A(z)$ and $B(z)$ is guaranteed, even though (2.14) below is not. Using specification (2.10)–(2.13), we find that

$$f(z) - A(z)B(z)^{-1} = \mathrm{O}(z^{\ell+m+1}), \tag{2.14}$$

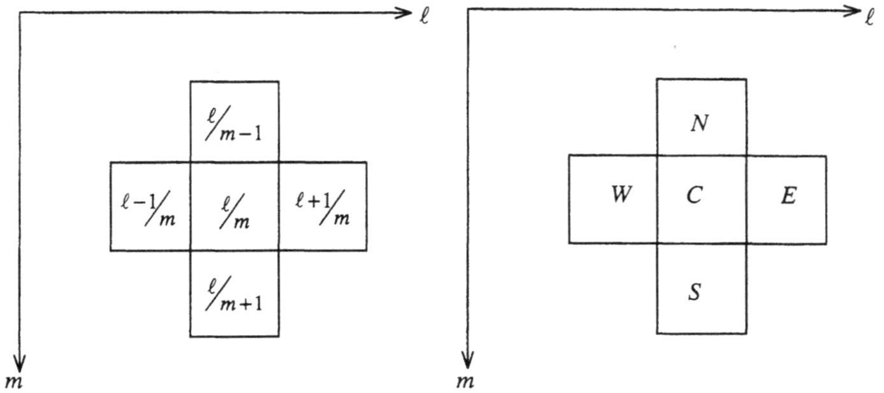

Fig. 3. Relative location of Padé approximants.

provided that a solution of (2.15) below can be found. To find $B(z)$, the linear equations corresponding to accuracy-through-orders $z^{\ell+1}, z^{\ell+2}, \ldots, z^{\ell+m}$ in (2.11) must be solved. They are

$$\begin{bmatrix} c_{\ell-m+1} & \cdots & c_\ell \\ \vdots & & \vdots \\ c_\ell & \cdots & c_{\ell+m-1} \end{bmatrix} \begin{bmatrix} b_m \\ \vdots \\ b_1 \end{bmatrix} = - \begin{bmatrix} c_{\ell+1} \\ \vdots \\ c_{\ell+m} \end{bmatrix}. \tag{2.15}$$

The coefficients of $B(z) = \sum_{i=0}^{m} b_i z^i$ are found using an accurate numerical solver of (2.15). By contrast, for purely theoretical purposes, Cramer's rule is applied to (2.15). We are led to define

$$q^{[\ell/m]}(z) = \begin{vmatrix} c_{\ell-m+1} & c_{\ell-m+2} & \cdots & c_{\ell+1} \\ c_{\ell-m+2} & c_{\ell-m+3} & \cdots & c_{\ell+2} \\ \vdots & \vdots & & \vdots \\ c_\ell & c_{\ell+1} & \cdots & c_{\ell+m} \\ z^m & z^{m-1} & \cdots & 1 \end{vmatrix} \tag{2.16}$$

and then we find that

$$B^{[\ell/m]}(z) = q^{[\ell/m]}(z)/q^{[\ell/m]}(0) \tag{2.17}$$

is the denominator polynomial for the Padé approximation problem (2.9)–(2.15) provided that $q^{[\ell/m]}(0) \neq 0$.

The collection of Padé approximants is called the Padé table, and in Fig. 3 we show five neighbouring approximants in the table.

These approximants satisfy a five-point star identity,

$$[N(z) - C(z)]^{-1} + [S(z) - C(z)]^{-1} = [E(z) - C(z)]^{-1} + [W(z) - C(z)]^{-1}, \tag{2.18}$$

called Wynn's identity or the compass identity. The proof of (2.18) is given in [1,2], and it is also a corollary (in the case $d = 1$) of the more general result (3.59) that we prove in the next section. Assuming (2.18) for the moment, the connection between Padé approximation and the epsilon

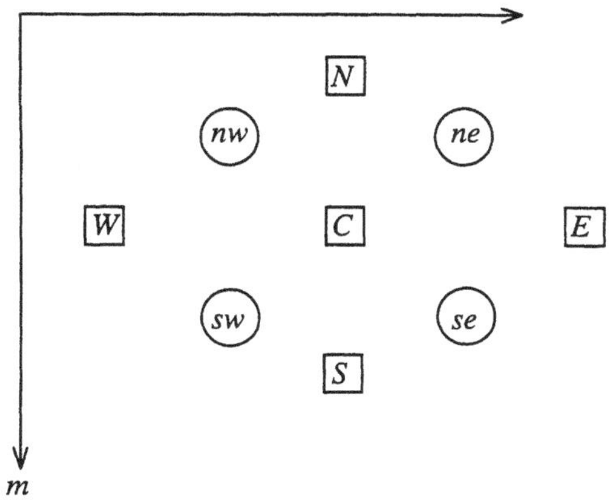

Fig. 4. Some artificial entries in the Padé table are shown circled.

algorithm is given by connecting the coefficients of $f(z)$ with those of S with

$$c_0 = s_0, \quad c_i = s_i - s_{i-1}, \quad i = 1, 2, 3, \ldots,$$

and by

Theorem 2.1. *The entries in columns of even index in the epsilon table are values of Padé approximants given by*

$$\varepsilon_{2k}^{(j)} = [j + k/k](1) \tag{2.19}$$

provided (i) *zero divisors do not occur in the construction of the epsilon table, and* (ii) *the corresponding Padé approximants identified by* (2.19) *exist.*

Proof. The entries W, C, E in the Padé table of Figs. 3 and 4 may be taken to correspond to entries $\varepsilon_{2k}^{(j-1)}, \varepsilon_{2k}^{(j)}, \varepsilon_{2k}^{(j+1)}$, respectively, in the epsilon table. They neighbour other elements in columns of odd index in the epsilon table, $\mathrm{nw} := \varepsilon_{2k-1}^{(j)}$, $\mathrm{ne} := \varepsilon_{2k-1}^{(j+1)}$, $\mathrm{se} := \varepsilon_{2k+1}^{(j)}$ and $\mathrm{sw} := \varepsilon_{2k+1}^{(j-1)}$. By re-pairing, we have

$$(\mathrm{nw} - \mathrm{sw}) - (\mathrm{ne} - \mathrm{se}) = (\mathrm{nw} - \mathrm{ne}) - (\mathrm{sw} - \mathrm{se}). \tag{2.20}$$

By applying the epsilon algorithm to each term in (2.20), we obtain the compass identity (2.18). □

With our conventions, the approximants of type $[\ell/0]$ lie in the first row $(m = 0)$ of the Padé table. This is quite natural when we regard these approximants as MacLaurin sections of $f(z)$. However, it must be noted that the row sequence $([\ell/m](1), \ell = m + j, m + j + 1, \ldots, m$ fixed) corresponds to the column sequence of entries $(\varepsilon_{2m}^{(j)}, j = 0, 1, 2, \ldots, m$ fixed); this identification follows from (2.19).

A key property of Padé approximants that is an axiom of their definition is that of accuracy-through-order, also called correspondence. Before Padé approximants were known as such, attention had rightly been focused on the particular sequence of rational fractions which are truncations of the continued fraction

$$f(z) = \frac{c_0}{1} - \frac{za_1}{1} - \frac{za_2}{1} - \frac{za_3}{1} - \cdots. \tag{2.21}$$

The right-hand side of (2.21) is called a C-fraction (for instance, see [36]), which is short for corresponding fraction, and its truncations are called its convergents. Normally, it can be constructed by successive reciprocation and re-expansion. The first stage of this process is

$$\frac{1 - c_0/f(z)}{z} = \frac{a_1}{1} - \frac{za_2}{1} - \frac{za_3}{1} - \cdots. \tag{2.22}$$

By undoing this process, we see that the convergents of the C-fraction are rational fractions in the variable z.

By construction, we see that these convergents agree order by order with $f(z)$, provided all $a_i \neq 0$, and this property is called correspondence.

Example 2.2. We truncate (2.21) after a_2 and obtain

$$\frac{A_2(z)}{B_2(z)} = \frac{c_0}{1} - \frac{za_1}{1 - za_2}. \tag{2.23}$$

This is a rational fraction of type [1/1], and we take

$$A_2(z) = c_0(1 - za_2), \quad B_2(z) = 1 - z(a_1 + a_2).$$

Provided all the $a_i \neq 0$, the convergents of (2.21) are well defined. The equality in (2.21) is not to be understood in the sense of pointwise convergence for each value of z, but in the sense of correspondence order by order in powers of z.

The numerators and denominators of the convergents of (2.21) are usually constructed using Euler's recursion. It is initialised, partly artificially, by

$$A_{-1}(z) = 0, \quad A_0(z) = c_0, \quad B_{-1}(z) = 1, \quad B_0(z) = 1 \tag{2.24}$$

and the recursion is

$$A_{i+1}(z) = A_i(z) - a_{i+1}zA_{i-1}(z), \quad i = 0, 1, 2, \ldots, \tag{2.25}$$

$$B_{i+1}(z) = B_i(z) - a_{i+1}zB_{i-1}(z), \quad i = 0, 1, 2, \ldots. \tag{2.26}$$

Euler's formula is proved in many texts, for example, [1,2,36]. From (2.24) to (2.26), it follows by induction that

$$\ell = \deg\{A_i(z)\} \leqslant \left[\frac{i}{2}\right], \quad m = \deg\{B_i(z)\} \leqslant \left[\frac{i+1}{2}\right], \tag{2.27}$$

where $[\,\cdot\,]$ represents the integer part function and the Baker normalisation is built in:

$$B_i(0) = 1, \quad i = 0, 1, 2, \ldots. \tag{2.28}$$

The sequence of approximants generated by (2.24)–(2.26) is shown in Fig. 5.

From (2.19) and (2.27), we see that the convergents of even index $i = 2k$ correspond to Padé approximants of type $[k/k]$; when they are evaluated at $z = 1$, they are values of $\varepsilon_{2k}^{(0)}$ on the leading diagonal of the epsilon table.

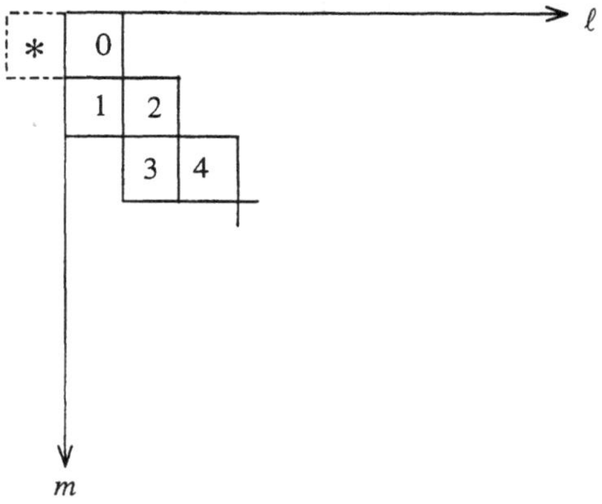

Fig. 5. A staircase sequence of approximants indexed by i, as in (2.27).

The epsilon algorithm was introduced in (2.1)–(2.4) as a numerical algorithm. Eq. (2.19) states its connection with values of certain Padé approximants. However, the epsilon algorithm can be given a symbolic interpretation if it is initialised with

$$\varepsilon_{-1}^{(j)} = 0, \quad \varepsilon_0^{(j)} = \sum_{i=0}^{j} c_i z^i \tag{2.29}$$

instead of (2.2) and (2.3). In this case, (2.19) would become

$$\varepsilon_{2k}^{(j)}(z) = [j + k/k](z). \tag{2.30}$$

The symbolic implementation of the iterative process (2.4) involves considerable cancellation of polynomial factors, and so we regard this procedure as being primarily of conceptual value.

We have avoided detailed discussions of normality and degeneracy [1,2,25] in this paper so as to focus on the algorithmic aspects. The case of numerical breakdown associated with zero divisors is treated by Cordellier [14,15] for example. Refs. [1,2] contain formulae for the difference between Padé approximants occupying neighbouring positions in the Padé table. Using these formulae, one can show that condition (i) of Theorem 2.1 implies that condition (ii) holds, and so conditions (ii) can be omitted.

It is always worthwhile to consider the case in which an approximation method gives exact results at an intermediate stage so that the algorithm is terminated at that stage. For example, let

$$f(z) = v_0 + \sum_{\kappa=1}^{k} \frac{v_\kappa}{1 - z\theta_\kappa} \tag{2.31}$$

with $v_\kappa, \theta_\kappa \in \mathbb{C}$, each $|\theta_\kappa| < 1$, each $v_\kappa \neq 0$ and all θ_κ distinct. Then $f(z)$ is a rational function of precise type $[k/k]$. It is the generating function of the generalised geometric sequence S with elements

$$s_j = v_0 + \sum_{\kappa=1}^{k} v_\kappa \frac{1 - \theta_\kappa^{j+1}}{1 - \theta_\kappa}, \quad j = 0, 1, 2, \ldots . \tag{2.32}$$

This sequence is sometimes called a Dirichlet series and it converges to $s_\infty = f(1)$ as $j \to \infty$. Its elements can also be expressed as

$$s_j = s_\infty - \sum_{\kappa=1}^{k} w_\kappa \theta_\kappa^j \tag{2.33}$$

if

$$s_\infty = \sum_{\kappa=0}^{k} v_\kappa + \sum_{\kappa=1}^{k} w_\kappa \quad \text{and} \quad w_\kappa = \theta_\kappa v_\kappa (1-\theta_\kappa)^{-1}.$$

Then (2.33) expresses the fact that S is composed of exactly k non-trivial, distinct geometric components. Theorem 2.1 shows that the epsilon algorithm yields

$$\varepsilon_{2k}^{(j)} = s_\infty, \quad j = 0, 1, 2, \ldots$$

which is the 'exact result' in each row of the column of index $2k$, provided that zero divisors have not occurred before this column is constructed. The algorithm should be terminated at this stage via a consistency test, because zero divisors necessarily occur at the next step. Remarkably, the epsilon algorithm has some smoothing properties [59], which may (or may not) disguise this problem when rounding errors occur.

In the next sections, these results will be generalised to the vector case. To do that, we will also need to consider the paradiagonal sequences of Padé approximants given by $([m+J/m](z),\ m = 0,1,2,\ldots, J \geqslant 0,\ J$ fixed). After evaluation at $z=1$, we find that this is a diagonal sequence $(\varepsilon_{2m}^{(J)},\ m = 0,1,2,\ldots, J \geqslant 0,\ J$ fixed) in the epsilon table.

3. The vector epsilon algorithm

The epsilon algorithm acquired greater interest when Wynn [57,58] showed that it has a useful and immediate generalisation to the vector case. Given a sequence

$$\boldsymbol{S} = (\boldsymbol{s}_0, \boldsymbol{s}_1, \boldsymbol{s}_2, \ldots : \boldsymbol{s}_i \in \mathbb{C}^d), \tag{3.1}$$

the standard implementation of the vector epsilon algorithm (VEA) consists of the following initialisation from $\boldsymbol{S}$ followed by its iteration phase:

Initialisation: For $j = 0, 1, 2, \ldots,$

$$\varepsilon_{-1}^{(j)} = \mathbf{0} \quad \text{(artificially)}, \tag{3.2}$$

$$\varepsilon_0^{(j)} = \boldsymbol{s}_j. \tag{3.3}$$

Iteration: For $j, k = 0, 1, 2, \ldots,$

$$\varepsilon_{k+1}^{(j)} = \varepsilon_{k-1}^{(j+1)} + [\varepsilon_k^{(j+1)} - \varepsilon_k^{(j)}]^{-1}. \tag{3.4}$$

The iteration formula (3.4) is identical to (2.4) for the scalar case, except that it requires the specification of an inverse (reciprocal) of a vector. Usually, the Moore–Penrose (or Samelson) inverse

$$\boldsymbol{v}^{-1} = \boldsymbol{v}^* / (\boldsymbol{v}^{\mathrm{H}} \boldsymbol{v}) = \boldsymbol{v}^* \Big/ \sum_{i=1}^{d} |v_i|^2 \tag{3.5}$$

−0·10 1·50		
0·59 2·35	0·96 1·58	
1·43 2·08	0·57 1·44	1·00 1·00
1·80 1·11	0·40 0·76	1·00 1·00
1·54 0·26	0·90 0·57	1·00 1·00
0·95 0·09	1·16 0·64	⋮
0·51 0·60	⋮	
⋮		

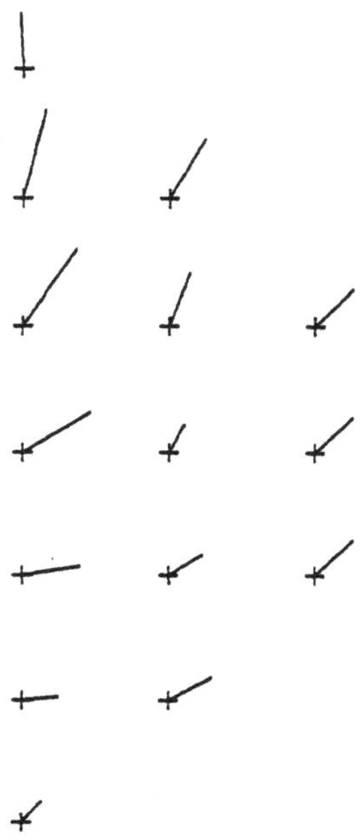

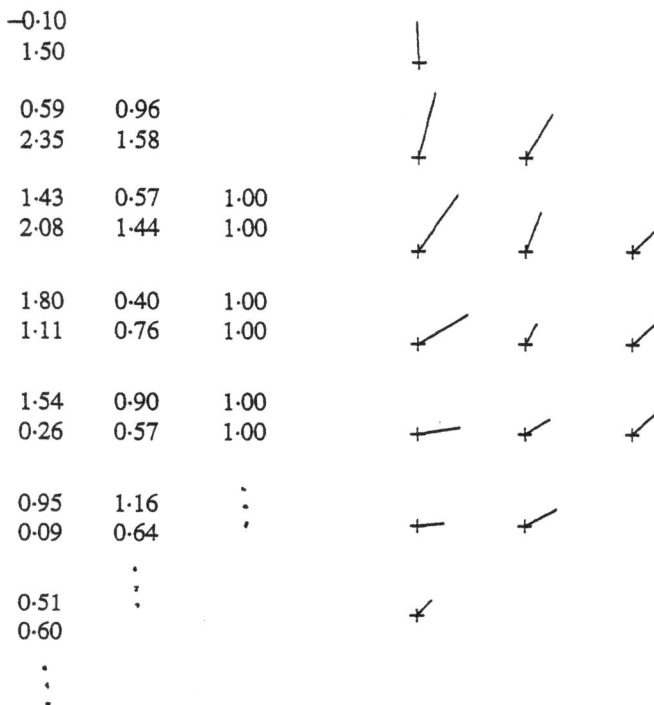

Fig. 6. Columns $k=0,2$ and 4 of the vector epsilon table for Example 3.1 are shown numerically and graphically.

(where the asterisk denotes the complex conjugate and H the Hermitian conjugate) is the most useful, but there are exceptions [39]. In this paper, the vector inverse is defined by (3.5). The vector epsilon table can then be constructed column by column from (3.2) to (3.4), as in the scalar case, and as shown in Fig. 6.

Example 3.1. The sequence $\boldsymbol{S}$ is initialised by

$$s_0 := \boldsymbol{b} := (-0.1, 1.5)^{\mathrm{T}} \tag{3.6}$$

(where T denotes the transpose) and it is generated recursively by

$$\boldsymbol{s}_{j+1} := \boldsymbol{b} + G\boldsymbol{s}_j, \quad j = 0, 1, 2, \ldots \tag{3.7}$$

with

$$G = \begin{bmatrix} 0.6 & 0.5 \\ -1 & 0.5 \end{bmatrix}. \tag{3.8}$$

The fixed point of (3.7) is $\boldsymbol{x} = [1, 1]$, which is the solution of $A\boldsymbol{x} = \boldsymbol{b}$ with $A = I - G$.

Notice that

$$\varepsilon_4^{(j)} = \boldsymbol{x} \quad \text{for } j = 0, 1, 2$$

and this 'exact' result is clearly demonstrated in the right-hand columns of Fig. 6.

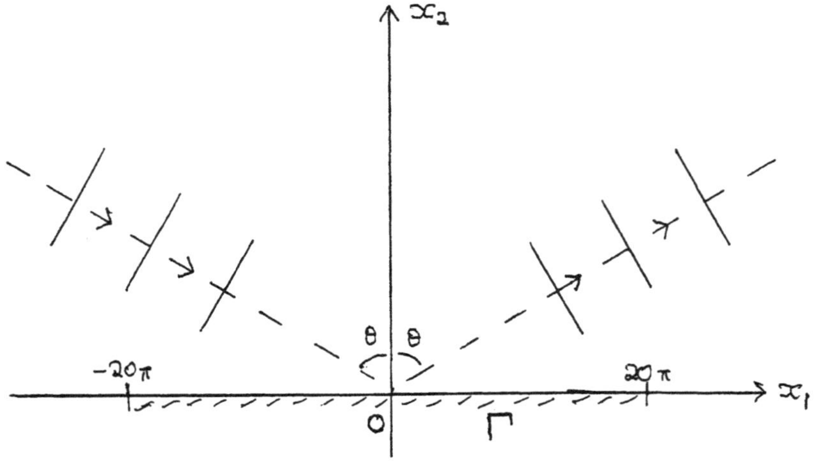

Fig. 7. Schematic view of the two components of $u_1(\boldsymbol{x})$ and the boundary Γ on the x_1-axis.

This elementary example demonstrates how the VEA can be a powerful convergence accelerator in an ideal situation. With the same rationale as was explained in the scalar case, the vector epsilon algorithm is used for sequences of vectors when their convergence is too slow. Likewise, the VEA can find an accurate solution (as a fixed point of an associated matrix operator) even when the sequence of vectors is weakly divergent. In applications, these vector sequences usually arise as sequences of discretised functions, and the operator is a (possibly nonlinear) integral operator. An example of this kind of vector sequence is one that arises in a problem of current interest. We consider a problem in acoustics, which is based on a boundary integral equation derived from the Helmholtz equation [12]. Our particular example includes impedance boundary conditions (3.12) relevant to the design of noise barriers.

Example 3.2. This is an application of the VEA for the solution of

$$u(\boldsymbol{x}) = u_1(\boldsymbol{x}) + ik\int_\Gamma G(\boldsymbol{x},\boldsymbol{y})[\beta(\boldsymbol{y}) - 1]u(\boldsymbol{y})\,d\boldsymbol{y} \tag{3.9}$$

for the acoustic field $u(\boldsymbol{x})$ at the space point $\boldsymbol{x} = (x_1, x_2)$. This field is confined to the half-space $x_2 \geqslant 0$ by a barrier shown in Fig. 7. The inhomogeneous term in (3.9) is

$$u_1(\boldsymbol{x}) = e^{ik(x_1 \sin\theta - x_2\cos\theta)} + R.e^{ik(x_1\sin\theta + x_2\cos\theta)} \tag{3.10}$$

which represents an incoming plane wave and a "partially reflected" outgoing plane wave with wave number k. The reflection coefficient in (3.10) is given by

$$R = -\tan^2\left(\frac{\theta}{2}\right), \tag{3.11}$$

so that $u_1(\boldsymbol{x})$ and $u(\boldsymbol{x})$ satisfy the impedance boundary conditions

$$\frac{\partial u_1}{\partial x_2} = -iku_1 \quad \text{and} \quad \frac{\partial u}{\partial x_2} = -ik\beta u \quad \text{on } \Gamma. \tag{3.12}$$

Notice that $u(x_1,0)=u_1(x_1,0)$ if $\beta(x_1,0)\equiv 1$. Then a numerically useful form of the Green's function in (3.9) is [10]

$$G(\boldsymbol{x},\boldsymbol{y})=\frac{\mathrm{i}}{2}H_0^{(1)}(kr)+\frac{\mathrm{e}^{\mathrm{i}kr}}{\pi}\int_0^\infty\frac{t^{-1/2}\mathrm{e}^{-krt}(1+\gamma+\gamma\mathrm{i}t)}{\sqrt{t-2\mathrm{i}}(t-\mathrm{i}-\mathrm{i}\gamma)^2}\,\mathrm{d}t, \tag{3.13}$$

where $\boldsymbol{w}=\boldsymbol{x}-\boldsymbol{y}$, $r=|\boldsymbol{w}|$, $\gamma=w_2/r$ and $H_0^{(1)}(z)$ is a Hankel function of the first kind, as specified more fully in [10,11]. By taking $x_2=0$ in (3.9), we see from (3.13) that $u(x_1,0)$ satisfies an integral equation with Toeplitz structure, and the fast Fourier transform yields its iterative solution efficiently.

Without loss of generality, we use the scale determined by $k=1$ in (3.9)–(3.13). For this example, the impedance is taken to be $\beta=1.4\mathrm{e}^{\mathrm{i}\pi/4}$ on the interval $\Gamma=\{\boldsymbol{x}\colon -40\pi<x_1<40\pi,\ x_2=0\}$. At two sample points ($x_1\approx-20\pi$ and 20π) taken from a 400-point discretisation of Γ, we found the following results with the VEA using 16 decimal place (MATLAB) arithmetic

$$\begin{aligned}
\varepsilon_0^{(12)}&=[\ldots,-\mathbf{0.36}843+\mathbf{0.4}4072\mathrm{i},\ldots,-\mathbf{0.145}07+\mathbf{0.5}5796\mathrm{i},\ldots],\\
\varepsilon_2^{(10)}&=[\ldots,-\mathbf{0.363}33+\mathbf{0.45}614\mathrm{i},\ldots,-\mathbf{0.1456}5+\mathbf{0.563}42\mathrm{i},\ldots],\\
\varepsilon_4^{(8)}&=[\ldots,-\mathbf{0.36341}+\mathbf{0.4558}2\mathrm{i},\ldots,-\mathbf{0.1456}8+\mathbf{0.563}12\mathrm{i},\ldots],\\
\varepsilon_6^{(6)}&=[\ldots,-\mathbf{0.36341}+\mathbf{0.45583}\mathrm{i},\ldots,-\mathbf{0.14569}+\mathbf{0.56311}\mathrm{i},\ldots],\\
\varepsilon_8^{(4)}&=[\ldots,-\mathbf{0.36341}+\mathbf{0.45583}\mathrm{i},\ldots,-\mathbf{0.14569}+\mathbf{0.56311}\mathrm{i},\ldots],
\end{aligned}$$

where the converged figures are shown in bold face.

Each of these results, showing just two of the components of a particular $\varepsilon_\kappa^{(j)}$ in columns $\kappa=0,2,\ldots,8$ of the vector-epsilon table, needs 12 iterations of (3.9) for its construction. In this application, these results show that the VEA converges reasonably steadily, in contrast to Lanczos type methods, eventually yielding five decimal places of precision.

Example 3.2 was chosen partly to demonstrate the use of the vector epsilon algorithm for a weakly convergent sequence of complex-valued data, and partly because the problem is one which lends itself to iterative methods. In fact, the example also shows that the VEA has used up 11 of the 15 decimal places of accuracy of the data to extrapolate the sequence to its limit. If greater precision is required, other methods such as stabilised Lanczos or multigrid methods should be considered.

The success of the VEA in examples such as those given above is usually attributed to the fact that the entries $\{\varepsilon_{2k}^{(j)}, j=0,1,2,\ldots\}$ are the exact limit of a convergent sequence S if S is generated by precisely k nontrivial geometric components. This result is an immediate and direct generalisation of that for the scalar case given in Section 2. The given vector sequence is represented by

$$\boldsymbol{s}_j=\boldsymbol{v}_0+\sum_{\kappa=1}^{k}\boldsymbol{v}_\kappa\sum_{i=0}^{j}(\theta_\kappa)^i=\boldsymbol{s}_\infty-\sum_{\kappa=1}^{k}\boldsymbol{w}_\kappa(\theta_\kappa)^j,\quad j=0,1,2,\ldots, \tag{3.14}$$

where each $\boldsymbol{v}_\kappa,\boldsymbol{w}_\kappa\in\mathbb{C}^d$, $\theta_\kappa\in\mathbb{C}$, $|\theta_\kappa|<1$, and all the θ_κ are distinct. The two representations used in (3.14) are consistent if

$$\sum_{\kappa=0}^{k}\boldsymbol{v}_\kappa=\boldsymbol{s}_\infty-\sum_{\kappa=1}^{k}\boldsymbol{w}_\kappa \quad\text{and}\quad \boldsymbol{v}_\kappa=\boldsymbol{w}_\kappa(\theta_\kappa^{-1}-1).$$

To establish this convergence result, and its generalisations, we must set up a formalism which allows vectors to be treated algebraically.

From the given sequence $\boldsymbol{S} = (\boldsymbol{s}_i, i = 0, 1, 2, \ldots, : \boldsymbol{s}_i \in \mathbb{C}^d)$, we form the series coefficients

$$\boldsymbol{c}_0 := \boldsymbol{s}_0, \quad \boldsymbol{c}_i := \boldsymbol{s}_i - \boldsymbol{s}_{i-1}, \qquad i = 1, 2, 3, \ldots \tag{3.15}$$

and the associated generating function

$$\boldsymbol{f}(z) = \boldsymbol{c}_0 + \boldsymbol{c}_1 z + \boldsymbol{c}_2 z^2 + \cdots \in \mathbb{C}^d[[z]]. \tag{3.16}$$

Our first aim is to find an analogue of (2.15) which allows construction, at least in principle, of the denominator polynomials of a vector-valued Padé approximant for $\boldsymbol{f}(z)$. This generalisation is possible if the vectors $\boldsymbol{c}_j$ in (3.16) are put in one–one correspondence with operators c_j in a Clifford algebra $\mathscr{A}$. The details of how this is done using an explicit matrix representation were basically set out by McLeod [37]. We use his approach [26,27,38] and square matrices E_i, $i = 1, 2, \ldots, 2d+1$ of dimension 2^{2d+1} which obey the anticommutation relations

$$E_i E_j + E_j E_i = 2\delta_{ij} I, \tag{3.17}$$

where I is an identity matrix. The special matrix $J = E_{2d+1}$ is used to form the operator products

$$F_i = J E_{d+i}, \quad i = 1, 2, \ldots, d. \tag{3.18}$$

Then, to each vector $\boldsymbol{w} = \boldsymbol{x} + \mathrm{i}\boldsymbol{y} \in \mathbb{C}^d$ whose real and imaginary parts $\boldsymbol{x}, \boldsymbol{y} \in \mathbb{R}^d$, we associate the operator

$$w = \sum_{i=1}^{d} x_i E_i + \sum_{i=1}^{d} y_i F_i. \tag{3.19}$$

The real linear space $\mathscr{V}_\mathbb{C}$ is defined as the set of all elements of the form (3.19). If $w_1, w_2 \in \mathscr{V}_\mathbb{C}$ correspond to $\boldsymbol{w}_1, \boldsymbol{w}_2 \in \mathbb{C}^d$ and α, β are real, then

$$w_3 = \alpha w_1 + \beta w_2 \in \mathscr{V}_\mathbb{C} \tag{3.20}$$

corresponds uniquely to $\boldsymbol{w}_3 = \alpha \boldsymbol{w}_1 + \beta \boldsymbol{w}_2 \in \mathbb{C}^d$. Were α, β complex, the correspondence would not be one–one. We refer to the space $\mathscr{V}_\mathbb{C}$ as the isomorphic image of $\mathbb{C}^d$, where the isomorphism preserves linearity only in respect of real multipliers as shown in (3.20). Thus the image of $\boldsymbol{f}(z)$ is

$$f(z) = c_0 + c_1 z + c_2 z^2 + \cdots \in \mathscr{V}_\mathbb{C}[[z]]. \tag{3.21}$$

The elements E_i, $i = 1, 2, \ldots, 2d+1$ are often called the basis vectors of $\mathscr{A}$, and their linear combinations are called the vectors of $\mathscr{A}$. Notice that the F_i are not vectors of $\mathscr{A}$ and so the vectors of $\mathscr{A}$ do not form the space $\mathscr{V}_\mathbb{C}$. Products of the nonnull vectors of $\mathscr{A}$ are said to form the Lipschitz group [40]. The reversion operator, denoted by a tilde, is defined as the anti-automorphism which reverses the order of the vectors constituting any element of the Lipschitz group and the operation is extended to the whole algebra $\mathscr{A}$ by linearity. For example, if $\alpha, \beta \in \mathbb{R}$ and

$$D = \alpha E_1 + \beta E_4 E_5 E_6,$$

then

$$\tilde{D} = \alpha E_1 + \beta E_6 E_5 E_4.$$

Hence (3.18) and (3.19) imply that

$$\tilde{w} = \sum_{i=1}^{d} x_i E_i - \sum_{i=1}^{d} y_i F_i. \tag{3.22}$$

We notice that $\tilde{w}$ corresponds to $\boldsymbol{w}^*$, the complex conjugate of $\boldsymbol{w}$, and that

$$\tilde{w}w = \sum_{i=1}^{d} (x_i^2 + y_i^2)I = ||\boldsymbol{w}||_2^2 I \tag{3.23}$$

is a real scalar in $\mathscr{A}$. The linear space of real scalars in $\mathscr{A}$ is defined as $\mathscr{S} := \{\alpha I,\ \alpha \in \mathbb{R}\}$. Using (3.23) we can form reciprocals, and

$$w^{-1} = \tilde{w}/|w|^2, \tag{3.24}$$

where

$$|w| := ||\boldsymbol{w}||, \tag{3.25}$$

so that w^{-1} is the image of $\boldsymbol{w}^{-1}$ as defined by (3.5). Thus (3.19) specifies an isomorphism between
(i) the space $\mathbb{C}^d$, having representative element

$$\boldsymbol{w} = \boldsymbol{x} + \mathrm{i}\boldsymbol{y} \quad \text{and an inverse} \quad \boldsymbol{w}^{-1} = \boldsymbol{w}^*/||\boldsymbol{w}||^2,$$

(ii) the real linear space $\mathscr{V}_\mathbb{C}$ with a representative element

$$w = \sum_{i=1}^{d} x_i E_i + \sum_{i=1}^{d} y_i F_i \quad \text{and its inverse given by} \quad w^{-1} = \tilde{w}/|w|^2.$$

The isomorphism preserves inverses and linearity with respect to real multipliers, as shown in (3.20). Using this formalism, we proceed to form the polynomial $q_{2j+1}(z)$ analogously to (2.15). The equations for its coefficients are

$$\begin{bmatrix} c_0 & \cdots & c_j \\ \vdots & & \vdots \\ c_j & \cdots & c_{2j} \end{bmatrix} \begin{bmatrix} q_{j+1}^{(2j+1)} \\ \vdots \\ q_1^{(2j+1)} \end{bmatrix} = \begin{bmatrix} -c_{j+1} \\ \vdots \\ -c_{2j+1} \end{bmatrix} \tag{3.26}$$

which represent the accuracy-through-order conditions; we assume that $q_0^{(2j+1)} = q_{2j+1}(0) = I$. In principle, we can eliminate the variables $q_{j+1}^{(2j+1)}, q_j^{(2j+1)}, \ldots, q_2^{(2j+1)}$ sequentially, find $q_1^{(2j+1)}$ and then the rest of the variables of (3.26) by back-substitution. However, the resulting $q_i^{(2j+1)}$ turn out to be higher grade quantities in the Clifford algebra, meaning that they involve higher-order outer products of the fundamental vectors. Numerical representation of these quantities uses up computer storage and is undesirable. For practical purposes, we prefer to work with low-grade quantities such as scalars and vectors [42].

The previous remarks reflect the fact that, in general, the product $w_1, w_2, w_3 \notin \mathscr{V}_\mathbb{C}$ when $w_1, w_2, w_3 \in \mathscr{V}_\mathbb{C}$. However, there is an important exception to this rule, which we formulate as follows [26], see Eqs. (6.3) and (6.4) in [40].

Lemma 3.3. *Let $w, t \in \mathscr{V}_\mathbb{C}$ be the images of $\boldsymbol{w} = \boldsymbol{x} + \mathrm{i}\boldsymbol{y}$, $\boldsymbol{t} = \boldsymbol{u} + \mathrm{i}\boldsymbol{v} \in \mathbb{C}^d$. Then*

(i) $t\tilde{w} + w\tilde{t} = 2\,\mathrm{Re}(\boldsymbol{w}^{\mathrm{H}}\boldsymbol{t})I \in \mathscr{S}$, (3.27)

(ii) $w\tilde{t}w = 2w\,\mathrm{Re}(\boldsymbol{w}^{\mathrm{H}}\boldsymbol{t}) - t||\boldsymbol{w}||^2 \in \mathscr{V}_\mathbb{C}$. (3.28)

Proof. Using (3.17), (3.18) and (3.22), we have

$$t\tilde{w} + w\tilde{t} = \sum_{i=1}^{d}\sum_{j=1}^{d}(u_iE_i + v_iF_i)(x_jE_j - y_jF_j) + (x_jE_j + y_jF_j)(u_iE_i - v_iF_i)$$
$$= (\boldsymbol{u}^{\mathrm{T}}\boldsymbol{x} + \boldsymbol{v}^{\mathrm{T}}\boldsymbol{y})I = 2\,\mathrm{Re}(\boldsymbol{w}^{\mathrm{H}}\boldsymbol{t})I$$

because, for $i,j = 1,2,\ldots,d$,

$$F_iE_j - E_jF_i = 0, \quad F_iF_j + F_jF_i = -2\delta_{ij}I.$$

For part (ii), we simply note that

$$w\tilde{t}w = w(\tilde{t}w + \tilde{w}t) - w\tilde{w}t. \qquad \Box$$

We have noted that, as j increases, the coefficients of $q_{2j+1}(z)$ are increasingly difficult to store. Economical approximations to $q_{2j+1}{}^{(z)}$ are given in [42]. Here we proceed with

$$\begin{bmatrix} c_0 & \cdots & c_{j+1} \\ \vdots & & \vdots \\ c_{j+1} & \cdots & c_{2j+2} \end{bmatrix} \begin{bmatrix} q_{j+1}^{(2j+1)} \\ \vdots \\ q_1^{(2j+1)} \\ I \end{bmatrix} = \begin{bmatrix} 0 \\ \vdots \\ 0 \\ e_{2j+1} \end{bmatrix} \tag{3.29}$$

which are the accuracy-through-order conditions for a right-handed operator Padé approximant (OPA) $p_{2j+1}(z)[q_{2j+1}(z)]^{-1}$ for $f(z)$ arising from

$$f(z)q_{2j+1}(z) = p_{2j+1}(z) + e_{2j+1}z^{2j+2} + \mathrm{O}(z^{2j+3}). \tag{3.30}$$

The left-hand side of (3.29) contains a general square Hankel matrix with elements that are operators from $\mathcal{V}_{\mathbb{C}}$. A remarkable fact, by no means obvious from (3.29) but proved in the next theorem, is that

$$e_{2j+1} \in \mathcal{V}_{\mathbb{C}}. \tag{3.31}$$

This result enables us to use OPAs of $f(z)$ without constructing the denominator polynomials. A quantity such as e_{2j+1} in (3.29) is called the left-designant of the operator matrix and it is denoted by

$$e_{2j+1} = \begin{vmatrix} c_0 & \cdots & c_{j+1} \\ \vdots & & \vdots \\ c_{j+1} & \cdots & c_{2j+2} \end{vmatrix}_{\ell}. \tag{3.31b}$$

The subscript ℓ (for left) distinguishes designants from determinants, which are very different constructs. Designants were introduced by Heyting [32] and in this context by Salam [43]. For present purposes, we regard them as being defined by the elimination process following (3.26).

Example 3.4. The denominator of the OPA of type [0/1] is constructed using

$$\begin{bmatrix} c_0 & c_1 \\ c_1 & c_2 \end{bmatrix}\begin{bmatrix} q_1^{(1)} \\ I \end{bmatrix} = \begin{bmatrix} 0 \\ e_1 \end{bmatrix}.$$

We eliminate $q_1^{(1)}$ as described above following (3.26) and find that

$$e_1 = \begin{vmatrix} c_2 & c_1 \\ c_1 & c_0 \end{vmatrix}_\ell = c_2 - c_1 c_0^{-1} c_1 \in \mathrm{span}\{c_0, c_1, c_2\}. \tag{3.32}$$

Proceeding with the elimination in (3.29), we obtain

$$\begin{bmatrix} c_2 - c_1 c_0^{-1} c_1 & \cdots & c_{j+2} - c_1 c_0^{-1} c_{j+1} \\ \vdots & & \vdots \\ c_{j+2} - c_{j+1} c_0^{-1} c_1 & \cdots & c_{2j+2} - c_{j+1} c_0^{-1} c_{j+1} \end{bmatrix} \begin{bmatrix} q_{j+2}^{(2j+1)} \\ \vdots \\ q_1^{(2j+1)} \\ I \end{bmatrix} = \begin{bmatrix} 0 \\ \vdots \\ 0 \\ e_{2j+1} \end{bmatrix}. \tag{3.33}$$

Not all the elements of the matrix in (3.33) are vectors. An inductive proof that e_{2j+1} is a vector (at least in the case when the c_j are real vectors and the algebra is a division ring) was given by Salam [43,44] and Roberts [41] using the designant forms of Sylvester's and Schweins' identities.

We next construct the numerator and denominator polynomials of the OPAs of $f(z)$ and prove (3.31) using Berlekamp's method [3], which leads on to the construction of vector Padé approximants.

Definitions. Given the series expansion (3.22) of $f(z)$, numerator and denominator polynomials $A_j(z)$, $B_j(z) \in A[z]$ of degrees ℓ_j, m_j are defined sequentially for $j = 0, 1, 2, \ldots$, by

$$A_{j+1}(z) = A_j(z) - z A_{j-1}(z) e_{j-1}^{-1} e_j, \tag{3.34}$$

$$B_{j+1}(z) = B_j(z) - z B_{j-1}(z) e_{j-1}^{-1} e_j \tag{3.35}$$

in terms of the error coefficients e_j and auxiliary polynomials $D_j(z)$ which are defined for $j=0, 1, 2, \ldots$ by

$$e_j := [f(z) B_j(z) \tilde{B}_j(z)]_{j+1}, \tag{3.36}$$

$$D_j(z) := \tilde{B}_j(z) B_{j-1}(z) e_{j-1}^{-1}. \tag{3.37}$$

These definitions are initialised with

$$\begin{aligned} &A_0(z) = c_0, \quad B_0(z) = I, \quad e_0 = c_1, \\ &A_{-1}(z) = 0, \quad B_{-1}(z) = I, \quad e_{-1} = c_0. \end{aligned} \tag{3.38}$$

Example 3.5.

$$\begin{aligned} &A_1(z) = c_0, \quad B_1(z) = I - z c_0^{-1} c_1, \quad e_1 = c_2 - c_1 c_0^{-1} c_1, \\ &D_1(z) = c_1^{-1} - z \tilde{c}_1 \tilde{c}_0^{-1} c_1^{-1}. \end{aligned} \tag{3.39}$$

Lemma 3.6.

$$B_j(0) = I, \quad j = 0, 1, 2, \ldots . \tag{3.40}$$

Proof. See (3.35) and (3.38). □

Theorem 3.7. *With the definitions above, for* $j = 0, 1, 2, \dots,$

(i) $f(z)B_j(z) - A_j(z) = O(z^{j+1}).$ (3.41)

(ii) $\ell_j := \deg\{A_j(z)\} = [j/2], \quad m_j := \deg\{B_j(z)\} = [(j+1)/2], \quad \deg\{A_j(z)\tilde{B}_j(z)\} = j.$ (3.42)

(iii) $B_j(z)\tilde{B}_j(z) = \tilde{B}_j(z)B_j(z) \in \mathscr{S}[z].$ (3.43)

(iv) $e_j \in \mathscr{V}_C.$ (3.44)

(v) $D_j(z), \ A_j(z)\tilde{B}_j(z) \in \mathscr{V}_C[z].$ (3.45)

(vi) $f(z)B_j(z) - A_j(z) = e_j z^{j+1} + O(z^{j+2}).$ (3.46)

Proof. Cases $j=0,1$ are verified explicitly using (3.38) and (3.39). We make the inductive hypothesis that (i)–(vi) hold for index j as stated, and for index $j-1$.

Part (i): Using (3.34), (3.35) and the inductive hypothesis (vi),

$$f(z)B_{j+1}(z) - A_{j+1}(z) = f(z)B_j(z) - A_j(z) - z(f(z)B_{j-1}(z) - A_{j-1}(z))e_{j-1}^{-1}e_j = O(z^{j+2}).$$

Part (ii): This follows from (3.34), (3.35) and the inductive hypothesis (ii).

Part (iii): Using (3.27) and (3.35), and hypotheses (iii)–(iv) inductively,

$$\tilde{B}_{j+1}(z)B_{j+1}(z) = \tilde{B}_j(z)B_j(z) + z^2\tilde{B}_{j-1}(z)B_{j-1}(z)|e_j|^2|e_{j-1}|^{-2} - z[D_j(z)e_j + \tilde{e}_j\tilde{D}_j(z)] \in \mathscr{S}[z]$$

and (iii) follows after postmultiplication by $\tilde{B}_{j+1}(z)$ and premultiplication by $[\tilde{B}_{j+1}(z)]^{-1}$, see [37, p. 45].

Part (iv): By definition (3.36),

$$e_{j+1} = \sum_{i=0}^{2m_{j+1}} c_{j+2-i}\beta_i,$$

where each $\beta_i = [B_{j+1}(z)\tilde{B}_{j+1}(z)]_i \in \mathscr{S}$ is real. Hence

$$e_{j+1} \in \mathscr{V}_C.$$

Part (v): From (3.35) and (3.37),

$$D_{j+1}(z) = [\tilde{B}_j(z)B_j(z)]e_j^{-1} - z[\tilde{e}_j\tilde{D}_j(z)e_j^{-1}].$$

Using part (v) inductively, parts (iii), (iv) and Lemma 3.3, it follows that $D_{j+1}(z) \in \mathscr{V}_C[z]$.

Using part (i), (3.40) and the method of proof of part (iv), we have

$$A_{j+1}(z)\tilde{B}_{j+1}(z) = [f(z)B_{j+1}(z)\tilde{B}_{j+1}(z)]_0^{j+1} \in \mathscr{V}_C[z].$$

Part (vi): From part (i), we have

$$f(z)B_{j+1}(z) - A_{j+1}(z) = \gamma_{j+1}z^{j+2} + O(z^{j+3})$$

for some $\gamma_{j+1} \in \mathscr{A}$. Hence,

$$f(z)B_{j+1}(z)\tilde{B}_{j+1}(z) - A_{j+1}(z)\tilde{B}_{j+1}(z) = \gamma_{j+1}z^{j+2}\tilde{B}_{j+1}(z) + O(z^{j+3}).$$

Using (ii) and (3.40), we obtain $\gamma_{j+1} = e_{j+1}$, as required. □

Corollary. *The designant of a Hankel matrix of real (or complex) vectors is a real (or complex) vector.*

Proof. Any designant of this type is expressed by e_{2j+1} in (3.31b), and (3.44) completes the proof. □

The implications of the previous theorem are extensive. From part (iii) we see that

$$Q_j(z).I := B_j(z)\tilde{B}_j(z) \tag{3.47}$$

defines a real polynomial $Q_j(z)$. Part (iv) shows that the e_j are images of vectors $\boldsymbol{e}_j \in \mathbb{C}^d$; part (vi) justifies calling them error vectors but they are also closely related to the residuals $\boldsymbol{b} - A\varepsilon_{2j}^{(0)}$ of Example 3.1. Part (v) shows that $A_j(z)\tilde{B}_j(z)$ is the image of some $\boldsymbol{P}_j(z) \in \mathbb{C}^d[z]$, so that

$$A_j(z)\tilde{B}_j(z) = \sum_{i=1}^{d} [\mathrm{Re}\{\boldsymbol{P}_j\}(z)]_i E_i + \sum_{i=1}^{d} [\mathrm{Im}\{\boldsymbol{P}_j\}(z)]_i F_i. \tag{3.48}$$

From (3.17) and (3.18), it follows that

$$\boldsymbol{P}_j(z) \cdot \boldsymbol{P}_j^*(z) = Q_j(z)\hat{Q}_j(z), \tag{3.49}$$

where $\hat{Q}_j(z)$ is a real scalar polynomial determined by $\hat{Q}_j(z)I = A_j(z)\tilde{A}_j(z)$. Property (3.49) will later be used to characterise certain VPAs independently of their origins in $\mathscr{A}$. Operator Padé approximants were introduced in (3.34) and (3.35) so as to satisfy the accuracy-through-order property (3.41) for $f(z)$. To generalise to the full table of approximants, only initialisation (3.38) and the degree specifications (3.42) need to be changed.

For $J > 0$, we use

$$A_0^{(J)}(z) = \sum_{i=0}^{J} c_i z^i, \quad B_0^{(J)}(z) = I, \quad e_0^{(J)} = c_{J+1},$$

$$A_{-1}^{(J)}(z) = \sum_{i=0}^{J-1} c_i z^i, \quad B_{-1}^{(J)}(z) = I, \quad e_{-1}^{(J)} = c_J, \tag{3.50}$$

$$\ell_j^{(J)} := \deg\{A_j^{(J)}(z)\} = J + [j/2],$$

$$m_j^{(J)} := \deg\{B_j^{(J)}(z)\} = [(j+1)/2] \tag{3.51}$$

and then (3.38) and (3.42) correspond to the case of $J = 0$.

For $J < 0$, we assume that $c_0 \neq 0$, and define

$$g(z) = [f(z)]^{-1} = \tilde{f}(z)[f(z)\tilde{f}(z)]^{-1} \tag{3.52}$$

corresponding to

$$\boldsymbol{g}(z) = [\boldsymbol{f}(z)]^{-1} = \boldsymbol{f}^*(z)[\boldsymbol{f}(z).\boldsymbol{f}^*(z)]^{-1}. \tag{3.53}$$

(If $c_0 = 0$, we would remove a maximal factor of z^ν from $f(z)$ and reformulate the problem.)

Then, for $J < 0$,

$$A_0^{(J)}(z) = I, \quad B_0^{(J)}(z) = \sum_{i=0}^{-J} g_i z^i, \quad e_0^{(J)} = [f(z)B_0^{(J)}(z)]_{1-J},$$

$$A_1^{(J)}(z) = I, \quad B_1^{(J)}(z) = \sum_{i=0}^{1-J} g_i z^i, \quad e_1^{(J)} = [f(z)B_1^{(J)}(z)]_{2-J},$$

$$\ell_j^{(J)} := \deg\{A_j^{(J)}(z)\} = [j/2],$$

$$m_j^{(J)} := \deg\{B_j^{(J)}(z)\} = [(j+1)/2] - J. \tag{3.54}$$

If an approximant of given type $[\ell/m]$ is required, there are usually two different staircase sequences of the form

$$S^{(J)} = (A_j^{(J)}(z)[B_j^{(J)}(z)]^{-1}, \quad j = 0, 1, 2, \ldots) \tag{3.55}$$

which contain the approximant, corresponding to two values of J for which $\ell = \ell_j^{(J)}$ and $m = m_j^{(J)}$. For ease of notation, we use $p^{[\ell/m]}(z) \equiv A_j^{(J)}(z)$ and $q^{[\ell/m]}(z) \equiv B_j^{(J)}(z)$. The construction based on (3.41) is for right-handed OPAs, as in

$$f(z) = p^{[\ell/m]}(z)[q^{[\ell/m]}(z)]^{-1} + \mathrm{O}(z^{\ell+m+1}), \tag{3.56}$$

but the construction can easily be adapted to that for left-handed OPAs for which

$$f(z) = [\check{q}^{[\ell/m]}(z)]^{-1}\check{p}^{[\ell/m]}(z) + \mathrm{O}(z^{\ell+m+1}). \tag{3.57}$$

Although the left- and right-handed numerator and denominator polynomials usually are different, the actual OPAs of given type are equal:

Theorem 3.8 (Uniqueness). *Left-handed and right-handed OPAs, as specified by* (3.56) *and* (3.57) *are identical*:

$$[\ell/m](z) := p^{[\ell/m]}(z)[q^{[\ell/m]}(z)]^{-1} = [\check{q}^{[\ell/m]}(z)]^{-1}\check{p}^{[\ell/m]}(z) \in \mathscr{V}_{\mathbb{C}} \tag{3.58}$$

and the OPA of type $[\ell/m]$ *for* $f(z)$ *is unique.*

Proof. Cross-multiply (3.58), use (3.56), (3.57) and then (3.40) to establish the formula in (3.58). Uniqueness of $[\ell/m](z)$ follows from this formula too, and its vector character follows from (3.43) and (3.45). □

The OPAs and the corresponding VPAs satisfy the compass (five-point star) identity amongst approximants of the type shown in the same format as Fig. 3.

Theorem 3.9 (Wynn's compass identity [57,58]).

$$[\boldsymbol{N}(z) - \boldsymbol{C}(z)]^{-1} + [\boldsymbol{S}(z) - \boldsymbol{C}(z)]^{-1} = [\boldsymbol{E}(z) - \boldsymbol{C}(z)]^{-1} + [\boldsymbol{W}(z) - \boldsymbol{C}(z)]^{-1}. \tag{3.59}$$

Proof. We consider the accuracy-through-order equations for the operators:

$$\breve{p}_N(z)q_C(z) - \breve{q}_N(z)p_C(z) = z^{\ell+m}\dot{\breve{p}}_N\dot{q}_C,$$

$$\breve{p}_C(z)q_W(z) - \breve{q}_C(z)p_W(z) = z^{\ell+m}\dot{\breve{p}}_C\dot{q}_W,$$

$$\breve{p}_N(z)q_W(z) - \breve{q}_N(z)p_W(z) = z^{\ell+m}\dot{\breve{p}}_N\dot{q}_W,$$

where $\dot{q}_\Omega$, $\dot{p}_\Omega$ denote the leading coefficients of $p_\Omega(z)$, $q_\Omega(z)$, and care has been taken to respect noncommutativity. Hence

$$\begin{aligned}&[N(z) - C(z)]^{-1} - [W(z) - C(z)]^{-1}\\ &\quad = [N(z) - C(z)]^{-1}(W(z) - N(z))[W(z) - C(z)]^{-1}\\ &\quad = q_C[\breve{p}_N q_C - \breve{q}_N p_C]^{-1}(\breve{q}_N p_W - \breve{p}_N q_W)[\breve{q}_C p_W - \breve{p}_C q_W]^{-1}\breve{q}_C\\ &\quad = z^{-\ell-m}q_C(z)\dot{q}_C^{-1}\dot{\breve{p}}_C^{-1}\breve{q}_C(z).\end{aligned}$$

Similarly, we find that

$$[E(z) - C(z)]^{-1} - [S(z) - C(z)]^{-1} = z^{-\ell-m}q_C(z)\dot{q}_C^{-1}\dot{\breve{p}}_C^{-1}\breve{q}_C(z)$$

and hence (3.59) is established in its operator form. Complex multipliers are not used in it, and so (3.59) holds as stated. □

An important consequence of the compass identity is that, with $z = 1$, it becomes equivalent to the vector epsilon algorithm for the construction of $\boldsymbol{E}(1)$ as we saw in the scalar case. If the elements $\boldsymbol{s}_j \in \boldsymbol{S}$ have representation (3.14), there exists a scalar polynomial $b(z)$ of degree k such that

$$\boldsymbol{f}(z) = \boldsymbol{a}(z)/b(z) \in \mathbb{C}^d[[z]]. \tag{3.60}$$

If the coefficients of $b(z)$ are real, we can uniquely associate an operator $f(z)$ with $\boldsymbol{f}(z)$ in (3.60), and then the uniqueness theorem implies that

$$\varepsilon_{2k}^{(j)} = \boldsymbol{f}(1) \tag{3.61}$$

and we are apt to say that column $2k$ of the epsilon table is exact in this case. However, Example 3.2 indicates that the condition that $b(z)$ must have real coefficients is not necessary. For greater generality in this respect, generalised inverse, vector-valued Padé approximants (GIPAs) were introduced [22]. The existence of a vector numerator polynomial $\boldsymbol{P}^{[n/2k]}(z) \in \mathbb{C}^d[z]$ and a real scalar denominator polynomial $Q^{[n/2k]}(z)$ having the following properties is normally established by (3.47) and (3.48):

$$\text{(i)}\quad \deg\{\boldsymbol{P}^{[n/2k]}(z)\} = n, \quad \deg\{Q^{[n/2k]}(z)\} = 2k, \tag{3.62}$$

$$\text{(ii)}\quad Q^{[n/2k]}(z) \text{ is a factor of } \boldsymbol{P}^{[n/2k]}(z).\boldsymbol{P}^{[n/2k]*}(z), \tag{3.63}$$

$$\text{(iii)}\quad Q^{[n/2k]}(0) = 1, \tag{3.64}$$

$$\text{(iv)}\quad \boldsymbol{f}(z) - \boldsymbol{P}^{[n/2k]}(z)/Q^{[n/2k]}(z) = O(z^{n+1}), \tag{3.65}$$

where the star in (3.63) denotes the functional complex-conjugate. These axioms suffice to prove the following result.

Theorem 3.10 (Uniqueness [24]). *If the vector-valued Padé approximant*

$$\boldsymbol{R}^{[n/2k]}(z) := \boldsymbol{P}^{[n/2k]}(z)/Q^{[n/2k]}(z) \tag{3.66}$$

of type $[n/2k]$ *for* $\boldsymbol{f}(z)$ *exists, then it is unique.*

Proof. Suppose that

$$\boldsymbol{R}(z) = \boldsymbol{P}(z)/Q(z), \quad \hat{\boldsymbol{R}}(z) = \hat{\boldsymbol{P}}(z)/\hat{Q}(z)$$

are two different vector-valued Padé approximants having the same specification as (3.62)–(3.66). Let $Q_{\text{gcd}}(z)$ be the greatest common divisor of $Q(z)$, $\hat{Q}(z)$ and define reduced and coprime polynomials by

$$Q_r(z) = Q(z)/Q_{\text{gcd}}(z), \quad \hat{Q}_r(z) = \hat{Q}(z)/Q_{\text{gcd}}(z).$$

From (3.63) and (3.65) we find that

$$z^{2n+2}Q_r(z)\hat{Q}_r(z) \text{ is a factor of } [\boldsymbol{P}(z)\hat{Q}_r(z) - \hat{\boldsymbol{P}}(z)Q_r(z)] \cdot [\boldsymbol{P}^*(z)\hat{Q}_r(z) - \hat{\boldsymbol{P}}^*(z)Q_r(z)]. \tag{3.67}$$

The left-hand expression of (3.67) is of degree $2n+4k-2.\deg\{Q_{\text{gcd}}(z)\}+2$. The right-hand expression of (3.67) is of degree $2n+4k-2.\deg\{Q_{\text{gcd}}(z)\}$. Therefore the right-hand expression of (3.67) is identically zero. □

By taking $\hat{Q}^{[n/2m]}(z) = b(z).b^*(z)$ and $\hat{\boldsymbol{P}}^{[n/2m]}(z) = \boldsymbol{a}(z)b^*(z)$, the uniqueness theorem shows that the generalised inverse vector-valued Padé approximant constructed using the compass identity yields

$$\boldsymbol{f}(z) = \boldsymbol{a}(z)b^*(z)/b(z)b^*(z)$$

exactly. On putting $z=1$, it follows that the sequence $\boldsymbol{S}$, such as the one given by (3.14), is summed exactly by the vector epsilon algorithm in the column of index $2k$. For normal cases, we have now outlined the proof of a principal result [37,2].

Theorem 3.11 (McLeod's theorem). *Suppose that the vector sequence* $\boldsymbol{S}$ *satisfies a nontrivial recursion relation*

$$\sum_{i=0}^{k} \beta_i \boldsymbol{s}_{i+j} = \left(\sum_{i=0}^{k} \beta_i \right) \boldsymbol{s}_\infty, \quad j = 0, 1, 2, \ldots \tag{3.68}$$

with $\beta_i \in \mathbb{C}$. *Then the vector epsilon algorithm leads to*

$$\varepsilon_{2k}^{(j)} = \boldsymbol{s}_\infty, \quad j = 0, 1, 2, \ldots \tag{3.69}$$

provided that zero divisors are not encountered in the construction.

The previous theorem is a statement about exact results in the column of index $2k$ in the vector epsilon table. This column corresponds to the row sequence of GIPAs of type $[n/2k]$ for $\boldsymbol{f}(z)$, evaluated at $z=1$. If the given vector sequence $\boldsymbol{S}$ is nearly, but not exactly, generalized geometric, we model this situation by supposing that its generating function $\boldsymbol{f}(z)$ is analytic in the closed unit disk $\bar{D}$, except for k poles in $D := \{z\colon |z| < 1\}$. This hypothesis ensures that $\boldsymbol{f}(z)$ is analytic at $z=1$, and it is sufficiently strong to guarantee convergence of the column of index $2k$ in the vector epsilon table. There are several convergence theorems of this type [28–30,39]. It is important to note that any row convergence theorem for generalised inverse vector-valued Padé approximants has immediate consequences as a convergence result for a column of the vector epsilon table.

A determinantal formula for $Q^{[n/2k]}(z)$ can be derived [24,25] by exploiting the factorisation property (3.63). The formula is

$$Q^{[n/2k]}(z) = \begin{vmatrix} 0 & M_{01} & M_{02} & \cdots & M_{0,2k} \\ M_{10} & 0 & M_{12} & \cdots & M_{1,2k} \\ \vdots & \vdots & \vdots & & \vdots \\ M_{2k-1,0} & M_{2k-1,1} & M_{2k-1,2} & \cdots & M_{2k-1,2k} \\ z^{2k} & z^{2k-1} & z^{2k-2} & \cdots & 1 \end{vmatrix}, \tag{3.70}$$

where the constant entries M_{ij} are those in the first $2k$ rows of an anti-symmetric matrix $M \in \mathbb{R}^{(2k+1)\times(2k+1)}$ defined by

$$M_{ij} = \begin{cases} \sum_{l=0}^{j-i-1} \boldsymbol{c}^{\mathrm{H}}_{\ell+i+n-2k+1} \cdot \boldsymbol{c}_{j-\ell+n-2k} & \text{for } j > i, \\ -M_{ji} & \text{for } i < j, \\ 0 & \text{for } i = j. \end{cases}$$

As a consequence of the compass identity (Theorem 3.9) and expansion (3.16), we see that entries in the vector epsilon table are given by

$$\varepsilon_{2k}^{(j)} = \boldsymbol{P}^{[j+2k/2k]}(1)/Q^{[j+2k/2k]}(1), \quad j,k \geqslant 0,$$

From this result, it readily follows that each entry in the columns of even index in the vector epsilon table is normally given succinctly by a ratio of determinants:

$$\varepsilon_{2k}^{(j)} = \begin{vmatrix} 0 & M_{01} & \cdots & M_{0,2k} \\ M_{10} & 0 & \cdots & M_{1,2k} \\ \vdots & \vdots & & \vdots \\ M_{2k-1,0} & M_{2k-1,1} & \cdots & M_{2k-1,2k} \\ \boldsymbol{s}_j & \boldsymbol{s}_{j+1} & \cdots & \boldsymbol{s}_{2k+j} \end{vmatrix} \div \begin{vmatrix} 0 & M_{01} & \cdots & M_{0,2k} \\ M_{10} & 0 & \cdots & M_{1,2k} \\ \vdots & \vdots & & \vdots \\ M_{2k-1,0} & M_{2k-1,1} & \cdots & M_{2k-1,2k} \\ 1 & 1 & \cdots & 1 \end{vmatrix}.$$

For computation, it is best to obtain numerical results from (3.4). The coefficients of $Q^{[n/2k]}(z) = \sum_{i=0}^{2k} Q_i^{[n/2k]} z^i$ should be found by solving the homogeneous, anti-symmetric (and therefore consistent) linear system equivalent to (3.70), namely

$$M\boldsymbol{q} = \boldsymbol{0},$$

where $\boldsymbol{q}^{\mathrm{T}} = (Q_{2k-i}^{[n/2k]},\ i = 0, 1, \ldots, 2k)$.

4. Vector-valued continued fractions and vector orthogonal polynomials

The elements $\varepsilon_{2k}^{(0)}$ lying at the head of each column of even index in the vector epsilon table are values of the convergents of a corresponding continued fraction. In Section 3, we noted that the entries in the vector epsilon table are values of vector Padé approximants of

$$\boldsymbol{f}(z) = \boldsymbol{c}_0 + \boldsymbol{c}_1 z + \boldsymbol{c}_2 z^2 + \cdots \tag{4.1}$$

as defined by (3.16). To obtain the continued fraction corresponding to (4.1), we use Viskovatov's algorithm, which is an ingenious rule for efficiently performing successive reciprocation and re-expansion of a series [2]. Because algebraic operations are required, we use the image of (4.1) in $\mathscr{A}$, which is

$$f(z) = c_0 + c_1 z + c_2 z^2 + \cdots \tag{4.2}$$

with $c_i \in \mathscr{V}_C$. Using reciprocation and re-expansion, we find

$$f(z) = \sum_{i=0}^{J-1} c_i z^i + \frac{z^J c_J}{1} - \frac{z\alpha_1^{(J)}}{1} - \frac{z\beta_1^{(J)}}{1} - \frac{z\alpha_2^{(J)}}{1} - \frac{z\beta_2^{(J)}}{1} - \cdots \tag{4.3}$$

with $\alpha_i^{(J)}, \beta_i^{(J)} \in \mathscr{A}$ and provided all $\alpha_i^{(J)} \neq 0$, $\beta_i^{(J)} \neq 0$. By definition, all the inverses implied in (4.3) are to be taken as right-handed inverses. For example, the second convergent of (4.3) is

$$[J+1/1](z) = \sum_{i=0}^{J-1} c_i z^i + z^J c_J [1 - z\alpha_1^{(J)}[1 - z\beta_1^{(J)}]^{-1}]^{-1}$$

and the corresponding element of the vector epsilon table is

$$\varepsilon_2^{(J)} = [J+1/1](1),$$

where the type refers to the allowed degrees of the numerator and denominator operator polynomials. The next algorithm is used to construct the elements of (4.3).

Theorem 4.1 (The vector qd algorithm [40]). *With the initialisation*

$$\beta_0^{(J)} = 0, \quad J = 1, 2, 3, \ldots, \tag{4.4}$$

$$\alpha_1^{(J)} = c_J^{-1} c_{J+1}, \quad J = 0, 1, 2, \ldots, \tag{4.5}$$

the remaining $\alpha_i^{(J)}, \beta_i^{(J)}$ *can be constructed using*

$$\alpha_m^{(J)} + \beta_m^{(J)} = \alpha_m^{(J+1)} + \beta_{m-1}^{(J+1)}, \tag{4.6}$$

$$\beta_m^{(J)} \alpha_{m+1}^{(J)} = \alpha_m^{(J+1)} \beta_m^{(J+1)} \tag{4.7}$$

for $J = 0, 1, 2, \ldots$ *and* $m = 1, 2, 3, \ldots$.

Remark. The elements connected by these rules form lozenges in the $\alpha - \beta$ array, as in Fig. 8. Rule (4.7) requires multiplications which are noncommutative except in the scalar case.

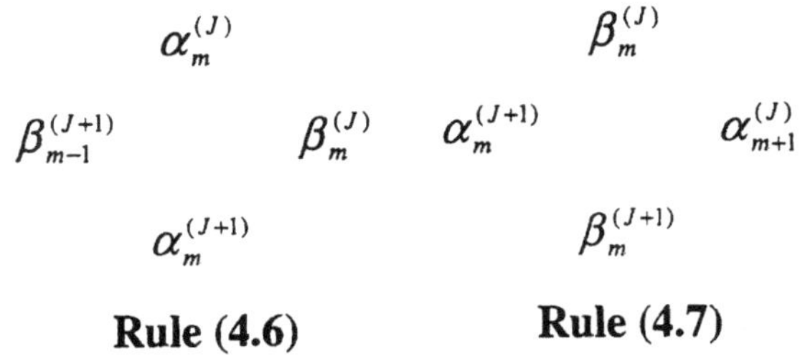

Fig. 8.

Proof. First, the identity

$$C + z\alpha[1 + z\beta D^{-1}]^{-1} = C + z\alpha - z^2\alpha\beta[z\beta + D]^{-1} \tag{4.8}$$

is applied to (4.3) with $\alpha = c_J$, $\beta = -\alpha_1^{(J)}$, then with $\alpha = -\beta_1^{(J)}$, $\beta = -\alpha_2^{(J)}$, etc. We obtain

$$f(z) = \sum_{i=0}^{J} c_i z^i + \frac{z^{J+1} c_J \alpha_1^{(J)}}{1 - z(\alpha_1^{(J)} + \beta_1^{(J)})} - \frac{z^2 \beta_1^{(J)} \alpha_2^{(J)}}{1 - z(\alpha_2^{(J)} + \beta_2^{(J)})} - \cdots . \tag{4.9}$$

Secondly, let $J \to J+1$ in (4.3), and then apply (4.8) with $\alpha = -\alpha_1^{(J+1)}$, $\beta = -\beta_1^{(J+1)}$, then with $\alpha = -\alpha_2^{(J+1)}$, $\beta = -\beta_2^{(J+1)}$, etc., to obtain

$$f(z) = \sum_{i=0}^{J} c_i z^i + \frac{z^{J+1} c_{J+1}}{1 - z\alpha_1^{(J+1)}} - \frac{z^2 \alpha_1^{(J+1)} \beta_1^{(J+1)}}{1 - z(\beta_1^{(J+1)} + \alpha_2^{(J+1)})} - \cdots . \tag{4.10}$$

These expansions (4.9) and (4.10) of $f(z)$ must be identical, and so (4.4)–(4.7) follow by identification of the coefficients. □

The purpose of this algorithm is the iterative construction of the elements of the C-fraction (4.3) starting from the coefficients $\mathbf{c}_i$ of (4.1). However, the elements $\alpha_i^{(J)}, \beta_i^{(J)}$ are not vectors in the algebra. Our next task is to reformulate this algorithm using vector quantities which are amenable for computational purposes.

The recursion for the numerator and denominator polynomials was derived in (3.34) and (3.35) for case of $J = 0$, and the more general sequence of approximants labelled by $J \geqslant 0$ was introduced in (3.50) and (3.51). For them, the recursions are

$$A_{j+1}^{(J)}(z) = A_j^{(J)}(z) - zA_{j-1}^{(J)}(z) e_{j-1}^{(J)-1} e_j^{(J)}, \tag{4.11}$$

$$B_{j+1}^{(J)}(z) = B_j^{(J)}(z) - zB_{j-1}^{(J)}(z) e_{j-1}^{(J)-1} e_j^{(J)} \tag{4.12}$$

and accuracy-through-order is expressed by

$$f(z) B_j^{(J)}(z) = A_j^{(J)}(z) + e_j^{(J)} z^{j+J+1} + \mathrm{O}(z^{j+J+2}) \tag{4.13}$$

for $j = 0, 1, 2, \ldots$ and $J \geqslant 0$. Euler's formula shows that (4.11) and (4.12) are the recursions associated with

$$f(z) = \sum_{i=0}^{J-1} c_i z^i + \frac{c_J z^J}{1} - \frac{e_0^{(J)} z}{1} - \frac{e_0^{(J)-1} e_1^{(J)} z}{1} - \frac{e_1^{(J)-1} e_2^{(J)} z}{1} - \cdots . \tag{4.14}$$

As was noted for (3.55), the approximant of (operator) type $[J+m/m]$ arising from (4.14) is also a convergent of (4.14) with $J \to J+1$. We find that

$$A_{2m}^{(J)}(z)[B_{2m}^{(J)}(z)]^{-1} = [J+m/m](z) = A_{2m-1}^{(J+1)}[B_{2m-1}^{(J+1)}(z)]^{-1} \tag{4.15}$$

and their error coefficients in (4.13) are also the same:

$$e_{2m}^{(J)} = e_{2m-1}^{(J+1)}, \quad m, J = 0, 1, 2, \dots . \tag{4.16}$$

These error vectors $e_i^{(J)} \in \mathscr{V}_C$ obey the following identity.

Theorem 4.2 (The cross-rule [27,40,41,46]). *With the partly artificial initialisation*

$$e_{-2}^{(J+1)} = \infty, \quad e_0^{(J)} = c_{J+1} \quad for\ J = 0, 1, 2, \dots, \tag{4.17}$$

the error vectors obey the identity

$$e_{i+2}^{(J-1)} = e_i^{(J+1)} + e_i^{(J)}[e_{i-2}^{(J+1)-1} - e_i^{(J-1)-1}]e_i^{(J)} \tag{4.18}$$

for $J \geqslant 0$ *and* $i \geqslant 0$.

Remark. These entries are displayed in Fig. 9 at positions corresponding to their associated approximants (see (4.13)) which satisfy the compass rule.

Proof. We identify the elements of (4.3) and (4.14) and obtain

$$\alpha_{j+1}^{(J)} = e_{2j-1}^{(J)-1} e_{2j}^{(J)}, \quad \beta_{j+1}^{(J)} = e_{2j}^{(J)-1} e_{2j+1}^{(J)}. \tag{4.19}$$

We use (4.16) to standardise on even-valued subscripts for the error vectors in (4.19):

$$\alpha_{j+1}^{(J)} = e_{2j}^{(J-1)-1} e_{2j}^{(J)}, \quad \beta_{j+1}^{(J)} = e_{2j}^{(J)-1} e_{2j+2}^{(J-1)}. \tag{4.20}$$

Substitute (4.20) in (4.6) with $m = j+1$ and $i = 2j$, giving

$$e_i^{(J-1)-1} e_i^{(J)} + e_i^{(J)-1} e_{i+2}^{(J-1)} = e_i^{(J)-1} e_i^{(J+1)} + e_{i-2}^{(J+1)-1} e_i^{(J)}. \tag{4.21}$$

Result (4.18) follows from (4.21) directly if i is even, but from (4.16) and (4.20) if i is odd. Initialisation (4.17) follows from (3.50). □

From Fig. 9, we note that the cross-rule can be informally expressed as

$$e_S = e_E + e_C(e_N^{-1} - e_W^{-1})e_C \tag{4.22}$$

where $e_\Omega \in V_C$ for $\Omega = N, S, E, W$ and C. Because these error vectors are designants (see (3.31b)), Eq. (4.22) is clearly a fundamental compass identity amongst designants.

In fact, this identity has also been established for the leading coefficients $\dot{p}_\Omega$ of the numerator polynomials [23]. If we were to use monic normalisation for the denominators

$$\dot{Q}_\Omega(z) = 1, \quad \dot{B}_j^{(J)}(z) = I, \quad \dot{p}_\Omega := \dot{A}_j^{(J)}(z) \tag{4.23}$$

(where the dot denotes that the leading coefficient of the polynomial beneath the dot is required), we would find that

$$\dot{p}_S = \dot{p}_E + \dot{p}_C(\dot{p}_N^{-1} - \dot{p}_W^{-1})\dot{p}_C, \tag{4.24}$$

corresponding to the same compass identity amongst designants.

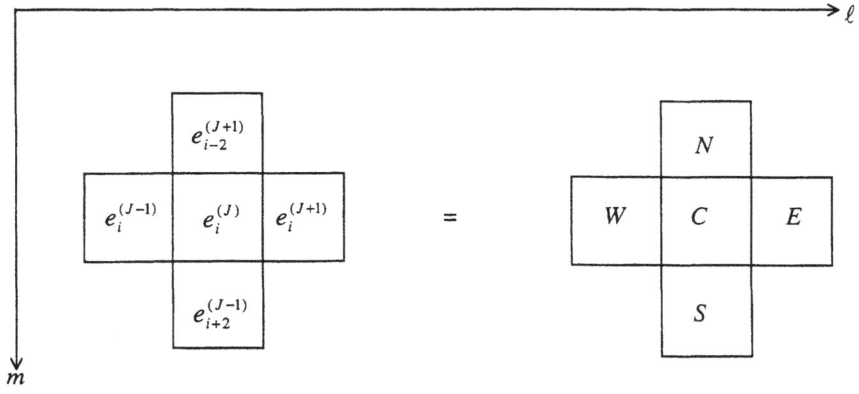

Fig. 9. Position of error vectors obeying the cross-rule.

Reverting to the normalisation of (3.64) with $q_\Omega(0) = I$ and $Q_\Omega(0) = 1$, we note that formula (3.28) is required to convert (4.22) to a usable relation amongst vectors $\boldsymbol{e}_\Omega \in \mathbb{C}^d$. We find that

$$\boldsymbol{e}_S = \boldsymbol{e}_E - |\boldsymbol{e}_C|^2 \left[\frac{\boldsymbol{e}_N}{|\boldsymbol{e}_N|^2} - \frac{\boldsymbol{e}_W}{|\boldsymbol{e}_W|^2} \right] + 2\boldsymbol{e}_C \operatorname{Re} \left[\boldsymbol{e}_C^{\mathrm{H}} \left(\frac{\boldsymbol{e}_N}{|\boldsymbol{e}_N|^2} - \frac{\boldsymbol{e}_W}{|\boldsymbol{e}_W|^2} \right) \right]$$

and this formula is computationally executable.

Implementation of this formula enables the calculation of the vectors $\boldsymbol{e}_\Omega$ in $\mathbb{C}^d$ in a rowwise fashion (see Fig. 9). For the case of vector-valued meromorphic functions of the type described following (3.69) it is shown in [40] that asymptotic (i.e., as J tends to infinity) results similar to the scalar case are valid, with an interesting interpretation for the behaviour of the vectors $\boldsymbol{e}_i^{(J)}$ as J tends to infinity. It is also shown in [40] that, as in the scalar case, the above procedure is numerically unstable, while a column-by-column computation retains stability – i.e., (4.22) is used to evaluate $\boldsymbol{e}_E$. There are also considerations of underflow and overflow which can be dealt with by a mild adaptation of the cross-rule.

Orthogonal polynomials lie at the heart of many approximation methods. In this context, the orthogonal polynomials are operators $\pi_i(\xi) \in \mathscr{A}[\xi]$, and they are defined using the functionals $c\{\cdot\}$ and $\boldsymbol{c}\{\cdot\}$. These functionals are defined by their action on monomials:

$$c\{\xi^i\} = c_i, \quad \boldsymbol{c}\{\xi^i\} = \boldsymbol{c}_i. \tag{4.25}$$

By linearity, we can normally define monic vector orthogonal polynomials by $\pi_0(\xi) = I$ and, for $i = 1, 2, 3, \ldots$, by

$$c\{\pi_i(\xi)\xi^j\} = 0, \quad j = 0, 1, \ldots, i-1. \tag{4.26}$$

The connection with the denominator polynomials (3.35) is

Theorem 4.3. *For* $i = 0, 1, 2, \ldots$

$$\pi_i(\xi) = \xi^i B_{2i-1}(\xi^{-1}).$$

Proof. Since $B_{2i-1}(z)$ is an operator polynomial of degree i, so is $\pi_i(\xi)$. Moreover, for $j=0,1,\ldots,i-1$,

$$c\{\xi^j\pi_i(\xi)\}=c\{\xi^{i+j}B_{2i-1}(\xi^{-1})\}=\sum_{\ell=0}^{i}c\{\xi^{i+j-\ell}B_\ell^{(2i-1)}\}=\sum_{\ell=0}^{i}c_{i+j-\ell}B_\ell^{(2i-1)}$$

$$=[f(z)B_{2i-1}(z)]_{i+j}=0$$

as is required for (4.26). □

This theorem establishes an equivalence between approximation methods based on vector orthogonal polynomials and those based on vector Padé approximation. To take account of noncommutativity, more care is needed over the issue of linearity with respect to multipliers from $\mathscr{A}$ than is shown in (4.26). Much fuller accounts, using variants of (4.26), are given by Roberts [41] and Salam [44,45].

In this section, we have focussed on the construction and properties of the continued fractions associated with the leading diagonal sequence of vector Padé approximants. When these approximants are evaluated at $z=1$, they equal $\varepsilon_{2k}^{(0)}$, the entries on the leading diagonal of the vector epsilon table. These entries are our natural first choice for use in the acceleration of convergence of a sequence of vectors.

Acknowledgements

Peter Graves-Morris is grateful to Dr. Simon Chandler-Wilde for making his computer programs available to us, and to Professor Ernst Weniger for his helpful review of the manuscript.

References

[1] G.A. Baker, Essentials of Padé Approximants, Academic Press, New York, 1975.
[2] G.A. Baker Jr., P.R. Graves-Morris, Padé approximants, Encyclopedia of Mathematics and its Applications, 2nd Edition, Vol. 59, Cambridge University Press, New York, 1996.
[3] E.R. Berlekamp, Algebraic Coding Theory, McGraw-Hill, New York, 1968.
[4] C. Brezinski, Etude sur les ε-et ρ-algorithmes, Numer. Math. 17 (1971) 153–162.
[5] C. Brezinski, Généralisations de la transformation de Shanks, de la table de Padé et de l'ε-algorithme, Calcolo 12 (1975) 317–360.
[6] C. Brezinski, Accélération de la Convergence en Analyse Numérique, Lecture Notes in Mathematics, Vol. 584, Springer, Berlin, 1977.
[7] C. Brezinski, Convergence acceleration of some sequences by the ε-algorithm, Numer. Math. 29 (1978) 173–177.
[8] C. Brezinski, Padé-Type Approximation and General Orthogonal Polynomials, Birkhäuser, Basel, 1980.
[9] C. Brezinski, M. Redivo-Zaglia, Extrapolation Methods, Theory and Practice, North-Holland, Amsterdam, 1991.
[10] S.N. Chandler-Wilde, D. Hothersall, Efficient calculation of the Green function for acoustic propagation above a homogeneous impedance plane, J. Sound Vibr. 180 (1995) 705–724.
[11] S.N. Chandler-Wilde, M. Rahman, C.R. Ross, A fast, two-grid method for the impedance problem in a half-plane, Proceedings of the Fourth International Conference on Mathematical Aspects of Wave Propagation, SIAM, Philadelphia, PA, 1998.
[12] D. Colton, R. Kress, Integral Equations Methods in Scattering Theory, Wiley, New York, 1983.
[13] F. Cordellier, L'ε-algorithme vectoriel, interprétation géometrique et règles singulières, Exposé au Colloque d'Analyse Numérique de Gourette, 1974.

[14] F. Cordellier, Démonstration algébrique de l'extension de l'identité de Wynn aux tables de Padé non-normales, in: L. Wuytack (Ed.), Padé Approximation and its Applications, Springer, Berlin, Lecture Notes in Mathematics, Vol. 765, 1979, pp. 36–60.
[15] F. Cordellier, Utilisation de l'invariance homographique dans les algorithmes de losange, in: H. Werner, H.J. Bünger (Eds.), Padé Approximation and its Applications, Bad Honnef 1983, Lecture Notes in Mathematics, Vol. 1071, Springer, Berlin, 1984, pp. 62–94.
[16] F. Cordellier, Thesis, University of Lille, 1989.
[17] A. Cuyt, L. Wuytack, Nonlinear Methods in Numerical Analysis, North-Holland, Amsterdam, 1987.
[18] J.-P. Delahaye, B. Germain-Bonne, The set of logarithmically convergent sequences cannot be accelerated, SIAM J. Numer. Anal. 19 (1982) 840–844.
[19] J.-P. Delahaye, Sequence Transformations, Springer, Berlin, 1988.
[20] W. Gander, E.H. Golub, D. Gruntz, Solving linear systems by extrapolation in Supercomputing, Trondheim, Computer Systems Science, Vol. 62, Springer, Berlin, 1989, pp. 279–293.
[21] S. Graffi, V. Grecchi, Borel summability and indeterminancy of the Stieltjes moment problem: Application to anharmonic oscillators, J. Math. Phys. 19 (1978) 1002–1006.
[22] P.R. Graves-Morris, Vector valued rational interpolants I, Numer. Math. 42 (1983) 331–348.
[23] P.R. Graves-Morris, B. Beckermann, The compass (star) identity for vector-valued rational interpolants, Adv. Comput. Math. 7 (1997) 279–294.
[24] P.R. Graves-Morris, C.D. Jenkins, Generalised inverse vector-valued rational interpolation, in: H. Werner, H.J. Bünger (Eds.), Padé Approximation and its Applications, Vol. 1071, Springer, Berlin, 1984, pp. 144–156.
[25] P.R. Graves-Morris, C.D. Jenkins, Vector-valued rational interpolants III, Constr. Approx. 2 (1986) 263–289.
[26] P.R. Graves-Morris, D.E. Roberts, From matrix to vector Padé approximants, J. Comput. Appl. Math. 51 (1994) 205–236.
[27] P.R. Graves-Morris, D.E. Roberts, Problems and progress in vector Padé approximation, J. Comput. Appl. Math. 77 (1997) 173–200.
[28] P.R. Graves-Morris, E.B. Saff, Row convergence theorems for generalised inverse vector-valued Padé approximants, J. Comput. Appl. Math. 23 (1988) 63–85.
[29] P.R. Graves-Morris, E.B. Saff, An extension of a row convergence theorem for vector Padé approximants, J. Comput. Appl. Math. 34 (1991) 315–324.
[30] P.R. Graves-Morris, J. Van Iseghem, Row convergence theorems for vector-valued Padé approximants, J. Approx. Theory 90 (1997) 153–173.
[31] M.H. Gutknecht, Lanczos type solvers for non-symmetric systems of linear equations, Acta Numer. 6 (1997) 271–397.
[32] A. Heyting, Die Theorie der linear Gleichungen in einer Zahlenspezies mit nichtkommutatives Multiplikation, Math. Ann. 98 (1927) 465–490.
[33] H.H.H. Homeier, Scalar Levin-type sequence transformations, this volume, J. Comput. Appl. Math. 122 (2000) 81–147.
[34] U.C. Jentschura, P.J. Mohr, G. Soff, E.J. Weniger, Convergence acceleration via combined nonlinear-condensation transformations, Comput. Phys. Comm. 116 (1999) 28–54.
[35] K. Jbilou, H. Sadok, Vector extrapolation methods, Applications and numerical comparison, this volume, J. Comput. Appl. Math. 122 (2000) 149–165.
[36] W.B. Jones, W. Thron, in: G.-C. Rota (Ed.), Continued Fractions, Encyclopedia of Mathematics and its Applications, Vol. 11, Addison-Wesley, Reading, MA, USA, 1980.
[37] J.B. McLeod, A note on the ε-algorithm, Computing 7 (1972) 17–24.
[38] D.E. Roberts, Clifford algebras and vector-valued rational forms I, Proc. Roy. Soc. London A 431 (1990) 285–300.
[39] D.E. Roberts, On the convergence of rows of vector Padé approximants, J. Comput. Appl. Math. 70 (1996) 95–109.
[40] D.E. Roberts, On a vector q-d algorithm, Adv. Comput. Math. 8 (1998) 193–219.
[41] D.E. Roberts, A vector Chebyshev algorithm, Numer. Algorithms 17 (1998) 33–50.
[42] D.E. Roberts, On a representation of vector continued fractions, J. Comput. Appl. Math. 105 (1999) 453–466.
[43] A. Salam, An algebraic approach to the vector ε-algorithm, Numer. Algorithms 11 (1996) 327–337.
[44] A. Salam, Formal vector orthogonal polynomials, Adv. Comput. Math. 8 (1998) 267–289.
[45] A. Salam, What is a vector Hankel determinant? Linear Algebra Appl. 278 (1998) 147–161.

[46] A. Salam, Padé-type approximants and vector Padé approximants, J. Approx. Theory 97 (1999) 92–112.
[47] D. Shanks, Non-linear transformations of divergent and slowly convergent sequences, J. Math. Phys. 34 (1955) 1–42.
[48] A. Sidi, W.F. Ford, D.A. Smith, SIAM J. Numer. Anal. 23 (1986) 178–196.
[49] R.C.E. Tan, Implementation of the topological epsilon algorithm, SIAM J. Sci. Statist. Comput. 9 (1988) 839–848.
[50] E.J. Weniger, Nonlinear sequence transformations for the acceleration of convergence and the summation of divergent series, Comput. Phys. Rep. 10 (1989) 371–1809.
[51] E.J. Weniger, A convergent, renormalised strong coupling perturbation expansion for the ground state energy of the quartic, sextic and octic anharmonic oscillator, Ann. Phys. 246 (1996) 133–165.
[52] E.J. Weniger, Prediction properties of Aitken's iterated Δ^2 process, of Wynn's epsilon algorithm and of Brezinski's iterated theta algorithm, this volume, J. Comp. Appl. Math. 122 (2000) 329–356.
[53] J. Wimp, Sequence Transformations and Their Applications, Academic Press, New York, 1981.
[54] P. Wynn, On a device for calculating the $e_m(S_n)$ transformations, Math. Tables Automat. Comp. 10 (1956) 91–96.
[55] P. Wynn, The epsilon algorithm and operational formulas of numerical analysis, Math. Comp. 15 (1961) 151–158.
[56] P. Wynn, L'ε-algoritmo e la tavola di Padé, Rendi. Mat. Roma 20 (1961) 403–408.
[57] P. Wynn, Acceleration techniques for iterative vector problems, Math. Comp. 16 (1962) 301–322.
[58] P. Wynn, Continued fractions whose coefficients obey a non-commutative law of multiplication, Arch. Rational Mech. Anal. 12 (1963) 273–312.
[59] P. Wynn, On the convergence and stability of the epsilon algorithm, SIAM J. Numer. Anal. 3 (1966) 91–122.

ELSEVIER

Journal of Computational and Applied Mathematics 122 (2000) 81–147

JOURNAL OF COMPUTATIONAL AND APPLIED MATHEMATICS

www.elsevier.nl/locate/cam

Scalar Levin-type sequence transformations

Herbert H.H. Homeier *, [1]

Institut für Physikalische und Theoretische Chemie, Universität Regensburg, D-93040 Regensburg, Germany

Received 7 June 1999; received in revised form 15 January 2000

Abstract

Sequence transformations are important tools for the convergence acceleration of slowly convergent scalar sequences or series and for the summation of divergent series. The basic idea is to construct from a given sequence $\{\{s_n\}\}$ a new sequence $\{\{s'_n\}\} = \mathscr{T}(\{\{s_n\}\})$ where each s'_n depends on a finite number of elements $s_{n_1}, \ldots, s_{n_m}$. Often, the s_n are the partial sums of an infinite series. The aim is to find transformations such that $\{\{s'_n\}\}$ converges faster than (or sums) $\{\{s_n\}\}$. Transformations $\mathscr{T}(\{\{s_n\}\}, \{\{\omega_n\}\})$ that depend not only on the sequence elements or partial sums s_n but also on an auxiliary sequence of the so-called remainder estimates ω_n are of Levin-type if they are linear in the s_n, and nonlinear in the ω_n. Such remainder estimates provide an easy-to-use possibility to use asymptotic information on the problem sequence for the construction of highly efficient sequence transformations. As shown first by Levin, it is possible to obtain such asymptotic information easily for large classes of sequences in such a way that the ω_n are simple functions of a few sequence elements s_n. Then, nonlinear sequence transformations are obtained. Special cases of such Levin-type transformations belong to the most powerful currently known extrapolation methods for scalar sequences and series. Here, we review known Levin-type sequence transformations and put them in a common theoretical framework. It is discussed how such transformations may be constructed by either a model sequence approach or by iteration of simple transformations. As illustration, two new sequence transformations are derived. Common properties and results on convergence acceleration and stability are given. For important special cases, extensions of the general results are presented. Also, guidelines for the application of Levin-type sequence transformations are discussed, and a few numerical examples are given. © 2000 Elsevier Science B.V. All rights reserved.

MSC: 65B05; 65B10; 65B15; 40A05; 40A25; 42C15

Keywords: Convergence acceleration; Extrapolation; Summation of divergent series; Stability analysis; Hierarchical consistency; Iterative sequence transformation; Levin-type transformations; Algorithm; Linear convergence; Logarithmic convergence; Fourier series; Power series; Rational approximation

* Fax: +49-941-943-4719.

E-mail address: herbert.homeier@na-net.ornl.gov (H.H.H. Homeier)

[1] WWW: http://www.chemie.uni-regensburg.de/ ~hoh05008

PII: S0377-0427(00)00359-9

1. Introduction

In applied mathematics and the numerate sciences, extrapolation methods are often used for the convergence acceleration of slowly convergent sequences or series and for the summation of divergent series. For an introduction to such methods, and also further information that cannot be covered here, see the books of Brezinski and Redivo Zaglia [14] and Wimp [102] and the work of Weniger [84,88] and Homeier [40], but also the books of Baker [3], Baker and Graves-Morris [5], Brezinski [7,8,10–12], Graves-Morris [24,25], Graves-Morris, Saff and Varga [26], Khovanskii [52], Lorentzen and Waadeland [56], Nikishin and Sorokin [62], Petrushev and Popov [66], Ross [67], Saff and Varga [68], Wall [83], Werner and Buenger [101] and Wuytack [103].

For the discussion of extrapolation methods, one considers a sequence $\{\{s_n\}\} = \{\{s_0, s_1, \ldots\}\}$ with elements s_n or the terms $a_n = s_n - s_{n-1}$ of a series $\sum_{j=0}^{\infty} a_j$ with partial sums $s_n = \sum_{j=0}^{n} a_j$ for large n. A common approach is to rewrite s_n as

$$s_n = s + R_n, \tag{1}$$

where s is the limit (or antilimit in the case of divergence) and R_n is the remainder or tail. The aim then is to find a new sequence $\{\{s'_n\}\}$ such that

$$s'_n = s + R'_n, \qquad R'_n/R_n \to 0 \text{ for } n \to \infty. \tag{2}$$

Thus, the sequence $\{\{s'_n\}\}$ converges faster to the limit s (or diverges less violently) than $\{\{s_n\}\}$.

To find the sequence $\{\{s'_n\}\}$, i.e., to construct a sequence transformation $\{\{s'_n\}\} = \mathscr{T}(\{\{s_n\}\})$, one needs asymptotic information about the s_n or the terms a_n for large n, and hence about the R_n. This information then allows to eliminate the remainder at least asymptotically, for instance by substracting the dominant part of the remainder. Either such information is obtained by a careful mathematical analysis of the behavior of the s_n and/or a_n, or it has to be extracted numerically from the values of a finite number of the s_n and/or a_n by some method that ideally can be proven to work for a large class of problems.

Suppose that one knows quantities ω_n such that $R_n/\omega_n = \mathrm{O}(1)$ for $n \to \infty$, for instance

$$\lim_{n\to\infty} R_n/\omega_n = c \neq 0, \tag{3}$$

where c is a constant. Such quantities are called remainder estimates. Quite often, such remainder estimates can be found with relatively low effort but the exact value of c is often quite hard to calculate. Then, it is rather natural to rewrite the rest as $R_n = \omega_n \mu_n$ where $\mu_n \to c$. The problem is how to describe or model the μ_n. Suppose that one has a system of known functions $\psi_j(n)$ such that $\psi_0(n) = 1$ and $\psi_{j+1} = \mathrm{o}(\psi_j(n))$ for $j \in \mathbb{N}_0$. An example of such a system is $\psi_j(n) = (n+\beta)^{-j}$ for some $\beta \in \mathbb{R}_+$. Then, one may model μ_n as a linear combination of the $\psi_j(n)$ according to

$$\mu_n \sim \sum_{j=0}^{\infty} c_j \psi_j(n) \qquad \text{for } n \to \infty, \tag{4}$$

whence the problem sequence is modelled according to

$$s_n \sim s + \omega_n \sum_{j=0}^{\infty} c_j \psi_j(n). \tag{5}$$

The idea now is to eliminate the leading terms of the remainder with the unknown constants c_j up to $j = k-1$, say. Thus, one uses a model sequence with elements

$$\sigma_m = \sigma + \omega_m \sum_{j=0}^{k-1} c_j \psi_j(m), \qquad m \in \mathbb{N}_0 \tag{6}$$

and calculates σ exactly by solving the system of $k+1$ equations resulting for $m = n, n+1, \ldots, n+k$ for the unknowns σ and c_j, $j = 0, \ldots, k-1$. The solution for σ is a ratio of determinants (see below) and may be denoted symbolically as

$$\sigma = T(\sigma_n, \ldots, \sigma_{n+k}; \omega_n, \ldots, \omega_{n+k}; \psi_j(n), \ldots, \psi_j(n+k)). \tag{7}$$

The resulting sequence transformation is

$$\mathscr{T}(\{\{s_n\}\}, \{\{\omega_n\}\}) = \{\{\mathscr{T}_n^{(k)}(\{\{s_n\}\}, \{\{\omega_n\}\})\}\} \tag{8}$$

with

$$\mathscr{T}_n^{(k)}(\{\{s_n\}\}, \{\{\omega_n\}\}) = T(s_n, \ldots, s_{n+k}; \omega_n, \ldots, \omega_{n+k}; \psi_j(n), \ldots, \psi_j(n+k)). \tag{9}$$

It eliminates the leading terms of the asymptotic expansion (5). The model sequences (6) are in the kernel of the sequence transformation $\mathscr{T}$, defined as the set of all sequences such that $\mathscr{T}$ reproduces their (anti)limit exactly.

A somewhat more general approach is based on model sequences of the form

$$\sigma_n = \sigma + \sum_{j=1}^{k} c_j g_j(n), \qquad n \in \mathbb{N}_0, \; k \in \mathbb{N}. \tag{10}$$

Virtually all known sequence transformations can be derived using such model sequences. This leads to the **E** algorithm as described below in Section 3.1. Also, some further important examples of sequence transformations are described in Section 3.

However, the introduction of remainder estimates proved to be an important theoretical step since it allows to make use of asymptotic information of the remainder easily. The most prominent of the resulting sequence transformations $\mathscr{T}(\{\{s_n\}\}, \{\{\omega_n\}\})$ is the Levin transformation [53] that corresponds to the asymptotic system of functions given by $\psi_j(n) = (n+\beta)^{-j}$, and thus, to Poincare-type expansions of the μ_n. But also other systems are of importance, like $\psi_j(n) = 1/(n+\beta)_j$ leading to factorial series, or $\psi_j(n) = t_n^j$ corresponding to Taylor expansions of t-dependent functions at the abscissae t_n that tend to zero for large n. The question which asymptotic system is best, cannot be decided generally. The answer to this question depends on the extrapolation problem. To obtain efficient extrapolation procedures for large classes of problems requires to use various asymptotic systems, and thus, a larger number of different sequence transformations. Also, different choices of ω_n lead to different variants of such transformations. Levin [53] has pioneered this question and introduced three variants that are both simple and rather successful for large classes of problems. These variants and some further ones will be discussed. The question which variant is best, also cannot be decided generally. There are, however, a number of results that favor certain variants for certain problems. For example, for Stieltjes series, the choice $\omega_n = a_{n+1}$ can be theoretically justified (see Appendix A).

Thus, we will focus on sequence transformations that involve an auxiliary sequence $\{\{\omega_n\}\}$. To be more specific, we consider transformations of the form $\mathscr{T}(\{\{s_n\}\}, \{\{\omega_n\}\}) = \{\{\mathscr{T}_n^{(k)}\}\}$ with

$$\mathscr{T}_n^{(k)} = \frac{\sum_{j=0}^{k} \lambda_{n,j}^{(k)} s_{n+j}/\omega_{n+j}}{\sum_{j=0}^{k} \lambda_{n,j}^{(k)}/\omega_{n+j}}. \tag{11}$$

This will be called a Levin-type transformations. The known sequence transformations that involve remainder estimates, for instance the $\mathscr{C}$, $\mathscr{S}$, and $\mathscr{M}$ transformations of Weniger [84], the W algorithm of Sidi [73], and the $\mathscr{J}$ transformation of Homeier with its many special cases like the important ${}_p\mathbf{J}$ transformations [35,36,38–40,46], are all of this type. Interestingly, also the $\mathscr{H}$, $\mathscr{I}$, and $\mathscr{K}$ transformations of Homeier [34,35,37,40–44] for the extrapolation of orthogonal expansions are of this type although the ω_n in some sense cease to be remainder estimates as defined in Eq. (3).

The Levin transformation was also generalized in a different way by Levin and Sidi [54] who introduced the $d^{(m)}$ transformations. This is an important class of transformations that would deserve a thorough review itself. This, however, is outside the scope of the present review. We collect some important facts regarding this class of transformations in Section 3.2.

Levin-type transformations as defined in Eq. (11) have been used for the solution of a large variety of problems. For instance, Levin-type sequence transformations have been applied for the convergence acceleration of infinite series representations of molecular integrals [28,29, 33,65,82,98–100], for the calculation of the lineshape of spectral holes [49], for the extrapolation of cluster- and crystal-orbital calculations of one-dimensional polymer chains to infinite chain length [16,88,97], for the calculation of special functions [28,40,82,88,89,94,100], for the summation of divergent and acceleration of convergent quantum mechanical perturbation series [17,18,27,85,90–93,95,96], for the evaluation of semiinfinite integrals with oscillating integrands and Sommerfeld integral tails [60,61,75,81], and for the convergence acceleration of multipolar and orthogonal expansions and Fourier series [34,35,37,40–45,63,77,80]. This list is clearly not complete but sufficient to demonstrate the possibility of successful application of these transformations.

The outline of this survey is as follows: After listing some definitions and notations, we discuss some basic sequence transformations in order to provide some background information. Then, special definitions relevant for Levin-type sequence transformations are given, including variants obtained by choosing specific remainder estimates ω_n. After this, important examples of Levin-type sequence transformations are introduced. In Section 5, we will discuss approaches for the construction of Levin-type sequence transformations, including model sequences, kernels and annihilation operators, and also the concept of hierarchical consistency. In Section 6, we derive basic properties, those of limiting transformations and discuss the application to power series. In Section 7, results on convergence acceleration are presented, while in Section 8, results on the numerical stability of the transformations are provided. Finally, we discuss guidelines for the application of the transformations and some numerical examples in Section 9.

2. Definitions and notations

2.1. General definitions

2.1.1. Sets

Natural numbers:

$$\mathbb{N} = \{1, 2, 3, \ldots\}, \qquad \mathbb{N}_0 = \mathbb{N} \cup \{0\}. \tag{12}$$

Integer numbers:

$$\mathbb{Z} = \mathbb{N} \cup \{0, -1, -2, -3, \ldots\}. \tag{13}$$

Real numbers and vectors:

$$\begin{aligned} &\mathbb{R} = \{x : x \text{ real}\}, \\ &\mathbb{R}_+ = \{x \in \mathbb{R} : x > 0\} \\ &\mathbb{R}^n = \{(x_1, \ldots, x_n) \,|\, x_j \in \mathbb{R}, j = 1, \ldots, n\}. \end{aligned} \tag{14}$$

Complex numbers:

$$\begin{aligned} &\mathbb{C} = \{z = x + \mathrm{i}y : x \in \mathbb{R}, y \in \mathbb{R}, \ \mathrm{i}^2 = -1\}, \\ &\mathbb{C}^n = \{(z_1, \ldots, z_n) \,|\, z_j \in \mathbb{C}, j = 1, \ldots, n\}. \end{aligned} \tag{15}$$

For $z = x + \mathrm{i}y$, real and imaginary parts are denoted as $x = \Re(z)$, $y = \Im(z)$. We use $\mathbb{K}$ to denote $\mathbb{R}$ or $\mathbb{C}$.

Vectors with nonvanishing components:

$$\mathbb{F}^n = \{(z_1, \ldots, z_n) \,|\, z_j \in \mathbb{C}, z_j \neq 0, \ j = 1, \ldots, n\}. \tag{16}$$

Polynomials:

$$\mathbb{P}^k = \left\{ P : z \mapsto \sum_{j=0}^{k} c_j z^j \,|\, z \in \mathbb{C}, (c_0, \ldots, c_k) \in \mathbb{K}^{k+1} \right\}. \tag{17}$$

Sequences:

$$\mathbb{S}^{\mathbb{K}} = \{\{\{s_0, s_1, \ldots, s_n, \ldots\}\} \,|\, s_n \in \mathbb{K}, \ n \in \mathbb{N}_0\}. \tag{18}$$

Sequences with nonvanishing terms:

$$\mathbb{O}^{\mathbb{K}} = \{\{\{s_0, s_1, \ldots, s_n, \ldots\}\} \,|\, s_n \neq 0, \ s_n \in \mathbb{K}, \ n \in \mathbb{N}_0\}. \tag{19}$$

2.1.2. Special functions and symbols

Gamma function [58, p. 1]:

$$\Gamma(z) = \int_0^\infty t^{z-1} \exp(-t)\, \mathrm{d}t \quad (z \in \mathbb{R}_+). \tag{20}$$

Factorial:

$$n! = \Gamma(n+1) = \prod_{j=1}^{n} j. \tag{21}$$

Pochhammer symbol [58, p. 2]:

$$(a)_n = \frac{\Gamma(a+n)}{\Gamma(a)} = \prod_{j=1}^{n}(a+j-1). \tag{22}$$

Binomial coefficients [1, p. 256, Eq. (6.1.21)]:

$$\binom{z}{w} = \frac{\Gamma(z+1)}{\Gamma(w+1)\Gamma(z-w+1)}. \tag{23}$$

Entier function:

$$[\![x]\!] = \max\{j \in \mathbb{Z}: j \leqslant x,\ x \in \mathbb{R}\}. \tag{24}$$

2.2. Sequences, series and operators

2.2.1. Sequences and series

For Stieltjes series see Appendix A.

Scalar sequences with elements s_n, tail R_n, and limit s:

$$\{\{s_n\}\} = \{\{s_n\}\}_{n=0}^{\infty} = \{\{s_0, s_1, s_2, \ldots\}\} \in \mathbb{S}^{\mathbb{K}}, \quad R_n = s_n - s, \quad \lim_{n\to\infty} s_n = s. \tag{25}$$

If the sequence is not convergent but summable to s, s is called the antilimit. The nth element s_n of a sequence $\sigma = \{\{s_n\}\} \in \mathbb{S}^{\mathbb{K}}$ is also denoted by $\langle\sigma\rangle_n$. A sequence is called a *constant sequence*, if all elements are constant, i.e., if there is a $c \in \mathbb{K}$ such that $s_n = c$ for all $n \in \mathbb{N}_0$, in which case it is denoted by $\{\{c\}\}$. The constant sequence $\{\{0\}\}$ is called the *zero sequence*.

Scalar series with terms $a_j \in \mathbb{K}$, partial sums s_n, tail R_n, and limit/antilimit s:

$$s = \sum_{j=0}^{\infty} a_j, \quad s_n = \sum_{j=0}^{n} a_j, \quad R_n = -\sum_{j=n+1}^{\infty} a_j = s_n - s. \tag{26}$$

We say that $\hat{a}_n$ are Kummer-related to the a_n with limit or antilimit $\hat{s}$ if $\hat{a}_n = \Delta\hat{s}_{n-1}$ satisfy $a_n \sim \hat{a}_n$ for $n \to \infty$ and $\hat{s}$ is the limit (or antilimit) of $\hat{s}_n = \sum_{j=0}^{n} \hat{a}_j$.

Scalar power series in $z \in \mathbb{C}$ with coefficients $c_j \in \mathbb{K}$, partial sums $f_n(z)$, tail $R_n(z)$, and limit/antilimit $f(z)$:

$$f(z) = \sum_{j=0}^{\infty} c_j z^j, \quad f_n(z) = \sum_{j=0}^{n} c_j z^j, \quad R_n(z) = \sum_{j=n+1}^{\infty} c_j z^j = f(z) - f_n(z). \tag{27}$$

2.2.2. Types of convergence

Sequences $\{\{s_n\}\}$ satisfying the equation

$$\lim_{n\to\infty} (s_{n+1} - s)/(s_n - s) = \rho \tag{28}$$

are called *linearly convergent* if $0 < |\rho| < 1$, *logarithmically convergent* for $\rho = 1$ and *hyperlinearly convergent* for $\rho = 0$. For $|\rho| > 1$, the sequence diverges.

A sequence $\{\{u_n\}\}$ accelerates a sequence $\{\{v_n\}\}$ to s if

$$\lim_{n\to\infty} (u_n - s)/(v_n - s) = 0. \tag{29}$$

If $\{\{v_n\}\}$ converges to s then we also say that $\{\{u_n\}\}$ converges faster than $\{\{v_n\}\}$.

A sequence $\{\{u_n\}\}$ accelerates a sequence $\{\{v_n\}\}$ to s with order $\alpha > 0$ if

$$(u_n - s)/(v_n - s) = \mathrm{O}(n^{-\alpha}). \tag{30}$$

If $\{\{v_n\}\}$ converges to s then we also say that $\{\{u_n\}\}$ converges faster than $\{\{v_n\}\}$ with order α.

2.2.3. Operators

Annihilation operator: An operator $\mathscr{A}\colon \mathbb{S}^{\mathbb{K}} \to \mathbb{K}$ is called an annihilation operator for a given sequence $\{\{\tau_n\}\}$ if it satisfies

$$\begin{aligned} &\mathscr{A}(\{\{s_n + z t_n\}\}) = \mathscr{A}(\{\{s_n\}\}) + z\mathscr{A}(\{\{t_n\}\}) \quad \text{for all } \{\{s_n\}\} \in \mathbb{S}^{\mathbb{K}},\ \{\{t_n\}\} \in \mathbb{S}^{\mathbb{K}}, z \in \mathbb{K},\\ &\mathscr{A}(\{\{\tau_n\}\}) = 0. \end{aligned} \tag{31}$$

Forward difference operator.

$$\begin{aligned} &\triangle_m g(m) = g(m+1) - g(m), \quad \triangle_m g_m = g_{m+1} - g_m,\\ &\triangle_m^k = \triangle_m \triangle_m^{k-1},\\ &\triangle = \triangle_n,\\ &\triangle^k g_n = \sum_{j=0}^{k} (-1)^{k-j} \binom{k}{j} g_{n+j}. \end{aligned} \tag{32}$$

Generalized difference operator $\nabla_n^{(k)}$ for given quantities $\delta_n^{(k)} \neq 0$:

$$\nabla_n^{(k)} = (\delta_n^{(k)})^{-1} \triangle. \tag{33}$$

Generalized difference operator $\tilde{\nabla}_n^{(k)}$ for given quantities $\zeta_n^{(k)} \neq 0$:

$$\tilde{\nabla}_n^{(k)} = (\zeta_n^{(k)})^{-1} \triangle^2. \tag{34}$$

Generalized difference operator $\bigtriangledown_n^{(k)}[\alpha]$ for given quantities $\Delta_n^{(k)} \neq 0$:

$$\bigtriangledown_n^{(k)}[\alpha] f_n = (\Delta_n^{(k)})^{-1} (f_{n+2} - 2\cos\alpha f_{n+1} + f_n). \tag{35}$$

Generalized difference operator $\partial_n^{(k)}[\zeta]$ for given quantities $\tilde{\Delta}_n^{(k)} \neq 0$:

$$\partial_n^{(k)}[\zeta] f_n = (\tilde{\Delta}_n^{(k)})^{-1} (\zeta_{n+k}^{(2)} f_{n+2} + \zeta_{n+k}^{(1)} f_{n+1} + \zeta_{n+k}^{(0)} f_n). \tag{36}$$

Weighted difference operators for given $P^{(k-1)} \in \mathbb{P}^{k-1}$:

$$\mathscr{W}_n^{(k)} = \mathscr{W}_n^{(k)}[P^{(k-1)}] = \triangle^{(k)} P^{(k-1)}(n). \tag{37}$$

Polynomial operators $\mathscr{P}$ for given $P^{(k)} \in \mathbb{P}^{(k)}$: Let $P^{(k)}(x) = \sum_{j=0}^{k} p_j^{(k)} x^j$. Then put

$$\mathscr{P}[P^{(k)}] g_n = \sum_{j=0}^{k} p_j^{(k)} g_{n+j}. \tag{38}$$

Divided difference operator. For given $\{\{x_n\}\}$ and $k, n \in \mathbb{N}_0$, put

$$\square_n^{(k)}[\{\{x_n\}\}](f(x)) = \square_n^{(k)}(f(x)) = f[x_n, \ldots, x_{n+k}] = \sum_{j=0}^{k} f(x_{n+j}) \prod_{\substack{i=0\\ i\neq j}}^{k} \frac{1}{x_{n+j} - x_{n+i}},$$

$$\square_n^{(k)}[\{\{x_n\}\}]g_n = \square_n^{(k)}g_n = \sum_{j=0}^{k} g_{n+j} \prod_{\substack{i=0\\ i\neq j}}^{k} \frac{1}{x_{n+j}-x_{n+i}}. \tag{39}$$

3. Some basic sequence transformations

3.1. E Algorithm

Putting for sequences $\{\{y_n\}\}$ and $\{\{g_j(n)\}\}$, $j=1,\dots,k$

$$E_n^{(k)}[\{\{y_n\}\};\{\{g_j(n)\}\}] = \begin{vmatrix} y_n & \cdots & y_{n+k} \\ g_1(n) & \cdots & g_1(n+k) \\ \vdots & \ddots & \vdots \\ g_k(n) & \cdots & g_k(n+k) \end{vmatrix}, \tag{40}$$

one may define the sequence transformation

$$\mathbf{E}_n^{(k)}(\{\{s_n\}\}) = \frac{E_n^{(k)}[\{\{s_n\}\};\{\{g_j(n)\}\}]}{E_n^{(k)}[\{\{1\}\};\{\{g_j(n)\}\}]}. \tag{41}$$

As is plain using Cramer's rule, we have $\mathbf{E}_n^{(k)}(\{\{\sigma_n\}\})=\sigma$ if the σ_n satisfy Eq. (10). Thus, the sequence transformation yields the limit σ exactly for model sequences (10).

The sequence transformation $\mathbf{E}$ is known as the $\mathbf{E}$ algorithm or also as Brezinski–Håvie–Protocol [102, Section 10] after two of its main investigators, Håvie [32] and Brezinski [9]. A good introduction to this transformation is also given in the book of Brezinski and Redivo Zaglia [14, Section 2.1], cf. also Ref. [15].

Numerically, the computation of the $\mathbf{E}_n^{(k)}(\{\{s_n\}\})$ can be performed recursively using either the algorithm of Brezinski [14, p. 58f]

$$\mathbf{E}_n^{(0)}(\{\{s_n\}\}) = s_n, \quad g_{0,i}^{(n)} = g_i(n), \quad n\in\mathbb{N}_0,\ i\in\mathbb{N},$$

$$\mathbf{E}_n^{(k)}(\{\{s_n\}\}) = \mathbf{E}_n^{(k-1)}(\{\{s_n\}\}) - \frac{\mathbf{E}_{(n+1)}^{(k-1)}(\{\{s_n\}\}) - \mathbf{E}_n^{(k-1)}(\{\{s_n\}\})}{g_{k-1,k}^{(n+1)} - g_{k-1,k}^{(n)}} g_{k-1,k}^{(n)},$$

$$g_{k,i}^{(n)} = g_{k-1,i}^{(n)} - \frac{g_{k-1,i}^{(n+1)} - g_{k-1,i}^{(n)}}{g_{k-1,k}^{(n+1)} - g_{k-,1,k}^{(n)}} g_{k-,1,k}^{(n)}, \quad i=k+1,k+2,\dots \tag{42}$$

or the algorithm of Ford and Sidi [22] that requires additionally the quantities $g_{k+1}(n+j)$, $j=0,\dots,k$ for the computation of $\mathbf{E}_n^{(k)}(\{\{s_n\}\})$. The algorithm of Ford and Sidi involves the quantities

$$\Psi_{k,n}(u) = \frac{E_n^{(k)}[\{\{u_n\}\};\{\{g_j(n)\}\}]}{E_n^{(k)}[\{\{g_{k+1}(n)\}\};\{\{g_j(n)\}\}]} \tag{43}$$

for any sequence $\{\{u_0,u_1,\dots\}\}$, where the $g_i(n)$ are not changed even if they depend on the u_n and the u_n are changed. Then we have

$$\mathbf{E}_n^{(k)}(\{\{s_n\}\}) = \frac{\Psi_k^{(n)}(s)}{\Psi_k^{(n)}(1)} \tag{44}$$

and the Ψ are calculated recursively via

$$\Psi_{k,n}(u) = \frac{\Psi_{k-1,n+1}(u) - \Psi_{k-1,n}(u)}{\Psi_{k-1,n+1}(g_{k+1}) - \Psi_{k-1,n}(g_{k+1})}. \tag{45}$$

Of course, for $g_j(n) = \omega_n \psi_{j-1}(n)$, i.e., in the context of sequences modelled via expansion (5), the **E** algorithm may be used to obtain an explicit representation for any Levin-type sequence transformation of the form (cf. Eq. (9))

$$\mathscr{T}_n^{(k)} = T(s_n, \dots, s_{n+k}; \omega_n, \dots, \omega_{n+k}; \psi_j(n), \dots, \psi_j(n+k)) \tag{46}$$

as ratio of two determinants

$$\mathscr{T}_n^{(k)}(\{\{s_n\}\}, \{\{\omega_n\}\}) = \frac{E_n^{(k)}[\{\{s_n/\omega_n\}\}; \{\{\psi_{j-1}(n)\}\}]}{E_n^{(k)}[\{\{1/\omega_n\}\}; \{\{\psi_{j-1}(n)\}\}]}. \tag{47}$$

This follows from the identity [14]

$$\frac{E_n^{(k)}[\{\{s_n\}\}; \{\{\omega_n \psi_{j-1}(n)\}\}]}{E_n^{(k)}[\{\{1\}\}; \{\{\omega_n \psi_{j-1}(n)\}\}]} = \frac{E_n^{(k)}[\{\{s_n/\omega_n\}\}; \{\{\psi_{j-1}(n)\}\}]}{E_n^{(k)}[\{\{1/\omega_n\}\}; \{\{\psi_{j-1}(n)\}\}]}, \tag{48}$$

that is an easy consequence of usual algebraic manipulations of determinants.

3.2. The $d^{(m)}$ transformations

As noted in the introduction, the $d^{(m)}$ transformations were introduced by Levin and Sidi [54] as a generalization of the u variant of the Levin transformation [53]. We describe a slightly modified variant of the $d^{(m)}$ transformations [77]:

Let s_r, $r = 0, 1, \dots$ be a real or complex sequence with limit or antilimit s and terms $a_0 = s_0$ and $a_r = s_r - s_{r-1}$, $r = 1, 2, \dots$ such that $s_r = \sum_{r=0}^{r} a_j$, $r = 0, 1, \dots$. For given $m \in \mathbb{N}$ and $\xi_l \in \mathbb{N}_0$ with $l \in \mathbb{N}_0$ and $0 \leqslant \xi_0 < \xi_1 < \xi_2 < \cdots$ and $v = (n_1, \dots, n_m)$ with $n_j \in \mathbb{N}_0$ the $d^{(m)}$ transformation yields a table of approximations $s_v^{(m,j)}$ for the (anti-)limit s as solution of the linear system of equations

$$s_{\xi_l} = s_v^{(m,j)} + \sum_{k=1}^{m} (\xi_l + \alpha)^k [\Delta^{k-1} a_{\xi_l}] \sum_{i=0}^{n_k} \frac{\bar{\beta}_{ki}}{(\xi_l + \alpha)^i}, \quad j \leqslant l \leqslant j + N \tag{49}$$

with $\alpha > 0$, $N = \sum_{k=1}^{m} n_k$ and the $N+1$ unknowns $s_v^{(m,j)}$ and $\bar{\beta}_{k\,i}$. The $[\Delta^k a_j]$ are defined via $[\Delta^0 a_j] = a_j$ and $[\Delta^k a_j] = [\Delta^{k-1} a_{j+1}] - [\Delta^{k-1} a_j]$, $k = 1, 2, \dots$. In most cases, all n_k are chosen equal and one puts $v = (n, n, \dots, n)$. Apart from the value of α, only the input of m and of ξ_ℓ is required from the user. As transformed sequence, often one chooses the elements $s_{(n,\dots,n)}^{(m,0)}$ for $n = 0, 1, \dots$. The u variant of the Levin transformation is obtained for $m = 1$, $\alpha = \beta$ and $\xi_l = l$. The definition above differs slightly from the original one [54] and was given in Ref. [22] with $\alpha = 1$.

Ford and Sidi have shown, how these transformations can be calculated recursively with the $\mathbf{W}^{(m)}$ algorithms [22]. The $d^{(m)}$ transformations are the best known special cases of the *generalised Richardson Extrapolation process* (GREP) as defined by Sidi [72,73,78].

The $d^{(m)}$ transformations are derived by asymptotic analysis of the remainders $s_r - s$ for $r \to \infty$ for the family $\tilde{B}^{(m)}$ of sequences $\{\{a_r\}\}$ as defined in Ref. [54]. For such sequences, the a_r satisfy a difference equation of order m of the form

$$a_r = \sum_{k=1}^{m} p_k(r) \Delta^k a_r. \tag{50}$$

The $p_k(r)$ satisfy the asymptotic relation

$$p_k(r) \sim r^{i_k} \sum_{\ell=0}^{\infty} \frac{p_{k\ell}}{r^\ell} \quad \text{for } r \to \infty. \tag{51}$$

The i_k are integers satisfying $i_k \leqslant k$ for $k=1,\ldots,m$. This family of sequences is very large. But still, Levin and Sidi could prove [54, Theorem 2] that under mild additional assumptions, the remainders for such sequences satisfy

$$s_r - s \sim \sum_{k=1}^{m} r^{j_k} (\Delta^{k-1} a_r) \sum_{\ell=0}^{\infty} \frac{\beta_{k\ell}}{r^\ell} \quad \text{for } r \to \infty. \tag{52}$$

The j_k are integers satisfying $j_k \leqslant k$ for $k=1,\ldots,m$. A corresponding result for $m=1$ was proven by Sidi [71, Theorem 6.1].

System (49) now is obtained by truncation of the expansions at $\ell = n_n$, evaluation at $r = \xi_l$, and some further obvious substitutions.

The introduction of suitable ξ_l was shown to improve the accuracy and stability in difficult situations considerably [77].

3.3. Shanks transformation and epsilon algorithm

An important special case of the **E** algorithm is the choice $g_j(n) = \Delta s_{n+j-1}$ leading to the Shanks transformation [70]

$$e_k(s_n) = \frac{E_n^{(k)}[\{\{s_n\}\};\{\{\Delta s_{n+j-1}\}\}]}{E_n^{(k)}[\{\{1\}\};\{\{\Delta s_{n+j-1}\}\}]}. \tag{53}$$

Instead of using one of the recursive schemes for the **E** algorithms, the Shanks transformation may be implemented using the epsilon algorithm [104] that is defined by the recursive scheme

$$\begin{aligned} &\varepsilon_{-1}^{(n)} = 0, \quad \varepsilon_0^{(n)} = s_n, \\ &\varepsilon_{k+1}^{(n)} = \varepsilon_{k-1}^{(n+1)} + 1/[\varepsilon_k^{(n+1)} - \varepsilon_k^{(n)}]. \end{aligned} \tag{54}$$

The relations

$$\varepsilon_{2k}^{(n)} = e_k(s_n), \quad \varepsilon_{2k+1}^{(n)} = 1/e_k(\Delta s_n) \tag{55}$$

hold and show that the elements $\varepsilon_{2k+1}^{(n)}$ are only auxiliary quantities.

The kernel of the Shanks transformation e_k is given by sequences of the form

$$s_n = s + \sum_{j=0}^{k-1} c_j \, \Delta \, s_{n+j}. \tag{56}$$

See also [14, Theorem 2.18].

Additionally, one can use the Shanks transformation – and hence the epsilon algorithm – to compute the upper-half of the Padé table according to [70,104]

$$e_k(f_n(z)) = [n+k/k]_f(z) \quad (k \geqslant 0, n \geqslant 0), \tag{57}$$

where

$$f_n(z) = \sum_{j=0}^{n} c_j z^j \tag{58}$$

are the partial sums of a power series of a function $f(z)$. Padé approximants of $f(z)$ are rational functions in z given as ratio of two polynomials $p_\ell \in \mathbb{P}^{(\ell)}$ and $q_m \in \mathbb{P}^{(m)}$ according to

$$[\ell/m]_f(z) = p_\ell(z)/q_m(z), \tag{59}$$

where the Taylor series of f and $[\ell/m]_f$ are identical to the highest possible power of z, i.e.,

$$f(z) - p_\ell(z)/q_m(z) = \mathrm{O}(z^{\ell+m+1}). \tag{60}$$

Methods for the extrapolation of power series will be treated later.

3.4. Aitken process

The special case $\varepsilon_2^{(n)} = e_1(s_n)$ is identical to the famous Δ^2 method of Aitken [2]

$$s_n^{(1)} = s_n - \frac{(s_{n+1} - s_n)^2}{s_{n+2} - 2s_{n+1} + s_n} \tag{61}$$

with kernel

$$s_n = s + c\,(s_{n+1} - s_n), \quad n \in \mathbb{N}_0. \tag{62}$$

Iteration of the Δ^2 method yields the iterated Aitken process [14,84,102]

$$\mathbf{A}_n^{(0)} = s_n,$$

$$\mathbf{A}_n^{(k+1)} = \mathbf{A}_n^{(k)} - \frac{(\mathbf{A}_{n+1}^{(k)} - \mathbf{A}_n^{(k)})^2}{\mathbf{A}_{n+2}^{(k)} - 2\mathbf{A}_{n+1}^{(k)} + \mathbf{A}_n^{(k)}}. \tag{63}$$

The iterated Aitken process and the epsilon algorithm accelerate linear convergence and can sometimes be applied successfully for the summation of alternating divergent series.

3.5. Overholt process

The Overholt process is defined by the recursive scheme [64]

$$V_n^{(0)}(\{\{s_n\}\}) = s_n,$$

$$V_n^{(k)}(\{\{s_n\}\}) = \frac{(\Delta s_{n+k-1})^k V_{n+1}^{(k-1)}(\{\{s_n\}\}) - (\Delta s_{n+k})^k V_n^{(k-1)}(\{\{s_n\}\})}{(\Delta s_{n+k-1})^k - (\Delta s_{n+k})^k} \tag{64}$$

for $k \in \mathbb{N}$ and $n \in \mathbb{N}_0$. It is important for the convergence acceleration of fixed point iterations.

4. Levin-type sequence transformations

4.1. Definitions for Levin-type transformations

A set $\Lambda^{(k)} = \{\lambda_{n,j}^{(k)} \in \mathbb{K} \,|\, n \in \mathbb{N}_0, 0 \leqslant j \leqslant k\}$ is called a *coefficient set of order* k with $k \in \mathbb{N}$ if $\lambda_{n,k}^{(k)} \neq 0$ for all $n \in \mathbb{N}_0$. Also, $\Lambda = \{\Lambda^{(k)} \,|\, k \in \mathbb{N}\}$ is called *coefficient set*. Two coefficient sets

$\Lambda = \{\{\lambda_{n,j}^{(k)}\}\}$ and $\hat{\Lambda} = \{\{\hat{\lambda}_{n,j}^{(k)}\}\}$ are called *equivalent*, if for all n and k, there is a constant $c_n^{(k)} \neq 0$ such that $\hat{\lambda}_{n,j}^{(k)} = c_n^{(k)} \lambda_{n,j}^{(k)}$ for all j with $0 \leqslant j \leqslant k$.

For each coefficient set $\Lambda^{(k)} = \{\lambda_{n,j}^{(k)} | n \in \mathbb{N}_0, 0 \leqslant j \leqslant k\}$ of order k, one may define a *Levin-type sequence transformation of order k* by

$$\begin{aligned} \mathscr{T}[\Lambda^{(k)}] &: \mathbb{S}^{\mathbb{K}} \times \mathbb{Y}^{(k)} \to \mathbb{S}^{\mathbb{K}} \\ &: (\{\{s_n\}\}, \{\{\omega_n\}\}) \mapsto \{\{s_n'\}\} = \mathscr{T}[\Lambda^{(k)}](\{\{s_n\}\}, \{\{\omega_n\}\}) \end{aligned} \tag{65}$$

with

$$s_n' = \mathscr{T}_n^{(k)}(\{\{s_n\}\}, \{\{\omega_n\}\}) = \frac{\sum_{j=0}^{k} \lambda_{n,j}^{(k)} s_{n+j}/\omega_{n+j}}{\sum_{j=0}^{k} \lambda_{n,j}^{(k)}/\omega_{n+j}} \tag{66}$$

and

$$\mathbb{Y}^{(k)} = \left\{ \{\{\omega_n\}\} \in \mathbb{O}^{\mathbb{K}} \colon \sum_{j=0}^{k} \lambda_{n,j}^{(k)}/\omega_{n+j} \neq 0 \text{ for all } n \in \mathbb{N}_0 \right\}. \tag{67}$$

We call $\mathscr{T}[\Lambda] = \{\mathscr{T}[\Lambda^{(k)}] | \, k \in \mathbb{N}\}$ the Levin-type sequence transformation corresponding to the coefficient set $\Lambda = \{\Lambda^{(k)} | \, k \in \mathbb{N}\}$. We write $\mathscr{T}^{(k)}$ and $\mathscr{T}$ instead of $\mathscr{T}[\Lambda^{(k)}]$ and $\mathscr{T}[\Lambda]$, respectively, whenever the coefficients $\lambda_{n,j}^{(k)}$ are clear from the context. Also, if two coefficient sets Λ and $\hat{\Lambda}$ are equivalent, they give rise to the same sequence transformation, i.e., $\mathscr{T}[\Lambda] = \mathscr{T}[\hat{\Lambda}]$, since

$$\frac{\sum_{j=0}^{k} \hat{\lambda}_{n,j}^{(k)} s_{n+j}/\omega_{n+j}}{\sum_{j=0}^{k} \hat{\lambda}_{n,j}^{(k)}/\omega_{n+j}} = \frac{\sum_{j=0}^{k} \lambda_{n,j}^{(k)} s_{n+j}/\omega_{n+j}}{\sum_{j=0}^{k} \lambda_{n,j}^{(k)}/\omega_{n+j}} \quad \text{for } \hat{\lambda}_{n,j}^{(k)} = c_n^{(k)} \lambda_n^{(k)} \tag{68}$$

with arbitrary $c_n^{(k)} \neq 0$.

The number $\mathscr{T}_n^{(k)}$ are often arranged in a two-dimensional table

$$\begin{matrix} \mathscr{T}_0^{(0)} & \mathscr{T}_0^{(1)} & \mathscr{T}_0^{(2)} & \cdots \\ \mathscr{T}_1^{(0)} & \mathscr{T}_1^{(1)} & \mathscr{T}_1^{(2)} & \cdots \\ \mathscr{T}_2^{(0)} & \mathscr{T}_2^{(1)} & \mathscr{T}_2^{(2)} & \cdots \\ \vdots & \vdots & \vdots & \ddots \end{matrix} \tag{69}$$

that is called the $\mathscr{T}$ table. The transformations $\mathscr{T}^{(k)}$ thus correspond to columns, i.e., to following vertical paths in the table. The numerators and denominators such that $\mathscr{T}_n^{(k)} = N_n^{(k)}/D_n^{(k)}$ also are often arranged in analogous N and D tables.

Note that for fixed N, one may also define a transformation

$$\mathscr{T}_N : \{\{s_{n+N}\}\} \mapsto \{\{\mathscr{T}_N^{(k)}\}\}_{k=0}^{\infty}. \tag{70}$$

This corresponds to horizontal paths in the $\mathscr{T}$ table. These are sometimes called diagonals, because rearranging the table in such a way that elements with constant values of $n+k$ are members of the same row, $\mathscr{T}_N^{(k)}$ for fixed N correspond to diagonals of the rearranged table.

For a given coefficient set Λ define the *moduli* by

$$\mu_n^{(k)} = \max_{0 \leqslant j \leqslant k} \{|\lambda_{n,j}^{(k)}|\} \tag{71}$$

and the *characteristic polynomials* by

$$\Pi_n^{(k)} \in \mathbb{P}^{(k)}:\ \Pi_n^{(k)}(z) = \sum_{j=0}^{k} \lambda_{n,j}^{(k)} z^j \tag{72}$$

for $n \in \mathbb{N}_0$ and $k \in \mathbb{N}$.

Then, $\mathscr{T}[\Lambda]$ is said to be *in normalized form* if $\mu_n^{(k)} = 1$ for all $k \in \mathbb{N}$ and $n \in \mathbb{N}_0$. Is is said to be *in subnormalized form* if for all $k \in \mathbb{N}$ there is a constant $\tilde{\mu}^{(k)}$ such that $\mu_n^{(k)} \leqslant \tilde{\mu}^{(k)}$ for all $n \in \mathbb{N}_0$.

Any Levin-type sequence transformation $\mathscr{T}[\Lambda]$ can rewritten in normalized form. To see this, use

$$c_n^{(k)} = 1/\mu_n^{(k)} \tag{73}$$

in Eq. (68). Similarly, each Levin-type sequence transformation can be rewritten in (many different) subnormalized forms.

A Levin-type sequence transformation of order k is said to be *convex* if $\Pi_n^{(k)}(1) = 0$ for all n in $\mathbb{N}_0$. Equivalently, it is convex if $\{\{1\}\} \notin \mathbb{Y}^{(k)}$, i.e., if the transformation vanishes for $\{\{s_n\}\} = \{\{c\omega_n\}\}$, $c \in \mathbb{K}$. Also, $\mathscr{T}[\Lambda]$ is called convex, if $\mathscr{T}[\Lambda^{(k)}]$ is convex for all $k \in \mathbb{N}$. We will see that this property is important for ensuring convergence acceleration for linearly convergent sequences.

A given Levin-type transformation $\mathscr{T}$ can also be rewritten as

$$\mathscr{T}_n^{(k)}(\{\{s_n\}\}, \{\{\omega_n\}\}) = \sum_{j=0}^{k} \gamma_{n,j}^{(k)}(\boldsymbol{\omega}_n)\, s_{n+j}, \qquad \boldsymbol{\omega}_n = (\omega_n, \ldots, \omega_{n+k}) \tag{74}$$

with

$$\gamma_{n,j}^{(k)}(\boldsymbol{\omega}_n) = \frac{\lambda_{n,j}^{(k)}}{\omega_{n+j}} \left[\sum_{j'=0}^{k} \frac{\lambda_{n,j'}^{(k)}}{\omega_{n+j'}} \right]^{-1}, \qquad \sum_{j=0}^{k} \gamma_{n,j}^{(k)}(\boldsymbol{\omega}_n) = 1. \tag{75}$$

Then, one may define *stability indices* by

$$\Gamma_n^{(k)}(\mathscr{T}) = \sum_{j=0}^{k} |\gamma_{n,j}^{(k)}(\boldsymbol{\omega}_n)| \geqslant 1. \tag{76}$$

Note that any sequence transformation $\mathscr{Q}$

$$\mathscr{Q}_n^{(k)} = \sum_{j=0}^{k} q_{n,j}^{(k)} s_{n+j} \tag{77}$$

with

$$\sum_{j=0}^{k} q_{n,j}^{(k)} = 1 \tag{78}$$

can formally be rewritten as a Levin-type sequence transformation according to $\mathscr{Q}_n^{(k)} = \mathscr{T}_n^{(k)}(\{\{s_n\}\}, \{\{\omega_n\}\})$ with coefficients $\lambda_{n,j}^{(k)} = \omega_{n+j} q_{n,j}^{(k)} \rho_n^{(k)}$ where the validity of Eq. (78) requires to set

$$\rho_n^{(k)} = \sum_{j=0}^{k} \lambda_{n,j}^{(k)} / \omega_{n+j}. \tag{79}$$

If for given $k \in \mathbb{N}$ and for a transformation $\mathscr{T}[\Lambda^{(k)}]$ the following limits exist and have the values:

$$\lim_{n\to\infty} \lambda_{n,j}^{(k)} = \mathring{\lambda}_j^{(k)} \tag{80}$$

for all $0 \leqslant j \leqslant k$, and if $\mathring{\Lambda}^{(k)}$ is a coefficient set of order k which means that at least the limit $\mathring{\lambda}_k^{(k)}$ does not vanish, then a limiting transformation $\mathring{\mathscr{T}}\,[\,\mathring{\Lambda}^{(k)}]$ exists where $\mathring{\Lambda}^{(k)} = \{\mathring{\lambda}_j^{(k)}\}$. More explicitly, we have

$$\begin{aligned} \mathring{\mathscr{T}}\,[\Lambda^{(k)}] &: \mathbb{S}^{\mathbb{K}} \times \mathring{\mathbb{Y}}^{(k)} \to \mathbb{S}^{\mathbb{K}} \\ &: (\{\{s_n\}\}, \{\{\omega_n\}\}) \mapsto \{\{s'_n\}\} \end{aligned} \tag{81}$$

with

$$s'_n = \mathring{\mathscr{T}}^{(k)}(\{\{s_n\}\}, \{\{\omega_n\}\}) = \frac{\sum_{j=0}^{k} \mathring{\lambda}_j^{(k)} s_{n+j}/\omega_{n+j}}{\sum_{j=0}^{k} \mathring{\lambda}_j^{(k)}/\omega_{n+j}} \tag{82}$$

and

$$\mathring{\mathbb{Y}}^{(k)} = \left\{ \{\{\omega_n\}\} \in \mathbb{O}^{\mathbb{K}} : \sum_{j=0}^{k} \mathring{\lambda}_j^{(k)}/\omega_{n+j} \neq 0 \quad \text{for all } n \in \mathbb{N}_0 \right\}. \tag{83}$$

Obviously, this limiting transformation itself is a Levin-type sequence transformation and automatically is given in subnormalized form.

4.1.1. Variants of Levin-type transformations

For the following, assume that $\beta > 0$ is an arbitrary constant, $a_n = \Delta s_{n-1}$, and $\hat{a}_n$ are Kummer-related to the a_n with limit or antilimit $\hat{s}$ (cf. Section 2.2.1).

A *variant* of a Levin-type sequence transformation $\mathscr{T}$ is obtained by a particular choice ω_n. For $\omega_n = f_n(\{\{s_n\}\})$, the transformation $\mathscr{T}$ is nonlinear in the s_n. In particular, we have [50,53,79]:

t Variant:

$${}^{t}\omega_n = \Delta s_{n-1} = a_n : {}^{t}\mathscr{T}_n^{(k)}(\{\{s_n\}\}) = \mathscr{T}_n^{(k)}(\{\{s_n\}\}, \{\{{}^{t}\omega_n\}\}). \tag{84}$$

u Variant:

$${}^{u}\omega_n = (n+\beta)\,\Delta\, s_{n-1} = (n+\beta)a_n : {}^{u}\mathscr{T}_n^{(k)}(\beta, \{\{s_n\}\}) = \mathscr{T}_n^{(k)}(\{\{s_n\}\}, \{\{{}^{u}\omega_n\}\}). \tag{85}$$

v Variant:

$${}^{v}\omega_n = -\frac{\Delta s_{n-1}\,\Delta\, s_n}{\Delta^2 s_{n-1}} = \frac{a_n a_{n+1}}{a_n - a_{n+1}} : {}^{v}\mathscr{T}_n^{(k)}(\{\{s_n\}\}) = \mathscr{T}_n^{(k)}(\{\{s_n\}\}, \{\{{}^{v}\omega_n\}\}). \tag{86}$$

$\tilde{t}$ Variant:

$${}^{\tilde{t}}\omega_n = \Delta s_n = a_{n+1} : {}^{\tilde{t}}\mathscr{T}_n^{(k)}(\{\{s_n\}\}) = \mathscr{T}_n^{(k)}(\{\{s_n\}\}, \{\{{}^{\tilde{t}}\omega_n\}\}). \tag{87}$$

lt Variant:

$${}^{lt}\omega_n = \hat{a}_n : {}^{lt}\mathscr{T}_n^{(k)}(\{\{s_n\}\}) = \mathscr{T}_n^{(k)}(\{\{s_n\}\}, \{\{{}^{lt}\omega_n\}\}). \tag{88}$$

lu Variant:

$$ {}^{lu}\omega_n = (n+\beta)\hat{a}_n\colon\ {}^{lu}\mathscr{T}_n^{(k)}(\beta,\{\{s_n\}\}) = \mathscr{T}_n^{(k)}(\{\{s_n\}\},\{\{{}^{lu}\omega_n\}\}). \tag{89} $$

lv Variant:

$$ {}^{lv}\omega_n = \frac{\hat{a}_n\hat{a}_{n+1}}{\hat{a}_n-\hat{a}_{n+1}}\colon\ {}^{lv}\mathscr{T}_n^{(k)}(\{\{s_n\}\}) = \mathscr{T}_n^{(k)}(\{\{s_n\}\},\{\{{}^{lv}\omega_n\}\}). \tag{90} $$

$l\tilde{t}$ Variant:

$$ {}^{l\tilde{t}}\omega_n = \hat{a}_{n+1}\colon\ {}^{l\tilde{t}}\mathscr{T}_n^{(k)}(\{\{s_n\}\}) = \mathscr{T}_n^{(k)}(\{\{s_n\}\},\{\{{}^{l\tilde{t}}\omega_n\}\}). \tag{91} $$

K Variant:

$$ {}^{K}\omega_n = \hat{s}_n - \hat{s}\colon\ {}^{K}\mathscr{T}_n^{(k)}(\{\{s_n\}\}) = \mathscr{T}_n^{(k)}(\{\{s_n\}\},\{\{{}^{K}\omega_n\}\}). \tag{92} $$

The K variant of a Levin-type transformation $\mathscr{T}$ is linear in the s_n. This holds also for the lt, lu, lv and $l\tilde{t}$ variants.

4.2. *Important examples of Levin-type sequence transformations*

In this section, we present important Levin-type sequence transformations. For each transformation, we give the definition, recursive algorithms and some background information.

4.2.1. *$\mathscr{J}$ transformation*

The $\mathscr{J}$ transformation was derived and studied by Homeier [35,36,38–40,46]. Although the $\mathscr{J}$ transformation was derived by hierarchically consistent iteration of the simple transformation

$$ s_n' = s_{n+1} - \omega_{n+1}\frac{\triangle s_n}{\triangle \omega_n}, \tag{93} $$

it was possible to derive an explicit formula for its kernel as is discussed later. It may be defined via the recursive scheme

$$ \begin{aligned} &N_n^{(0)} = s_n/\omega_n, \quad D_n^{(0)} = 1/\omega_n,\\ &N_n^{(k)} = \nabla_n^{(k-1)}N_n^{(k-1)}, \quad D_n^{(k)} = \nabla_n^{(k-1)}D_n^{(k-1)},\\ &\mathscr{J}_n^{(k)}(\{\{s_n\}\},\{\{\omega_n\}\},\{\delta_n^{(k)}\}) = N_n^{(k)}/D_n^{(k)}, \end{aligned} \tag{94} $$

where the generalized difference operator defined in Eq. (33) involves quantities $\delta_n^{(k)} \neq 0$ for $k \in \mathbb{N}_0$. Special cases of the $\mathscr{J}$ transformation result from corresponding choices of the $\delta_n^{(k)}$. These are summarized in Table 1.

Using generalized difference operators $\nabla_n^{(k)}$, we also have the representation [36, Eq. (38)]

$$ \mathscr{J}_n^{(k)}(\{\{s_n\}\},\{\{\omega_n\}\},\{\{\delta_n^{(k)}\}\}) = \frac{\nabla_n^{(k-1)}\nabla_n^{(k-2)}\ldots\nabla_n^{(0)}[s_n/\omega_n]}{\nabla_n^{(k-1)}\nabla_n^{(k-2)}\ldots\nabla_n^{(0)}[1/\omega_n]}. \tag{95} $$

The $\mathscr{J}$ transformation may also be computed using the alternative recursive schemes [36,46]

$$ \begin{aligned} &\hat{D}_n^{(0)} = 1/\omega_n, \qquad \hat{N}_n^{(0)} = s_n/\omega_n,\\ &\hat{D}_n^{(k)} = \Phi_n^{(k-1)}\hat{D}_{n+1}^{(k-1)} - \hat{D}_n^{(k-1)}, \quad k\in\mathbb{N},\\ &\hat{N}_n^{(k)} = \Phi_n^{(k-1)}\hat{N}_{n+1}^{(k-1)} - \hat{N}_n^{(k-1)}, \quad k\in\mathbb{N}, \end{aligned} \tag{96} $$

Table 1
Special cases of the $\mathcal{J}$ transformation[a]

Case	$\psi_j(n)$[b]	$\delta_n^{(k)}$[c]
Drummond transformation $\mathscr{D}_n^{(k)}(\{\{s_n\}\},\{\{\omega_n\}\})$	n^j	1
Homeier $\mathscr{I}$ transformation $\mathscr{I}_n^{(k)}(\alpha,\{\{s_n\}\},\{\{\omega_n\}\},\{\Delta_n^{(k)}\})$ $=\mathscr{J}_n^{(2k)}(\{\{s_n\}\},\{\{e^{-i\alpha n}\omega_n\}\},\{\delta_n^{(k)}\})$	Eq. (231)	$\delta_n^{(2\ell)}=\exp(2\mathrm{i}\alpha n)$, $\delta_n^{(2\ell+1)}=\exp(-2\mathrm{i}\alpha n)\Delta_n^{(\ell)}$
Homeier $\mathscr{F}$ transformation $\mathscr{F}_n^{(k)}(\{\{s_n\}\},\{\{\omega_n\}\},\{\{x_n\}\})$	$1/(x_n)_j$	$\frac{x_{n+k+1}-x_n}{x_n+k-1}\prod_{j=0}^{n-1}\frac{(x_j+k)(x_{j+k+1}+k-1)}{(x_j+k-1)(x_{j+k+2}+k)}$
Homeier ${}_p\mathbf{J}$ transformation ${}_p\mathbf{J}_n^{(k)}(\beta,\{\{s_n\}\},\{\{\omega_n\}\})$	Eq. (231)	$\frac{1}{(n+\beta+(p-1)k)_2}$
Levin transformation $\mathscr{L}_n^{(k)}(\beta,\{\{s_n\}\},\{\{\omega_n\}\})$	$(n+\beta)^{-j}$	$\frac{1}{(n+\beta)(n+\beta+k+1)}$
generalized $\mathscr{L}$ transformation $\mathscr{L}_n^{(k)}(\alpha,\beta,\{\{s_n\}\},\{\{\omega_n\}\})$	$(n+\beta)^{-j\alpha}$	$\frac{(n+\beta+k+1)^\alpha-(n+\beta)^\alpha}{(n+\beta)^\alpha(n+\beta+k+1)^\alpha}$
Levin-Sidi $d^{(1)}$ transformation [22,54,77] $(d^{(1)})_n^{(k)}(\alpha,\{\{s_n\}\})$	$(R_n+\alpha)^{-j}$	$\frac{1}{R_{n+k+1}+\alpha}-\frac{1}{R_n+\alpha}$
Mosig–Michalski algorithm [60,61] $M_n^{(k)}(\{\{s_n\}\},\{\{\omega_n\}\},\{\{x_n\}\})$	Eq. (231)	$\frac{1}{x_n^2}\left(1-\frac{\omega_n x_{n+1}^{2k}}{\omega_{n+1}x_n^{2k}}\right)$
Sidi W algorithm (GREP[(1)]) [73,77,78] $W_n^{(k)}(\{\{s_n\}\},\{\{\omega_n\}\},\{\{t_n\}\})$	t_n^j	$t_{n+k+1}-t_n$
Weniger $\mathscr{C}$ transformation [87] $\mathscr{C}_n^{(k)}(\gamma,\beta/\gamma,\{\{s_n\}\},\{\{\omega_n\}\})$	$\frac{1}{(\gamma n+\beta)_j}$	$\frac{(n+1+(\beta+k-1)/\gamma)_k}{(n+(\beta+k)/\gamma)_{k+2}}$
Weniger $\mathscr{M}$ transformation $\mathscr{M}_n^{(k)}(\xi,\{\{s_n\}\},\{\{\omega_n\}\})$	$\frac{1}{(-n-\xi)_j}$	$\frac{(n+1+\xi-(k-1))_k}{(n+\xi-k)_{k+2}}$
Weniger $\mathscr{S}$ transformation $\mathscr{S}_n^{(k)}(\beta,\{\{s_n\}\},\{\{\omega_n\}\})$	$1/(n+\beta)_j$	$\frac{1}{(n+\beta+2k)_2}$
Iterated Aitken process [2,84] $\mathbf{A}_n^{(k)}(\{\{s_n\}\})$ $=\mathscr{J}_n^{(k)}(\{\{s_n\}\},\{\{\triangle s_n\}\},\{\delta_n^{(k)}\})$	Eq. (231)	$\frac{(\triangle\mathbf{A}_n^{(k+1)}(\{\{s_n\}\}))(\triangle^2\mathbf{A}_n^{(k)})(\{\{s_n\}\})}{(\triangle\mathbf{A}_n^{(k)}(\{\{s_n\}\}))(\triangle\mathbf{A}_{n+1}^{(k)}(\{\{s_n\}\}))}$
Overholt process [64] $V_n^{(k)}(\{\{s_n\}\})$ $=\mathscr{J}_n^{(k)}(\{\{s_n\}\},\{\{\triangle s_n\}\},\{\delta_n^{(k)}\})$	Eq. (231)	$\frac{(\Delta s_{n+k+1})\Delta[(\Delta s_{n+k})^{k+1}]}{(\Delta s_{n+k})^{k+1}}$

[a]Refs. [36,38,40].
[b]For the definition of the $\psi_{j,n}$ see Eq. (5).
[c]Factors independent of n are irrelevant.

$$\mathscr{J}_n^{(k)}(\{\{s_n\}\},\{\{\omega_n\}\},\{\delta_n^{(k)}\}) = \frac{\hat{N}_n^{(k)}}{\hat{D}_n^{(k)}}$$

with

$$\Phi_n^{(0)} = 1, \qquad \Phi_n^{(k)} = \frac{\delta_n^{(0)}\delta_n^{(1)}\cdots\delta_n^{(k-1)}}{\delta_{n+1}^{(0)}\delta_{n+1}^{(1)}\cdots\delta_{n+1}^{(k-1)}}, \quad k \in \mathbb{N} \tag{97}$$

and

$$\begin{aligned}
\tilde{D}_n^{(0)} &= 1/\omega_n, \quad \tilde{N}_n^{(0)} = s_n/\omega_n, \\
\tilde{D}_n^{(k)} &= \tilde{D}_{n+1}^{(k-1)} - \Psi_n^{(k-1)}\tilde{D}_n^{(k-1)}, \quad k \in \mathbb{N}, \\
\tilde{N}_n^{(k)} &= \tilde{N}_{n+1}^{(k-1)} - \Psi_n^{(k-1)}\tilde{N}_n^{(k-1)}, \quad k \in \mathbb{N},
\end{aligned} \tag{98}$$

$$\mathscr{J}_n^{(k)}(\{\{s_n\}\},\{\{\omega_n\}\},\{\delta_n^{(k)}\}) = \frac{\tilde{N}_n^{(k)}}{\tilde{D}_n^{(k)}}$$

with

$$\Psi_n^{(0)} = 1, \qquad \Psi_n^{(k)} = \frac{\delta_{n+k}^{(0)}\delta_{n+k-1}^{(1)}\cdots\delta_{n+1}^{(k-1)}}{\delta_{n+k-1}^{(0)}\delta_{n+k-2}^{(1)}\cdots\delta_n^{(k-1)}}, \quad k \in \mathbb{N}. \tag{99}$$

The quantities $\Psi_n^{(k)}$ should not be mixed up with the $\Psi_{k,n}(u)$ as defined in Eq. (43).

As shown in [46], the coefficients for the algorithm (96) that are defined via $\hat{D}_n^{(k)} = \sum_{j=0}^{k} \lambda_{n,j}^{(k)}/\omega_{n+j}$, satisfy the recursion

$$\lambda_{n,j}^{(k+1)} = \Phi_n^{(k)}\lambda_{n+1,j-1}^{(k)} - \lambda_{n,j}^{(k)} \tag{100}$$

with starting values $\lambda_{n,j}^{(0)} = 1$. This holds for all j if we define $\lambda_{n,j}^{(k)} = 0$ for $j < 0$ or $j > k$. Because $\Phi_n^{(k)} \neq 0$, we have $\lambda_{n,k}^{(k)} \neq 0$ such that $\{\lambda_{n,j}^{(k)}\}$ is a coefficient set for all $k \in \mathbb{N}_0$.

Similarly, the coefficients for algorithm (98) that are defined via $\tilde{D}_n^{(k)} = \sum_{j=0}^{k} \tilde{\lambda}_{n,j}^{(k)}/\omega_{n+j}$, satisfy the recursion

$$\tilde{\lambda}_{n,j}^{(k+1)} = \tilde{\lambda}_{n+1,j-1}^{(k)} - \Psi_n^{(k)}\tilde{\lambda}_{n,j}^{(k)} \tag{101}$$

with starting values $\tilde{\lambda}_{n,j}^{(0)} = 1$. This holds for all j if we define $\tilde{\lambda}_{n,j}^{(k)} = 0$ for $j < 0$ or $j > k$. In this case, we have $\tilde{\lambda}_{n,k}^{(k)} = 1$ such that $\{\tilde{\lambda}_{n,j}^{(k)}\}$ is a coefficient set for all $k \in \mathbb{N}_0$.

Since the $\mathscr{J}$ transformation vanishes for $\{\{s_n\}\} = \{\{c\omega_n\}\}$, $c \in \mathbb{K}$ according to Eq. (95) for all $k \in \mathbb{N}$, it is convex. This may also be shown by using induction in k using $\lambda_{n,1}^{(1)} = -\lambda_{n,0}^{(1)} = 1$ and the equation

$$\sum_{j=0}^{k+1} \lambda_{n,j}^{(k+1)} = \Phi_n^{(k)} \sum_{j=0}^{k} \lambda_{n+1,j}^{(k)} - \sum_{j=0}^{k} \lambda_{n,j}^{(k)} \tag{102}$$

that follows from Eq. (100).

Assuming that the limits $\Phi_k = \lim_{n\to\infty} \Phi_n^{(k)}$ exist for all $k \in \mathbb{N}$ and noting that for $k = 0$ always $\Phi_0 = 1$ holds, it follows that there exists a limiting transformation $\mathring{\mathcal{J}}[\mathring{\Lambda}]$ that can be considered as special variant of the $\mathcal{J}$ transformation and with coefficients given explicitly as [46, Eq. (16)]

$$\mathring{\lambda}_j^{(k)} = (-1)^{k-j} \sum_{\substack{j_0+j_1+\dots+j_{k-1}=j,\\ j_0\in\{0,1\},\dots,j_{k-1}\in\{0,1\}}} \prod_{m=0}^{k-1} (\Phi_m)^{j_m}. \tag{103}$$

As characteristic polynomial we obtain

$$\mathring{\Pi}^{(k)}(z) = \sum_{j=0}^{k} \mathring{\lambda}_j^{(k)} z^j = \prod_{j=0}^{k-1} (\Phi_j z - 1). \tag{104}$$

Hence, the $\mathring{\mathcal{J}}$ transformation is convex since $\mathring{\Pi}^{(k)}(1) = 0$ due to $\Phi_0 = 1$.

The ${}_p\mathbf{J}$ *Transformation*: This is the special case of the $\mathcal{J}$ transformation corresponding to

$$\delta_n^{(k)} = \frac{1}{(n+\beta+(p-1)k)_2} \tag{105}$$

or to [46, Eq. (18)][2]

$$\Phi_n^{(k)} = \begin{cases} \left(\dfrac{n+\beta+2}{p-1}\right)_k \Big/ \left(\dfrac{n+\beta}{p-1}\right)_k & \text{for } p \neq 1, \\[2ex] \left(\dfrac{n+\beta+2}{n+\beta}\right)^k & \text{for } p = 1 \end{cases} \tag{106}$$

or to

$$\Psi_n^{(k)} = \begin{cases} \left(\dfrac{n+\beta+k-1}{p-2}\right)_k \Big/ \left(\dfrac{n+\beta+k+1}{p-2}\right)_k & \text{for } p \neq 2, \\[2ex] \left(\dfrac{n+\beta+k-1}{n+\beta+k+1}\right)^k & \text{for } p = 2, \end{cases} \tag{107}$$

that is,

$${}_p\mathbf{J}_n^{(k)}(\beta, \{\{s_n\}\}, \{\{\omega_n\}\}) = \mathcal{J}_n^{(k)}(\{\{s_n\}\}, \{\{\omega_n\}\}, \{1/(n+\beta+(p-1)k)_2\}). \tag{108}$$

The limiting transformation ${}_p\mathring{\mathbf{J}}$ of the ${}_p\mathbf{J}$ transformation exists for all p and corresponds to the $\mathring{\mathcal{J}}$ transformation with $\Phi_k = 1$ for all k in $\mathbb{N}_0$. This is exactly the Drummond transformation discussed in Section 4.2.2, i.e., we have

$${}_p\mathring{\mathbf{J}}_n^{(k)}(\beta, \{\{s_n\}\}, \{\{\omega_n\}\}) = \mathcal{D}_n^{(k)}(\{\{s_n\}\}, \{\{\omega_n\}\}). \tag{109}$$

---

[2] The equation in [46] contains an error.

4.2.2. Drummond transformation

This transformation was given by Drummond [19]. It was also discussed by Weniger [84]. It may be defined as

$$\mathscr{D}_n^{(k)}(\{\{s_n\}\},\{\{\omega_n\}\}) = \frac{\triangle^k[s_n/\omega_n]}{\triangle^k[1/\omega_n]}. \tag{110}$$

Using the definition (32) of the forward difference operator, the coefficients may be taken as

$$\lambda_{n,j}^{(k)} = (-1)^j \binom{k}{j}, \tag{111}$$

i.e., independent of n. As moduli, one has $\mu_n^{(k)} = \binom{k}{[\![k/2]\!]} = \tilde{\mu}^{(k)}$. Consequently, the Drummond transformation is given in subnormalized form. As characteristic polynomial we obtain

$$\Pi_n^{(k)}(z) = \sum_{j=0}^{k}(-1)^j \binom{k}{j} z^j = (1-z)^k. \tag{112}$$

Hence, the Drummond transformation is convex since $\Pi_n^{(k)}(1) = 0$. Interestingly, the Drummond transformation is identical to its limiting transformation:

$$\mathring{\mathscr{D}}^{(k)}(\{\{s_n\}\},\{\{\omega_n\}\}) = \mathscr{D}_n^{(k)}(\{\{s_n\}\},\{\{\omega_n\}\}). \tag{113}$$

The Drummond transformation may be computed using the recursive scheme

$$\begin{aligned} &N_n^{(0)} = s_n/\omega_n, \quad D_n^{(0)} = 1/\omega_n\,, \\ &N_n^{(k)} = \triangle N_n^{(k-1)}, \quad D_n^{(k)} = \triangle D_n^{(k-1)}, \\ &\mathscr{D}_n^{(k)} = N_n^{(k)}/D_n^{(k)}. \end{aligned} \tag{114}$$

4.2.3. Levin transformation

This transformation was given by Levin [53]. It was also discussed by Weniger [84]. It may be defined as [3]

$$\mathscr{L}_n^{(k)}(\beta,\{\{s_n\}\},\{\{\omega_n\}\}) = \frac{(n+\beta+k)^{1-k}\,\triangle^k\,[(n+\beta)^{k-1}s_n/\omega_n]}{(n+\beta+k)^{1-k}\,\triangle^k\,[(n+\beta)^{k-1}/\omega_n]}. \tag{115}$$

Using the definition (32) of the forward difference operator, the coefficients may be taken as

$$\lambda_{n,j}^{(k)} = (-1)^j \binom{k}{j} (n+\beta+j)^{k-1}/(n+\beta+k)^{k-1}. \tag{116}$$

The moduli satisfy $\mu_n^{(k)} \leqslant \binom{k}{[\![k/2]\!]} = \tilde{\mu}^{(k)}$ for given k. Consequently, the Levin transformation is given in subnormalized form. As characteristic polynomial we obtain

$$\Pi_n^{(k)}(z) = \sum_{j=0}^{k}(-1)^j \binom{k}{j} z^j (n+\beta+j)^{k-1}/(n+\beta+k)^{k-1}. \tag{117}$$

---

[3] Note that the order of indices is different from that in the literature.

Since $\Pi_n^{(k)}(1)=0$ because $\triangle^k$ annihilates any polynomial in n with degree less than k, the Levin transformation is convex. The limiting transformation is identical to the Drummond transformation

$$\overset{\circ}{\mathscr{L}}{}^{(k)}(\{\{s_n\}\},\{\{\omega_n\}\}) = \mathscr{D}_n^{(k)}(\{\{s_n\}\},\{\{\omega_n\}\}). \tag{118}$$

The Levin transformation may be computed using the recursive scheme [21,55,84,14, Section 2.7]

$$\begin{aligned}
&N_n^{(0)} = s_n/\omega_n, \quad D_n^{(0)} = 1/\omega_n,\\
&N_n^{(k)} = N_{n+1}^{(k-1)} - \frac{(\beta+n)(\beta+n+k-1)^{k-2}}{(\beta+n+k)^{k-1}} N_n^{(k-1)},\\
&D_n^{(k)} = D_{n+1}^{(k-1)} - \frac{(\beta+n)(\beta+n+k-1)^{k-2}}{(\beta+n+k)^{k-1}} D_n^{(k-1)},\\
&\mathscr{L}_n^{(k)}(\beta,\{\{s_n\}\},\{\{\omega_n\}\}) = N_n^{(k)}/D_n^{(k)}.
\end{aligned} \tag{119}$$

This is essentially the same as the recursive scheme (98) for the $\mathscr{J}$ transformation with

$$\Psi_n^{(k)} = \frac{(\beta+n)(\beta+n+k)^{k-1}}{(\beta+n+k+1)^k}, \tag{120}$$

since the Levin transformation is a special case of the $\mathscr{J}$ transformation (see Table 1). Thus, the Levin transformation can also be computed recursively using scheme (94)

$$\delta_n^{(k)} = \frac{1}{(n+\beta)(n+\beta+k+1)} \tag{121}$$

or scheme (96) with [46]

$$\Phi_n^{(k)} = (n+\beta+k+1)\frac{(n+\beta+1)^{k-1}}{(n+\beta)^k}. \tag{122}$$

4.2.4. Weniger transformations

Weniger [84,87,88] derived sequence transformations related to factorial series. These may be regarded as special cases of the transformation

$$\mathscr{C}_n^{(k)}(\alpha,\zeta,\{\{s_n\}\},\{\{\omega_n\}\}) = \frac{((\alpha[n+\zeta+k])_{k-1})^{-1}\,\triangle^k\,[(\alpha[n+\zeta])_{k-1}s_n/\omega_n]}{((\alpha[n+\zeta+k])_{k-1})^{-1}\,\triangle^k\,[(\alpha[n+\zeta])_{k-1}/\omega_n]}. \tag{123}$$

In particular, the Weniger $\mathscr{S}$ transformation may be defined as

$$\mathscr{S}_n^{(k)}(\beta,\{\{s_n\}\},\{\{\omega_n\}\}) = \mathscr{C}_n^{(k)}(1,\beta,\{\{s_n\}\},\{\{\omega_n\}\}) \tag{124}$$

and the Weniger $\mathscr{M}$ transformation as

$$\mathscr{M}_n^{(k)}(\xi,\{\{s_n\}\},\{\{\omega_n\}\}) = \mathscr{C}_n^{(k)}(-1,\xi,\{\{s_n\}\},\{\{\omega_n\}\}). \tag{125}$$

The parameters β, ξ, and ζ are taken to be positive real numbers. Weniger considered the $\mathscr{C}$ transformation only for $\alpha>0$ [87,88] and thus, he was not considering the $\mathscr{M}$ transformation as a special case of the $\mathscr{C}$ transformation. He also found that one should choose $\xi \geqslant k-1$. In the u variant of the $\mathscr{M}$ transformation he proposed to choose $\omega_n = (-n-\xi)\triangle s_{n-1}$. This variant is denoted as ${}^u\mathscr{M}$ transformation in the present work.

Using the definition (32) of the forward difference operator, the coefficients may be taken as

$$\lambda_{n,j}^{(k)} = (-1)^j \binom{k}{j} (\alpha[n+\zeta+j])_{k-1}/(\alpha[n+\zeta+k])_{k-1} \tag{126}$$

in the case of the $\mathscr{C}$ transformation, as

$$\lambda_{n,j}^{(k)} = (-1)^j \binom{k}{j} (n+\beta+j)_{k-1}/(n+\beta+k)_{k-1} \tag{127}$$

in the case of the $\mathscr{S}$ transformation, and as

$$\lambda_{n,j}^{(k)} = (-1)^j \binom{k}{j} (-n-\xi-j)_{k-1}/(-n-\xi-k)_{k-1} \tag{128}$$

in the case of the $\mathscr{M}$ transformation.

The $\mathscr{S}$ transformation in (124) may be computed using the recursive scheme (98) with [84, Section 8.3]

$$\Psi_n^{(k)} = \frac{(\beta+n+k)(\beta+n+k-1)}{(\beta+n+2k)(\beta+n+2k-1)}. \tag{129}$$

The $\mathscr{M}$ transformation in (125) may be computed using the recursive scheme (98) with [84, Section 9.3]

$$\Psi_n^{(k)} = \frac{\xi+n-k+1}{\xi+n+k+1}. \tag{130}$$

The $\mathscr{C}$ transformation in (123) may be computed using the recursive scheme (98) with [87, Eq. (3.3)]

$$\Psi_n^{(k)} = (\alpha[\zeta+n]+k-2)\frac{(\alpha[n+\zeta+k-1])_{k-2}}{(\alpha[n+\zeta+k])_{k-1}}. \tag{131}$$

Since the operator $\triangle^k$ for $k \in \mathbb{N}$ annihilates all polynomials in n of degree smaller than k, the transformations $\mathscr{S}, \mathscr{M}$, and $\mathscr{C}$ are convex. The moduli satisfy $\mu_n^{(k)} \leqslant \binom{k}{[k/2]} = \tilde{\mu}^{(k)}$ for given k. Consequently, the three Weniger transformations are given in subnormalized form.

For $\alpha \to \infty$, the Levin transformation is obtained from the $\mathscr{C}$ transformation [87]. The $\mathscr{S}$ transformation is identical to the ${}_3\mathbf{J}$ transformation. It is also the special case $x_n = n + \beta$ of the $\mathscr{F}$ transformation. Analogously, the $\mathscr{C}$ transformation is obtained for $x_n = \alpha[\zeta + n]$. All these Weniger transformations are special cases of the $\mathscr{J}$ transformation (cf Table 1).

The limiting transformation of all these Weniger transformations is the Drummond transformation.

4.2.5. Levin–Sidi transformations and W algorithms

As noted above in Section 3.2, the $d^{(m)}$ transformations were introduced by Levin and Sidi [54] as a generalization of the u variant of the Levin transformation, and these transformations may implemented recursively using the $\mathbf{W}^{(m)}$ algorithms.

The case $m = 1$ corresponding to the $d^{(1)}$ transformation and the $\mathbf{W}^{(1)} = W$ algorithm is relevant for the present survey of Levin-type transformations. In the following, the kth-order transformation $\mathscr{T}^{(k)}$ of Levin-type transformation $\mathscr{T}$ as given by the W algorithm is denoted by $W^{(k)}$ which should not be confused with the $\mathbf{W}^{(m)}$ algorithms of Ford and Sidi [22].

The W algorithm [73] was also studied by other authors [84, Section 7.4], [14, p. 71f, 116f] and may be regarded as a special case of the $\mathscr{J}$ transformation [36]. It may be defined as (cf [78, Theorems 1.1 and 1.2])

$$
\begin{aligned}
N_n^{(0)} &= \frac{s_n}{\omega_n}, \quad D_n^{(0)} = \frac{1}{\omega_n},\\
N_n^{(k)} &= \frac{N_{n+1}^{(k-1)} - N_n^{(k-1)}}{t_{n+k} - t_n},\\
D_n^{(k)} &= \frac{D_{n+1}^{(k-1)} - D_n^{(k-1)}}{t_{n+k} - t_n},\\
W_n^{(k)}(\{\{s_n\}\}, \{\{\omega_n\}\}, \{\{t_n\}\}) &= N_n^{(k)}/D_n^{(k)}
\end{aligned}
\tag{132}
$$

and computes

$$
W_n^{(k)}(\{\{s_n\}\}, \{\{\omega_n\}\}, \{\{t_n\}\}) = \frac{\square_n^{(k)}(s_n/\omega_n)}{\square_n^{(k)}(1/\omega_n)}, \tag{133}
$$

where the divided difference operators $\square_n^{(k)} = \square_n^{(k)}[\{\{t_n\}\}]$ are used. The W algorithm may be used to calculate the Levin transformation on putting $t_n = 1/(n+\beta)$. Some authors call a linear variant of the W algorithm with $\omega_n = (-1)^{n+1}\mathrm{e}^{-n\zeta q} t_n^{\alpha}$ the W transformation, while the $\tilde{t}$ variant of the W algorithm [74,75] is sometimes called mW transformation [31,57,60].

If $t_{n+1}/t_n \to \tau$ for large n, one obtains as limiting transformation the $\mathring{\mathscr{J}}$ transformation with $\Phi_j = \tau^{-j}$ and characteristic polynomial

$$
\mathring{\Pi}^{(k)}(z) = \prod_{j=0}^{k-1} (z/\tau^j - 1). \tag{134}
$$

For the $d^{(1)}$ transformation, we write

$$
(d^{(1)})_n^{(k)}(\alpha, \{\{s_n\}\}, \{\{\xi_n\}\}) = W_n^{(k)}(\{\{s_{\xi_n}\}\}, \{\{(\xi_n+\alpha)(s_{\xi_n} - s_{\xi_n - 1})\}\}, \{\{1/(\xi_n+\alpha)\}\}). \tag{135}
$$

Thus, it corresponds to the variant of the W algorithm with remainder estimates chosen as $(\xi_n + \alpha)(s_{\xi_n} - s_{\xi_n - 1})$ operating on the subsequence $\{\{s_{\xi_n}\}\}$ of $\{\{s_n\}\}$ with $t_n = 1/(\xi_n + \alpha)$. It should be noted that this is not(!) identical to the u variant

$$
{}^{u}W_n^{(k)}(\{\{s_{\xi_n}\}\}, \{\{1/(\xi_n+\alpha)\}\}) = W_n^{(k)}(\{\{s_{\xi_n}\}\}, \{\{{}^{u}\omega_n\}\}, \{\{1/(\xi_n+\alpha)\}\}), \tag{136}
$$

neither for ${}^{u}\omega_n = (n+\alpha)(s_{\xi_n} - s_{\xi_{n-1}})$ nor for ${}^{u}\omega_n = (\xi_n+\alpha)(s_{\xi_n} - s_{\xi_{n-1}})$, since the remainder estimates are chosen differently in Eq. (135).

The $d^{(1)}$ transformation was thoroughly analyzed by Sidi (see [77,78] and references therein).

4.2.6. Mosig–Michalski transformation

The Mosig–Michalski transformation — also known as "weighted–averages algorithm" — was introduced by Mosig [61] and modified later by Michalski who gave the $\tilde{t}$ variant of the transformation the name $\mathscr{K}$ transformation (that is used for a different transformation in the present article(!)), and applied it to the computation of Sommerfeld integrals [60].

The Mosig–Michalski transformation M may be defined via the recursive scheme

$$s_n^{(0)} = s_n,$$
$$s_n^{(k+1)} = \frac{s_n^{(k)} + \eta_n^{(k)} s_{n+1}^{(k)}}{1+\eta_n^{(k)}},$$
$$M_n^{(k)}(\{\{s_n\}\},\{\{\omega_n\}\},\{\{x_n\}\}) = s_n^{(k)} \tag{137}$$

for $n \in \mathbb{N}_0$ and $k \in \mathbb{N}_0$ where $\{\{x_n\}\}$ is an auxiliary sequence with $\lim_{n\to\infty} 1/x_n = 0$ such that $x_{n+\ell} > x_n$ for $\ell \in \mathbb{N}_0$ and $x_0 > 1$, i.e., a diverging sequence of monotonously increasing positive numbers, and

$$\eta_n^{(k)} = -\frac{\omega_n}{\omega_{n+1}}\left(\frac{x_{n+1}}{x_n}\right)^{2k}. \tag{138}$$

Putting $\omega_n^{(k)} = \omega_n/x_n^{2k}$, $N_n^{(k)} = s_n^{(k)}/\omega_n^{(k)}$, and $D_n^{(k)} = 1/\omega_n^{(k)}$, it is easily seen that the recursive scheme (137) is equivalent to the scheme (94) with

$$\delta_n^{(k)} = \frac{1}{x_n^2}\left(1 - \frac{\omega_n x_{n+1}^{2k}}{\omega_{n+1} x_n^{2k}}\right). \tag{139}$$

Thus, the Mosig–Michalski transformation is a special case of the $\mathcal{J}$ transformation. Its character as a Levin-type transformation is somewhat formal since the $\delta_n^{(k)}$ and, hence, the coefficients $\lambda_{n,j}^{(k)}$ depend on the ω_n.

If $x_{n+1}/x_n \sim \xi > 1$ for large n, then a limiting transformation exists, namely $M(\{\{s_n\}\},\{\{\omega_n\}\},\{\{\xi^{n+1}\}\})$. It corresponds to the $\mathring{\mathcal{J}}$ transformation with $\Phi_k = \xi^{2k}$. This may be seen by putting $\widehat{D}_n^{(k)} = 1/\omega_n$, $\widehat{N}_n^{(k)} = s_n^{(k)} D_n^{(k)}$ and $\Phi_n^{(k)} = \xi^{2k}$ in Eq. (96).

4.2.7. $\mathcal{F}$ transformation

This transformation is seemingly new. It will be derived in a later section. It may be defined as

$$\mathcal{F}_n^{(k)}(\{\{s_n\}\},\{\{\omega_n\}\},\{\{x_n\}\}) = \frac{\square_n^{(k)}((x_n)_{k-1}s_n/\omega_n)}{\square_n^{(k)}((x_n)_{k-1}/\omega_n)} = \frac{x_n^k/(x_n)_{k-1}\square_n^{(k)}((x_n)_{k-1}s_n/\omega_n)}{x_n^k/(x_n)_{k-1}\square_n^{(k)}((x_n)_{k-1}/\omega_n)}, \tag{140}$$

where $\{\{x_n\}\}$ is an auxiliary sequence with $\lim_{n\to\infty} 1/x_n = 0$ such that $x_{n+\ell} > x_n$ for $\ell \in \mathbb{N}$ and $x_0 > 1$, i.e., a diverging sequence of monotonously increasing positive numbers. Using the definition (39) of the divided difference operator $\square_n^{(k)} = \square_n^{(k)}[\{\{x_n\}\}]$, the coefficients may be taken as

$$\lambda_{n,j}^{(k)} = \frac{(x_{n+j})_{k-1}}{(x_n)_{k-1}} \prod_{\substack{i=0\\ i\neq j}}^{k} \frac{x_n}{x_{n+j}-x_{n+i}} = \prod_{m=0}^{k-2} \frac{x_{n+j}+m}{x_n+m} \left(\frac{x_n}{x_{n+j}}\right)^k \prod_{\substack{i=0\\ i\neq j}}^{k} \frac{1}{1-x_{n+i}/x_{n+j}}. \tag{141}$$

Assuming that the following limit exists such that

$$\lim_{n\to\infty} \frac{x_{n+1}}{x_n} = \xi > 1 \tag{142}$$

holds, we see that one can define a limiting transformation $\mathring{\mathcal{F}}^{(k)}$ with coefficients

$$\overset{\circ}{\lambda}{}_j^{(k)} = \lim_{n\to\infty} \lambda_{n,j}^{(k)} = \frac{1}{\xi^j} \prod_{\substack{\ell=0\\ \ell\neq j}}^{k} \frac{1}{1-\xi^{\ell-j}} = (-1)^k \xi^{-k(k+1)/2} \prod_{\substack{\ell=0\\ \ell\neq j}}^{k} \frac{1}{\xi^{-j}-\xi^{-\ell}}, \tag{143}$$

since

$$\prod_{m=0}^{k-2} \frac{x_{n+j}+m}{x_n+m} \left(\frac{x_n}{x_{n+j}}\right)^k \prod_{\substack{\ell=0\\ \ell\neq j}}^{k} \frac{1}{1-x_{n+\ell}/x_{n+j}} \to \xi^{(k-1)j}\xi^{k(-j)} \prod_{\substack{\ell=0\\ \ell\neq j}}^{k} \frac{\xi^{-l}}{\xi^{-\ell}-\xi^{-j}} \tag{144}$$

for $n\to\infty$. Thus, the limiting transformation is given by

$$\overset{\circ}{\mathscr{F}}{}^{(k)}(\{\{s_n\}\},\{\{\omega_n\}\},\xi) = \frac{\sum_{j=0}^{k} s_{n+j}/\omega_{n+j} \prod_{\substack{\ell=0\\ \ell\neq j}}^{k} 1/(\xi^{-j}-\xi^{-\ell})}{\sum_{j=0}^{k} \frac{1}{\omega_{n+j}} \prod_{\substack{\ell=0\\ \ell\neq j}}^{k} 1/(\xi^{-j}-\xi^{-\ell})}. \tag{145}$$

Comparison with definition (39) of the divided difference operators reveals that the limiting transformation can be rewritten as

$$\overset{\circ}{\mathscr{F}}{}^{(k)}(\{\{s_n\}\},\{\{\omega_n\}\},\xi) = \frac{\square_n^{(k)}[\{\{\xi^{-n}\}\}](s_n/\omega_n)}{\square_n^{(k)}[\{\{\xi^{-n}\}\}](1/\omega_n)}. \tag{146}$$

Comparison to Eq. (133) shows that the limiting transformation is nothing but the W algorithm for $t_n=\xi^{-n}$. As characteristic polynomial we obtain

$$\overset{\circ}{\Pi}{}^{(k)}(z) = \sum_{j=0}^{k} z^j \prod_{\substack{\ell=0\\ \ell\neq j}}^{k} \frac{1}{\xi^{-j}-\xi^{-\ell}} = \xi^{k(k+1)/2} \prod_{j=0}^{k-1} \frac{1-z\xi^j}{\xi^{j+1}-1}. \tag{147}$$

The last equality is easily proved by induction. Hence, the $\overset{\circ}{\mathscr{F}}$ transformation is convex since $\overset{\circ}{\Pi}{}^{(k)}(1)=0$.

As shown in Appendix B, the $\mathscr{F}$ transformation may be computed using the recursive scheme

$$\begin{aligned}
N_n^{(0)} &= \frac{1}{x_n-1}\frac{s_n}{\omega_n}, \quad D_n^{(0)} = \frac{1}{x_n-1}\frac{1}{\omega_n},\\
N_n^{(k)} &= \frac{(x_{n+k}+k-2)N_{n+1}^{(k-1)} - (x_n+k-2)N_n^{(k-1)}}{x_{n+k}-x_n},\\
D_n^{(k)} &= \frac{(x_{n+k}+k-2)D_{n+1}^{(k-1)} - (x_n+k-2)D_n^{(k-1)}}{x_{n+k}-x_n},\\
\mathscr{F}_n^{(k)} &= N_n^{(k)}/D_n^{(k)}.
\end{aligned} \tag{148}$$

It follows directly from Eq. (146) and the recursion relation for divided differences that the limiting transformation can be computed via the recursive scheme

$$\begin{aligned}
\overset{\circ}{N}{}_n^{(0)} &= \frac{s_n}{\omega_n}, \quad \overset{\circ}{D}{}_n^{(0)} = \frac{1}{\omega_n},\\
\overset{\circ}{N}{}_n^{(k)} &= \frac{\overset{\circ}{N}{}_{n+1}^{(k-1)} - \overset{\circ}{N}{}_n^{(k-1)}}{\xi^{-(n+k)}-\xi^{-n}},
\end{aligned}$$

$$\mathring{D}_n^{(k)} = \frac{\mathring{D}_{n+1}^{(k-1)} - \mathring{D}_n^{(k-1)}}{\xi^{-(n+k)} - \xi^{-n}},$$

$$\mathring{\mathscr{F}}_n^{(k)} = \mathring{N}_n^{(k)} / \mathring{D}_n^{(k)}. \tag{149}$$

4.2.8. $\mathscr{J}\mathscr{D}$ transformation

This transformation is newly introduced in this article. In Section 5.2.1, it is derived via (asymptotically) hierarchically consistent iteration of the $\mathscr{D}^{(2)}$ transformation, i.e., of

$$s'_n = \frac{\triangle^2(s_n/\omega_n)}{\triangle^2(1/\omega_n)}. \tag{150}$$

The $\mathscr{J}\mathscr{D}$ transformation may be defined via the recursive scheme

$$N_n^{(0)} = s_n/\omega_n, \quad D_n^{(0)} = 1/\omega_n,$$

$$N_n^{(k)} = \tilde{\nabla}_n^{(k-1)} N_n^{(k-1)}, \quad D_n^{(k)} = \tilde{\nabla}_n^{(k-1)} D_n^{(k-1)},$$

$$\mathscr{J}\mathscr{D}_n^{(k)}(\{\{s_n\}\}, \{\{\omega_n\}\}, \{\zeta_n^{(k)}\}) = N_n^{(k)}/D_n^{(k)}, \tag{151}$$

where the generalized difference operator defined in Eq. (34) involves quantities $\zeta_n^{(k)} \neq 0$ for $k \in \mathbb{N}_0$. Special cases of the $\mathscr{J}\mathscr{D}$ transformation result from corresponding choices of the $\zeta_n^{(k)}$. From Eq. (151) one easily obtains the alternative representation

$$\mathscr{J}\mathscr{D}_n^{(k)}(\{\{s_n\}\}, \{\{\omega_n\}\}, \{\zeta_n^{(k)}\}) = \frac{\tilde{\nabla}_n^{(k-1)} \tilde{\nabla}_n^{(k-2)} \dots \tilde{\nabla}_n^{(0)}[s_n/\omega_n]}{\tilde{\nabla}_n^{(k-1)} \tilde{\nabla}_n^{(k-2)} \dots \tilde{\nabla}_n^{(0)}[1/\omega_n]}. \tag{152}$$

Thus, the $\mathscr{J}\mathscr{D}^{(k)}$ is a Levin-type sequence transformation of order $2k$.

4.2.9. $\mathscr{H}$ transformation and generalized $\mathscr{H}$ transformation

The $\mathscr{H}$ transformation was introduced by Homeier [34] and used or studied in a series of articles [35,41–44,63]. Target of the $\mathscr{H}$ transformation are Fourier series

$$s = A_0/2 + \sum_{j=1}^{\infty} (A_j \cos(j\alpha) + B_j \sin(j\alpha)) \tag{153}$$

with partial sums $s_n = A_0/2 + \sum_{j=1}^{n} (A_j \cos(j\alpha) + B_j \sin(j\alpha))$ where the Fourier coefficients A_n and B_n have asymptotic expansions of the form

$$C_n \sim \rho^n n^{\varepsilon} \sum_{j=0}^{\infty} c_j n^{-j} \tag{154}$$

for $n \to \infty$ with $\rho \in \mathbb{K}$, $\varepsilon \in \mathbb{K}$ and $c_0 \neq 0$.

The $\mathscr{H}$ transformation was critized by Sidi [77] as very unstable and useless near singularities of the Fourier series. However, Sidi failed to notice that – as in the case of the $d^{(1)}$ transformation with $\xi_n = \tau n$ – one can apply also the $\mathscr{H}$ transformation (and also most other Levin-type sequence transformations) to the subsequence $\{\{s_{\xi_n}\}\}$ of $\{\{s_n\}\}$. The new sequence elements $s_{\xi_n} = s_{\tau n}$ can be regarded as the partial sums of a Fourier series with τ–fold frequency. Using this τ-fold frequency approach, one can obtain stable and accurate convergence acceleration even in the vicinity of singularities [41–44].

The $\mathscr{H}$ transformation may be defined as

$$N_n^{(0)} = (n+\beta)^{-1} s_n/\omega_n, \quad D_n^{(0)} = (n+\beta)^{-1}/\omega_n,$$

$$N_n^{(k)} = (n+\beta)N_n^{(k-1)} + (n+2k+\beta)N_{n+2}^{(k-1)} - 2\cos(\alpha)(n+k+\beta)N_{n+1}^{(k-1)},$$

$$D_n^{(k)} = (n+\beta)D_n^{(k-1)} + (n+2k+\beta)D_{n+2}^{(k-1)} - 2\cos(\alpha)(n+k+\beta)D_{n+1}^{(k-1)},$$

$$\mathscr{H}_n^{(k)}(\alpha, \beta, \{\{s_n\}\}, \{\{\omega_n\}\}) = N_n^{(k)}/D_n^{(k)}, \tag{155}$$

where $\cos\alpha \neq \pm 1$ and $\beta \in \mathbb{R}_+$.

It can also be represented in the explicit form [34]

$$\mathscr{H}_n^{(k)}(\alpha, \beta, \{\{s_n\}\}, \{\{\omega_n\}\}) = \frac{\mathscr{P}[P^{(2k)}(\alpha)][(n+\beta)^{k-1} s_n/\omega_n]}{\mathscr{P}[P^{(2k)}(\alpha)][(n+\beta)^{k-1}/\omega_n]}, \tag{156}$$

where the $p_m^{(2k)}(\alpha)$ and the polynomial $P^{(2k)}(\alpha) \in \mathscr{P}^{(2k)}$ are defined via

$$P^{(2k)}(\alpha)(x) = (x^2 - 2x\cos\alpha + 1)^k = \sum_{m=0}^{2k} p_m^{(2k)}(\alpha) x^m \tag{157}$$

and $\mathscr{P}$ is the polynomial operator defined in Eq. (38). This shows that the $\mathscr{H}^{(k)}$ transformation is a Levin-type transformation of order $2k$. It is not convex.

A subnormalized form is

$$\mathscr{H}_n^{(k)}(\alpha, \beta, \{\{s_n\}\}, \{\{\omega_n\}\}) = \frac{\sum_{m=0}^{2k} p_m^{(2k)}(\alpha) \frac{(n+\beta+m)^{k-1}}{(n+\beta+2k)^{k-1}} \frac{s_{n+m}}{\omega_{n+m}}}{\sum_{m=0}^{2k} p_m^{(2k)}(\alpha) \frac{(n+\beta+m)^{k-1}}{(n+\beta+2k)^{k-1}} \frac{1}{\omega_{n+m}}}. \tag{158}$$

This relation shows that the limiting transformation

$$\mathring{\mathscr{H}}^{(k)} = \frac{\mathscr{P}[P^{(2k)}(\alpha)][s_n/\omega_n]}{\mathscr{P}[P^{(2k)}(\alpha)][1/\omega_n]} \tag{159}$$

exists, and has characteristic polynomial $P^{(2k)}(\alpha)$.

A generalized $\mathscr{H}$ transformation was defined by Homeier [40,43]. It is given in terms of the polynomial $P^{(k,M)}(\mathbf{e}) \in \mathbb{P}^{(kM)}$ with

$$P^{(k,M)}(\mathbf{e})(x) = \prod_{m=1}^{M} (x - e_m)^k = \sum_{\ell=0}^{kM} p_\ell^{(k,M)}(\mathbf{e}) x^\ell, \tag{160}$$

where $\mathbf{e} = (e_1, \ldots, e_M) \in \mathbb{K}^M$ is a vector of constant parameters. Then, the generalized $\mathscr{H}$ transformation is defined as

$$\mathscr{H}_n^{(k,M)}(\beta, \{\{s_n\}\}, \{\{\omega_n\}\}, \mathbf{e}) = \frac{\mathscr{P}[P^{(k,M)}(\mathbf{e})][(n+\beta)^{k-1} s_n/\omega_n]}{\mathscr{P}[P^{(k,M)}(\mathbf{e})][(n+\beta)^{k-1}/\omega_n]}. \tag{161}$$

This shows that the generalized $\mathscr{H}^{(k,M)}$ is a Levin-type sequence transformation of order kM. The generalized $\mathscr{H}$ transformation can be computed recursively using the scheme [40,43]

$$N_n^{(0)} = (n+\beta)^{-1}\, s_n/\omega_n, \quad D_n^{(0)} = (n+\beta)^{-1}/\omega_n,$$
$$N_n^{(k)} = \sum_{j=0}^{M} q_j\,(n+\beta+j\,k) N_{n+j}^{(k-1)},$$
$$D_n^{(k)} = \sum_{j=0}^{M} q_j(n+\beta+jk) D_{n+j}^{(k-1)}, \tag{162}$$

$$\mathscr{H}_n(k,M)(\beta,\{\{s_n\}\},\{\{\omega_n\}\},\mathbf{e}) = \frac{N_n^{(k)}}{D_n^{(k)}}.$$

Here, the q_j are defined by

$$\prod_{m=1}^{M}(x-e_m) = \sum_{j=0}^{M} q_j x^j. \tag{163}$$

Algorithm (155) is a special case of algorithm (162). To see this, one observes that $M=2$, $e_1 = \exp(\mathrm{i}\alpha)$ und $e_2 = \exp(-\mathrm{i}\alpha)$ imply $q_0 = q_2 = 1$ and $q_1 = -2\cos(\alpha)$.

For $M=1$ and $e_1 = 1$, the Levin transformation is recovered.

4.2.10. $\mathscr{I}$ transformation

The $\mathscr{I}$ transformation was in a slightly different form introduced by Homeier [35]. It was derived via (asymptotically) hierarchically consistent iteration of the $\mathscr{H}^{(1)}$ transformation, i.e., of

$$s_n' = \frac{s_{n+2}/\omega_{n+2} - 2\cos(\alpha)s_{n+1}/\omega_{n+1} + s_n/\omega_n}{1/\omega_{n+2} - 2\cos(\alpha)/\omega_{n+1} + 1/\omega_n}. \tag{164}$$

For the derivation and an analysis of the properties of the $\mathscr{I}$ transformation see [40,44]. The $\mathscr{I}$ transformation may be defined via the recursive scheme

$$N_n^{(0)} = s_n/\omega_n, \quad D_n^{(0)} = 1/\omega_n,$$
$$N_n^{(k+1)} = \nabla_n^{(k)}[\alpha] N_n^{(k)},$$
$$D_n^{(k+1)} = \nabla_n^{(k)}[\alpha] D_n^{(k)}, \tag{165}$$

$$\mathscr{I}_n^{(k)}(\alpha,\{\{s_n\}\},\{\{\omega_n\}\},\{\Delta_n^{(k)}\}) = \frac{N_n^{(k)}}{D_n^{(k)}},$$

where the generalized difference operator $\nabla_n^{(k)}[\alpha]$ defined in Eq. (35) involves quantities $\Delta_n^{(k)} \neq 0$ for $k \in \mathbb{N}_0$. Special cases of the $\mathscr{I}$ transformation result from corresponding choices of the $\Delta_n^{(k)}$. From Eq. (165) one easily obtains the alternative representation

$$\mathscr{I}_n^{(k)}(\{\{s_n\}\},\{\{\omega_n\}\},\{\Delta_n^{(k)}\}) = \frac{\nabla_n^{(k-1)}[\alpha]\,\nabla_n^{(k-2)}[\alpha]\ldots\nabla_n^{(0)}[\alpha][s_n/\omega_n]}{\nabla_n^{(k-1)}[\alpha]\,\nabla_n^{(k-2)}[\alpha]\ldots\nabla_n^{(0)}[\alpha][1/\omega_n]}. \tag{166}$$

Thus, $\mathscr{I}^{(k)}$ is a Levin-type sequence transformation of order $2k$. It is not convex.

Put $\Theta_n^{(0)} = 1$ and for $k > 0$ define

$$\Theta_n^{(k)} = \frac{\Delta_n^{(0)} \ldots \Delta_n^{(k-1)}}{\Delta_{n+1}^{(0)} \ldots \Delta_{n+1}^{(k-1)}}. \tag{167}$$

If for all $k \in \mathbb{N}$ the limits

$$\lim_{n\to\infty} \Theta_n^{(k)} = \Theta_k \tag{168}$$

exist (we have always $\Theta_0 = 1$), then one can define a limiting transformation $\mathring{\mathscr{I}}$ for large n. It is a special case of the $\mathscr{I}$ transformation according to [44]

$$\mathring{\mathscr{I}}_n^{(k)}(\alpha, \{\{s_n\}\}, \{\{\omega_n\}\}, \{\{\Theta_k\}\}) = \mathscr{I}_n^{(k)}(\alpha, \{\{s_n\}\}, \{\{\omega_n\}\}, \{\{(\Theta_k/\Theta_{k+1})^n\}\}). \tag{169}$$

This is a transformation of order $2k$. The characteristic polynomials of $\mathring{\mathscr{I}}$ are known [44] to be

$$Q^{(2k)}(\alpha) \in \mathbb{P}^{2k}: Q^{(2k)}(\alpha)(z) = \prod_{j=0}^{k-1} [(1 - z\Theta_j \exp(\mathrm{i}\alpha))(1 - z\Theta_j \exp(-\mathrm{i}\alpha))]. \tag{170}$$

4.2.11. $\mathscr{K}$ transformation

The $\mathscr{K}$ transformation was introduced by Homeier [37] in a slightly different form. It was obtained via iteration of the simple transformation

$$s_n' = \frac{\zeta_n^{(0)} s_n/\omega_n + \zeta_n^{(1)} s_{n+1}/\omega_{n+1} + \zeta_n^{(2)} s_{n+2}/\omega_{n+2}}{\zeta_n^{(0)} 1/\omega_n + \zeta_n^{(1)} 1/\omega_{n+1} + \zeta_n^{(2)} 1/\omega_{n+2}}, \tag{171}$$

that is exact for sequences of the form

$$s_n = s + \omega_n(cP_n + dQ_n), \tag{172}$$

where c and d are arbitrary constants, while P_n and Q_n are two linearly independent solutions of the three-term recurrence

$$\zeta_n^{(0)} v_n + \zeta_n^{(1)} v_{n+1} + \zeta_n^{(2)} v_{n+2} = 0. \tag{173}$$

The $\mathscr{K}$ transformation may be defined via the recursive scheme

$$\begin{aligned} &N_n^{(0)} = s_n/\omega_n, \quad D_n^{(0)} = 1/\omega_n, \\ &N_n^{(k+1)} = \partial_n^{(k)}[\zeta] N_n^{(k)}, \\ &D_n^{(k+1)} = \partial_n^{(k)}[\zeta] D_n^{(k)}, \end{aligned} \tag{174}$$

$$\mathscr{K}_n^{(k)}(\{\{s_n\}\}, \{\{\omega_n\}\}, \{\tilde{\Delta}_n^{(k)}\}, \{\zeta_n^{(j)}\}) = \frac{N_n^{(k)}}{D_n^{(k)}},$$

where the generalized difference operator $\partial_n^{(k)}[\zeta]$ defined in Eq. (36) involves recursion coefficients $\zeta_{n+k}^{(j)}$ with $j = 0, 1, 2$ and quantities $\tilde{\Delta}_n^{(k)} \neq 0$ for $k \in \mathbb{N}_0$. Special cases of the $\mathscr{K}$ transformation for given recursion, i.e., for given $\zeta_n^{(j)}$, result from corresponding choices of the $\tilde{\Delta}_n^{(k)}$. From Eq. (174) one easily obtains the alternative representation

$$\mathscr{K}_n^{(k)}(\{\{s_n\}\}, \{\{\omega_n\}\}, \{\tilde{\Delta}_n^{(k)}\}, \{\zeta_n^{(j)}\}) = \frac{\partial_n^{(k-1)}[\zeta]\partial_n^{(k-2)}[\zeta] \ldots \partial_n^{(0)}[\zeta][s_n/\omega_n]}{\partial_n^{(k-1)}[\zeta]\partial_n^{(k-2)}[\zeta] \ldots \partial_n^{(0)}[\zeta][1/\omega_n]}. \tag{175}$$

Thus, $\mathscr{K}^{(k)}$ is a Levin-type sequence transformation of order $2k$. It is not convex. For applications of the $\mathscr{K}$ transformation see [37,40,42,45].

5. Methods for the construction of Levin-type transformations

In this section, we discuss approaches for the construction of Levin-type sequence transformations and point out the relation to their kernel.

5.1. Model sequences and annihilation operators

As discussed in the introduction, the derivation of sequence transformations may be based on model sequences. These may be of the form (10) or of the form (6). Here, we consider model sequences of the latter type that involves remainder estimates ω_n. As described in Section 3.1, determinantal representations for the corresponding sequence transformations can be derived using Cramer's rule, and one of the recursive schemes of the **E** algorithm may be used for the computation. However, for important special choices of the functions $\psi_j(n)$, simpler recursive schemes and more explicit representations in the form (11) can be obtained using the annihilation operator approach of Weniger [84]. This approach was also studied by Brezinski and Matos [13] who showed that it leads to a unified derivation of many extrapolation algorithms and related devices and general results about their kernels. Further, we mention the work of Matos [59] who analysed the approach further and derived a number of convergence acceleration results for Levin-type sequence transformations.

In this approach, an annihilation operator $\mathscr{A}=\mathscr{A}_n^{(k)}$ as defined in Eq. (31) is needed that annihilates the sequences $\{\{\psi_j(n)\}\}$, i.e., such that

$$\mathscr{A}_n^{(k)}(\{\{\psi_j(n)\}\})=0 \quad \text{for } j=0,\ldots,k-1. \tag{176}$$

Rewriting Eq. (6) in the form

$$\frac{\sigma_n-\sigma}{\omega_n}=\sum_{j=0}^{k-1} c_j\psi_j(n) \tag{177}$$

and applying $\mathscr{A}$ to both sides of this equation, one sees that

$$\mathscr{A}_n^{(k)}\left\{\left\{\frac{\sigma_n-\sigma}{\omega_n}\right\}\right\}=0 \tag{178}$$

This equation may be solved for σ due to the linearity of $\mathscr{A}$. The result is

$$\sigma=\frac{\mathscr{A}_n^{(k)}(\{\{\sigma_n/\omega_n\}\})}{\mathscr{A}_n^{(k)}(\{\{1/\omega_n\}\})} \tag{179}$$

leading to a sequence transformation

$$\mathscr{T}_n^{(k)}(\{\{s_n\}\},\{\{\omega\}\})=\frac{\mathscr{A}_n^{(k)}(\{\{s_n/\omega_n\}\})}{\mathscr{A}_n^{(k)}(\{\{1/\omega_n\}\})}. \tag{180}$$

Since $\mathscr{A}$ is linear, this transformation can be rewritten in the form (11), i.e., a Levin-type transformation has been obtained.

We note that this process can be reversed, that is, for each Levin-type sequence transformation $\mathcal{T}[\Lambda^{(k)}]$ of order k there is an annihilation operator, namely the polynomial operator $\mathcal{P}[\Pi_n^{(k)}]$ as defined in Eq. (38) where $\Pi_n^{(k)}$ are the characteristic polynomials as defined in Eq. (72). Using this operator, the defining Eq. (66) can be rewritten as

$$\mathcal{T}_n^{(k)}(\{\{s_n\}\},\{\{\omega_n\}\}) = \frac{\mathcal{P}[\Pi_n^{(k)}](s_n/\omega_n)}{\mathcal{P}[\Pi_n^{(k)}](1/\omega_n)}. \tag{181}$$

Let $\phi_{n,m}(k)$ for $m=0,\ldots,k-1$ be k linearly independent solutions of the linear $(k+1)$–term recurrence

$$\sum_{j=0}^{k} \lambda_{n,j}^{(k)} v_{n+j} = 0. \tag{182}$$

Then $\mathcal{P}[\Pi_n^{(k)}]\phi_{n,m}(k)=0$ for $m=0,\ldots,k-1$, i.e., $\mathcal{P}[\Pi_n^{(k)}]$ is an annihilation operator for all solutions of Eq. (182). Thus, all sequences that are annihilated by this operator are linear combinations of the k sequences $\{\{\phi_{n,m}^{(k)}\}\}$.

If $\{\{\sigma_n\}\}$ is a sequence in the kernel of $\mathcal{T}^{(k)}$ with (anti)limit σ, we must have

$$\sigma = \frac{\mathcal{P}[\Pi_n^{(k)}](\sigma_n/\omega_n)}{\mathcal{P}[\Pi_n^{(k)}](1/\omega_n)} \tag{183}$$

or after some rearrangement using the linearity of $\mathcal{P}$

$$\mathcal{P}[\Pi_n^{(k)}]\left(\frac{\sigma_n-\sigma}{\omega_n}\right) = 0. \tag{184}$$

Hence, we must have

$$\frac{\sigma_n-\sigma}{\omega_n} = \sum_{m=0}^{k-1} c_m \phi_{n,m}^{(k)}, \tag{185}$$

or, equivalently

$$\sigma_n = \sigma + \omega_n \sum_{m=0}^{k-1} c_m \phi_{n,m}^{(k)} \tag{186}$$

for some constants c_m. Thus, we have determined the kernel of $\mathcal{T}^{(k)}$ that can also be considered as the set of model sequences for this transformation. Thus, we have proved the following theorem:

Theorem 1. *Let $\phi_{n,m}^{(k)}$ for $m=0,\ldots,k-1$ be the k linearly independent solutions of the linear $(k+1)$–term recurrence* (182). *The kernel of $\mathcal{T}[\Lambda^{(k)}](\{\{s_n\}\},\{\{\omega_n\}\})$ is given by all sequences $\{\{\sigma_n\}\}$ with (anti)limit σ and elements σ_n of the form* (186) *for arbitrary constants c_m.*

We note that the $\psi_j(n)$ for $j=0,\ldots,k-1$ can essentially be identified with the $\phi_{n,j}^{(k)}$. Thus, we have determinantal representations for known $\psi_j(n)$ as noted above in the context of the E algorithm. See also [38] for determinantal representations of the $\mathcal{J}$ transformations and the relation to its kernel.

Examples of annihilation operators and the functions $\psi_j(n)$ that are annihilated are given in Table 2. Examples for the Levin-type sequence transformations that have been derived using the approach of model sequences are discussed in Section 5.1.2.

Table 2
Examples of annihilation operators[a]

Type	Operator	$\psi_j(n),\ j=0,\ldots,k-1$
Differences	$\triangle^k$	$(n+\beta)^j$
		$(n+\beta)_j$
		$(\alpha[n+\zeta])^j$
		$(\alpha[n+\zeta])_j$
		$p_j(n),\ p_j \in \mathbb{P}^{(j)}$
Weighted differences	$\triangle^k(n+\beta)^{k-1}$	$1/(n+\beta)^j$
	$\triangle^k(n+\beta)_{k-1}$	$1/(n+\beta)_j$
	$\triangle^k(\alpha[n+\zeta])^{k-1}$	$1/(\alpha[n+\zeta])^j$
	$\triangle^k(\alpha[n+\zeta])_{k-1}$	$1/(\alpha[n+\zeta])_j$
Divided differences	$\square_n^{(k)}[\{\{t_n\}\}]$	t_n^j
	$\square_n^{(k)}[\{\{t_n\}\}]$	$p_j(t_n),\ p_j \in \mathbb{P}^{(j)}$
	$\square_n^{(k)}[\{\{x_n\}\}](x_n)_{k-1}$	$1/(x_n)_j$
Polynomial	$\mathscr{P}[P^{(2k)}(\alpha)]$	$\exp(+\mathrm{i}\alpha n)p_j(n),\ p_j \in \mathbb{P}^{(j)}$
		$\exp(-\mathrm{i}\alpha n)p_j(n),\ p_j \in \mathbb{P}^{(j)}$
	$\mathscr{P}[P^{(2k)}(\alpha)](n+\beta)^{k-1}$	$\exp(+\mathrm{i}\alpha n)/(n+\beta)^j$
		$\exp(-\mathrm{i}\alpha n)/(n+\beta)^j$
	$\mathscr{P}[P^{(2k)}(\alpha)](n+\beta)_{k-1}$	$\exp(+\mathrm{i}\alpha n)/(n+\beta)_j$
		$\exp(-\mathrm{i}\alpha n)/(n+\beta)_j$
	$\mathscr{P}[P^{(k)}]$	$\psi_j(n)$ is solution of
		$\sum_{m=0}^{k} p_n^{(k)} v_{n+j}=0$
	$\mathscr{P}[P^{(k)}](n+\beta)^m$	$(n+\beta)^m\psi_j(n)$ is solution of
		$\sum_{m=0}^{k} p_n^{(k)} v_{n+j}=0$
	L_1 (see (188))	$\frac{\alpha_j^{n+1}}{n!}$
	L_2 (see (189))	$\frac{n^j\alpha^{n+1}}{n!}$
	$\tilde{L}$ (see (191))	$\frac{\Gamma(n+\alpha_j+1)}{n!}$

[a]See also Section 5.1.1.

Note that the annihilation operators used by Weniger [84,87,88] were weighted difference operators $\mathscr{W}_n^{(k)}$ as defined in Eq. (37). Homeier [36,38,39] discussed operator representations for the $\mathscr{J}$ transformation that are equivalent to many of the annihilation operators and related sequence transformations as given by Brezinski and Matos [13]. The latter have been further discussed by Matos [59] who considered among others Levin-type sequence transformations with constant coefficients, $\lambda_{n,j}^{(k)}=\text{const.}$, and with polynomial coefficients $\lambda_{n,j}^{(k)}=\lambda_j(n+1)$, with $\lambda_j \in \mathbb{P}$, and $n \in \mathbb{N}_0$, in particular annihilation operators of the form

$$L(u_n)=(\Omega^l+\lambda_1\Omega^{l-1}+\cdots+\lambda_l)(u_n) \tag{187}$$

with the special cases

$$L_1(u_n)=(\Omega-\alpha_1)(\Omega-\alpha_2)\cdots(\Omega-\alpha_l)(u_n) \quad (\alpha_i \neq \alpha_j \quad \text{for all } i \neq j) \tag{188}$$

and

$$L_2(u_n)=(\Omega-\alpha)^l(u_n), \tag{189}$$

where

$$\Omega^r(u_n) = (n+1)_r u_{n+r}, \quad n \in \mathbb{N}_0 \tag{190}$$

and

$$\tilde{L}(u_n) = (\pi - \alpha_1)(\pi - \alpha_2)\cdots(\pi - \alpha_l)(u_n), \tag{191}$$

where

$$\pi(u_n) = (n+1)\,\Delta\, u_n, \quad \pi^r(u_n) = \pi(\pi^{r-1}(u_n)), \quad n \in \mathbb{N}_0 \tag{192}$$

and the λ's and α's are constants. Note that n is shifted in comparison to [59] where the convention $n \in \mathbb{N}$ was used. See also Table 2 for the corresponding annihilated functions $\psi_j(n)$.

Matos [59] also considered difference operators of the form

$$L(u_n) = \Delta^k + p_{k-1}(n)\,\Delta^{k-1} + \cdots + p_1(n)\,\Delta + p_0(n), \tag{193}$$

where the functions f_j given by $f_j(t) = p_j(1/t)t^{-k+j}$ for $j = 0,\ldots,k-1$ are analytic in the neighborhood of 0. For such operators, there is no explicit formula for the functions that are annihilated. However, the asymptotic behavior of such functions is known [6,59]. We will later return to such annihilation operators and state some convergence results.

5.1.1. Derivation of the $\mathscr{F}$ transformation

As an example for the application of the annihilation operator approach, we derive the $\mathscr{F}$ transformation. Consider the model sequence

$$\sigma_n = \sigma + \omega_n \sum_{j=0}^{k-1} c_j \frac{1}{(x_n)_j}, \tag{194}$$

that may be rewritten as

$$\frac{\sigma_n - \sigma}{\omega_n} = \sum_{j=0}^{k-1} c_j \frac{1}{(x_n)_j}. \tag{195}$$

We note that Eq. (194) corresponds to modeling $\mu_n = R_n/\omega_n$ as a truncated factorial series in x_n (instead as a truncated power series as in the case of the W algorithm). The x_n are elements of $\{\{x_n\}\}$ an auxiliary sequence $\{\{x_n\}\}$ such that $\lim_{n\to\infty} 1/x_n = 0$ and also $x_{n+\ell} > x_n$ for $\ell \in \mathbb{N}$ and $x_0 > 1$, i.e., a diverging sequence of monotonously increasing positive numbers. To find an annihilation operator for the $\psi_j(n) = 1/(x_n)_j$, we make use of the fact that the divided difference operator $\square_n^{(k)} = \square_n^{(k)}[\{\{x_n\}\}]$ annihilates polynomials in x_n of degree less than k. Also, we observe that the definition of the Pochhammer symbols entails that

$$(x_n)_{k-1}/(x_n)_j = (x_n + j)_{k-1-j} \tag{196}$$

is a polynomial of degree less than k in x_n for $0 \leqslant j \leqslant k-1$. Thus, the sought annihilation operator is $\mathscr{A} = \square_n^{(k)}(x_n)_{k-1}$ because

$$\square_n^{(k)}(x_n)_{k-1}\frac{1}{(x_n)_j} = 0, \quad 0 \leqslant j < k. \tag{197}$$

Hence, for the model sequence (194), one can calculate σ via

$$\sigma = \frac{\Box_n^{(k)}((x_n)_{k-1}\sigma_n/\omega_n)}{\Box_n^{(k)}((x_n)_{k-1}/\omega_n)} \tag{198}$$

and the $\mathscr{F}$ transformation (140) results by replacing σ_n by s_n in the right-hand side of Eq. (198).

5.1.2. Important special cases

Here, we collect model sequences and annihilation operators for some important Levin-type sequence transformations that were derived using the model sequence approach. For further examples see also [13]. The model sequences are the kernels by construction. In Section 5.2.2, kernels and annihilation operators are stated for important Levin-type transformation that were derived using iterative methods.

Levin transformation: The model sequence for $\mathscr{L}^{(k)}$ is

$$\sigma_n = \sigma + \omega_n \sum_{j=0}^{k-1} c_j/(n+\beta)^j. \tag{199}$$

The annihilation operator is

$$\mathscr{A}_n^{(k)} = \Delta^k (n+\beta)^{k-1}. \tag{200}$$

Weniger transformations: The model sequence for $\mathscr{S}^{(k)}$ is

$$\sigma_n = \sigma + \omega_n \sum_{j=0}^{k-1} c_j/(n+\beta)_j. \tag{201}$$

The annihilation operator is

$$\mathscr{A}_n^{(k)} = \Delta^k (n+\beta)_{k-1}. \tag{202}$$

The model sequence for $\mathscr{M}^{(k)}$ is

$$\sigma_n = \sigma + \omega_n \sum_{j=0}^{k-1} c_j/(-n-\xi)_j. \tag{203}$$

The annihilation operator is

$$\mathscr{A}_n^{(k)} = \Delta^k (-n-\xi)_{k-1}. \tag{204}$$

The model sequence for $\mathscr{C}^{(k)}$ is

$$\sigma_n = \sigma + \omega_n \sum_{j=0}^{k-1} c_j/(\alpha[n+\zeta])_j. \tag{205}$$

The annihilation operator is

$$\mathscr{A}_n^{(k)} = \Delta^k (\alpha[n+\zeta])_{k-1}. \tag{206}$$

W algorithm: The model sequence for $W^{(k)}$ is

$$\sigma_n = \sigma + \omega_n \sum_{j=0}^{k-1} c_j t_n^j. \tag{207}$$

The annihilation operator is

$$\mathscr{A}_n^{(k)} = \square_n^{(k)}[\{\{t_n\}\}]. \tag{208}$$

$\mathscr{H}$ transformation: The model sequence for $\mathscr{H}^{(k)}$ is

$$\sigma_n = \sigma + \omega_n \left(\exp(\mathrm{i}\alpha n) \sum_{j=0}^{k-1} c_j^+/(n+\beta)^j + \exp(-\mathrm{i}\alpha n) \sum_{j=0}^{k-1} c_j^-/(n+\beta)^j \right). \tag{209}$$

The annihilation operator is

$$\mathscr{A}_n^{(k)} = \mathscr{P}[P^{(2k)}(\alpha)](n+\beta)^{k-1}. \tag{210}$$

Generalized $\mathscr{H}$ transformation: The model sequence for $\mathscr{H}^{(k,m)}$ is

$$\sigma_n = \sigma + \omega_n \sum_{m=1}^{M} e_m^n \sum_{j=0}^{k-1} c_{m,j}(n+\beta)^{-j}. \tag{211}$$

The annihilation operator is

$$\mathscr{A}_n^{(k)} = \mathscr{P}[P^{(k,m)}(\mathbf{e})](n+\beta)^{k-1}. \tag{212}$$

5.2. Hierarchically consistent iteration

As alternative to the derivation of sequence transformations using model sequences and possibly annihilation operators, one may take some simple sequence transformation T and iterate it k times to obtain a transformation $T^{(k)} = T \circ \cdots \circ T$. For the iterated transformation, by construction one has a simple algorithm by construction, but the theoretical analysis is complicated since usually no kernel is known. See for instance the iterated Aitken process where the $\triangle^2$ method plays the role of the simple transformation. However, as is discussed at length in Refs. [36,86], there are usually several possibilities for the iteration. Both problems – unknown kernel and arbitrariness of iteration – are overcome using the concept of hierarchical consistency [36,40,44] that was shown to give rise to powerful algorithms like the $\mathscr{J}$ and the $\mathscr{I}$ transformations [39,40,44]. The basic idea of the concept is to provide a hierarchy of model sequences such that the simple transformation provides a mapping between neighboring levels of the hierarchy. To ensure the latter, normally one has to fix some parameters in the simple transformation to make the iteration consistent with the hierarchy.

A formal description of the concept is given in the following taken mainly from the literature [44]. As an example, the concept is later used to derive the $\mathscr{J}\mathscr{D}$ transformation in Section 5.2.1.

Let $\{\{\sigma_n(\boldsymbol{c},\boldsymbol{p})\}\}_{n=0}^{\infty}$ be a simple "basic" model sequence that depends on a vector $\boldsymbol{c} \in \mathbb{K}^a$ of constants, and further parameters $\boldsymbol{p}$. Assume that its (anti)limit $\sigma(\boldsymbol{p})$ exists and is independent of $\boldsymbol{c}$. Assume that the basic transformation $T = T(\boldsymbol{p})$ allows to compute the (anti)limit exactly according to

$$T(\boldsymbol{p}) : \{\{\sigma_n(\boldsymbol{c},\boldsymbol{p})\}\} \to \{\{\sigma(\boldsymbol{p})\}\}. \tag{213}$$

Let the hierarchy of model sequences be given by

$$\{\{\{\sigma_n^{(\ell)}(\boldsymbol{c}^{(\ell)},\boldsymbol{p}^{(\ell)})|\boldsymbol{c}^{(\ell)} \in \mathbb{K}^{a^{(\ell)}}\}\}\}_{\ell=0}^{L} \tag{214}$$

with $a^{(\ell)} > a^{(\ell')}$ for $\ell > \ell'$. Here, ℓ numbers the levels of the hierarchy. Each of the model sequences $\{\{\sigma_n^{(\ell)}(\boldsymbol{c}^{(\ell)},\boldsymbol{p}^{(\ell)})\}\}$ depends on an $a^{(\ell)}$-dimensional complex vector $\boldsymbol{c}^{(\ell)}$ and further parameters $\boldsymbol{p}^{(\ell)}$. Assume that the model sequences of lower levels are also contained in those of higher levels: For all $\ell < L$ and all $\ell' > \ell$ and $\ell' \leqslant L$, every sequence $\{\{\sigma_n^{(\ell)}(\boldsymbol{c}^{(\ell)},\boldsymbol{p}^{(\ell)})\}\}$ is assumed to be representable as a model sequence $\{\{\sigma_n^{(\ell')}(\boldsymbol{c}^{(\ell')},\boldsymbol{p}^{(\ell')})\}\}$ where $\boldsymbol{c}^{(\ell')}$ is obtained from $\boldsymbol{c}^{(\ell)}$ by the natural injection $\mathbb{K}^{a^{(\ell)}} \to \mathbb{K}^{a^{(\ell')}}$. Assume that for all ℓ with $0 < \ell \leqslant L$

$$T(\boldsymbol{p}^{(\ell)}) : \{\{\sigma_n^{(\ell)}(\boldsymbol{c}^{(\ell)},\boldsymbol{p}^{(\ell)})\}\} \to \{\{\sigma_n^{(\ell-1)}(\boldsymbol{c}^{(\ell-1)},\boldsymbol{p}^{(\ell-1)})\}\} \tag{215}$$

is a mapping between neighboring levels of the hierarchy. Composition yields an iterative transformation

$$T^{(L)} = T(\boldsymbol{p}^{(0)}) \circ T(\boldsymbol{p}^{(1)}) \circ \cdots \circ T(\boldsymbol{p}^{(L)}). \tag{216}$$

This transformation is called "hierarchically consistent" or "consistent with the hierarchy". It maps model sequences $\sigma_n^{(\ell)}(\boldsymbol{c}^{(\ell)},\boldsymbol{p}^{(\ell)})$ to constant sequences if Eq. (213) holds with

$$\{\{\sigma_n^{(0)}(\boldsymbol{c}^{(0)},\boldsymbol{p}^{(0)})\}\} = \{\{\sigma_n(\boldsymbol{c},\boldsymbol{p})\}\}. \tag{217}$$

If instead of Eq. (215) we have

$$T(\boldsymbol{p}^{(\ell)})(\{\{\sigma_n^{(\ell)}(\boldsymbol{c}^{(\ell)},\boldsymbol{p}^{(\ell)})\}\}) \sim \{\{\sigma_n^{(\ell-1)}(\boldsymbol{c}^{(\ell-1)},\boldsymbol{p}^{(\ell-1)})\}\} \tag{218}$$

for $n \to \infty$ for all $\ell > 0$ then the iterative transformation $T^{(L)}$ is called "asymptotically consistent with the hierarchy" or "asymptotically hierarchy-consistent".

5.2.1. Derivation of the $\mathcal{J}\mathcal{D}$ transformation

The simple transformation is the $\mathcal{D}^{(2)}$ transformation

$$s_n' = T(\{\{\omega_n\}\})(\{\{s_n\}\}) = \frac{\Delta^2(s_n/\omega_n)}{\Delta^2(1/\omega_n)} \tag{219}$$

depending on the "parameters" $\{\{\omega_n\}\}$, with basic model sequences

$$\frac{\sigma_n}{\omega_n} = \sigma\frac{1}{\omega_n} + (an + b). \tag{220}$$

The more complicated model sequences of the next level are taken to be

$$\frac{\sigma_n}{\omega_n} = \sigma\frac{1}{\omega_n} + (an + b + (a_1 n + b_1)r_n). \tag{221}$$

Application of $\triangle^2$ eliminates the terms involving a and b. The result is

$$\frac{\triangle^2 \sigma_n/\omega_n}{\triangle^2 r_n} = \sigma \frac{\triangle^2 1/\omega_n}{\triangle^2 r_n} + \left(a_1 n + b_1 + 2a_1 \frac{\triangle r_n}{\triangle^2 r_n}\right) \tag{222}$$

for $\triangle^2 r_n \neq 0$. Assuming that for large n

$$\frac{\triangle r_n}{\triangle^2 r_n} = An + B + \mathrm{o}(1) \tag{223}$$

holds, the result is asymptotically of the same *form* as the model sequence in Eq. (220), namely

$$\frac{\sigma'_n}{\omega'_n} = \sigma \frac{1}{\omega'_n} + (a'n + b' + \mathrm{o}(1)) \tag{224}$$

with renormalized "parameters"

$$1/\omega'_n = \frac{\triangle^2(1/\omega_n)}{\triangle^2 r_n} \tag{225}$$

and obvious identifications for a' and b'.

We now assume that this mapping between two neighboring levels of the hierarchy can be extended to any two neighboring levels, provided that one introduces ℓ-dependent quantities, especially $r_n \to r_n^{(\ell)}$ with $\zeta_n^{(\ell)} = \triangle^2 r_n^{(\ell)} \neq 0$, $s_n/\omega_n \to N_n^{(\ell)}$, $1/\omega_n \to D_n^{(\ell)}$ and $s'_n/\omega'_n \to N_n^{(\ell+1)}$, $1/\omega'_n \to D_n^{(\ell+1)}$.

Iterating in this way leads to algorithm (151).

Condition (223) or more generally

$$\frac{\triangle r_n^{(\ell)}}{\triangle^2 r_n^{(\ell)}} = A_\ell n + B_\ell + \mathrm{o}(1) \tag{226}$$

for given ℓ and for large n is satisfied in many cases. For instance, it is satisfied if there are constants $\beta_\ell \neq 0$, γ_ℓ and $\delta_\ell \neq 0$ such that

$$\triangle\, r_n^{(\ell)} \sim \beta_\ell \begin{cases} \left(\dfrac{\delta_\ell + 1}{\delta_\ell}\right)^n & \text{for } \gamma_\ell = 0, \\[2ex] \left(\dfrac{\delta_\ell + 1}{\gamma_\ell}\right)_n \Big/ \left(\dfrac{\delta_\ell}{\gamma_\ell}\right)_n & \text{otherwise.} \end{cases} \tag{227}$$

This is for instance the case for $r_n^{(\ell)} = n^{\zeta_\ell}$ with $\zeta_\ell(\zeta_\ell - 1) \neq 0$.

The kernel of $\mathcal{J}\mathcal{D}^{(k)}$ may be found inductively in the following way:

$$\begin{aligned} & N_n^{(k)} - \sigma D_n^{(k)} = 0 \\ & \Rightarrow \triangle^2(N_n^{(k-1)} - \sigma D_n^{(k-1)}) = 0 \\ & \Rightarrow N_n^{(k-1)} - \sigma D_n^{(k-1)} = a_{k-1} n + b_{k-1} \\ & \Rightarrow \triangle^2(N_n^{(k-2)} - \sigma D_n^{(k-2)}) = (a_{k-1} n + b_{k-1}) \zeta_n^{(k-2)} \\ & \Rightarrow N_n^{(k-2)} - \sigma D_n^{(k-2)}) = a_{k-2} n + b_{k-2} + \sum_{j=0}^{n-2} \sum_{n'=0}^{j} (a_{k-1} n' + b_{k-1}) \zeta_{n'}^{(k-2)} \end{aligned} \tag{228}$$

yielding the result

$$N_n^{(0)} - \sigma D_n^{(0)} = a_0 n + b_0 + \sum_{j=0}^{n-2} \sum_{n_1=0}^{j} \zeta_{n_1}^{(0)} (a_1 n_1 + b_1 + \cdots$$

$$+ (a_{k-2} n + b_{k-2} + \sum_{j_{k-2}=0}^{n_{k-2}-2} \sum_{n_{k-1}=0}^{j_{k-2}} \zeta_{n_{k-1}}^{(k-2)} (a_{k-1} n_{k-1} + b_{k-1}) \Bigg) . \tag{229}$$

Here, the definitions $N_n^{(0)} = \sigma_n/\omega_n$ and $D_n^{(0)} = 1/\omega_n$ may be used to obtain the model sequence $\{\{\sigma_n\}\}$ for $\mathcal{J}\mathcal{D}^{(k)}$, that may be identified as kernel of that transformation, and also may be regarded as model sequence of the kth level according to $\{\{\sigma_n^{(k)}(\boldsymbol{c}^{(k)}, \boldsymbol{p}^{(k)})\}\}$ with $\boldsymbol{c}^{(k)} = (a_0, b_0, \ldots, a_{k-1}, b_{k-1})$ and $\boldsymbol{p}^{(k)}$ corresponds to $\omega_n^{(k)} = 1/D_n^{(k)}$ and the $\{\zeta_n^{(\kappa)} | 0 \leqslant \kappa \leqslant k-2\}$.

We note this as a theorem:

Theorem 2. *The kernel of* $\mathcal{J}\mathcal{D}^{(k)}$ *is given by the set of sequences* $\{\{\sigma_n\}\}$ *such that Eq.* (229) *holds with* $N_n^{(0)} = \sigma_n/\omega_n$ *and* $D_n^{(0)} = 1/\omega_n$.

5.2.2. Important special cases

Here, we give the hierarchies of model sequences for sequence transformations derived via hierarchically consistent iteration.

$\mathcal{J}$ transformation: The most prominent example is the $\mathcal{J}$ transformation (actually a large class of transformations). The corresponding hierarchy of model sequences provided by the kernels that are explicitly known according to the following theorem:

Theorem 3 (Homeier [36]). *The kernel of the* $\mathcal{J}^{(k)}$ *transformation is given by the sequences* $\{\{\sigma_n\}\}$ *with elements of the form*

$$\sigma_n = \sigma + \omega_n \sum_{j=0}^{k-1} c_j \psi_j(n) \tag{230}$$

with

$$\psi_0(n) = 1,$$

$$\psi_1(n) = \sum_{n_1=0}^{n-1} \delta_{n_1}^{(0)},$$

$$\psi_2(n) = \sum_{n_1=0}^{n-1} \delta_{n_1}^{(0)} \sum_{n_2=0}^{n_1-1} \delta_{n_2}^{(1)}, \tag{231}$$

$$\vdots$$

$$\psi_{k-1}(n) = \sum_{n > n_1 > n_2 > \cdots > n_{k-1}} \delta_{n_1}^{(0)} \delta_{n_2}^{(1)} \cdots \delta_{n_{k-1}}^{(k-2)}$$

with arbitrary constants $c_0, \ldots, c_{k-1}$.

$\mathscr{I}$ *transformation*: Since the $\mathscr{I}$ transformation is a special case of the $\mathscr{J}$ transformation (cf. Table 1) and [44], its kernels (corresponding to the hierarchy of model sequences) are explicitly known according to the following theorem:

Theorem 4 (Homeier [44, Theorem 8]). *The kernel of the* $\mathscr{I}^{(k)}$ *transformation is given by the sequences* $\{\{\sigma_n\}\}$ *with elements of the form*

$$\sigma_n = \sigma + \exp(-\mathrm{i}\alpha n)\omega_n \Bigg[d_0 + d_1 \exp(2\mathrm{i}\alpha n)$$

$$+ \sum_{n_1=0}^{n-1} \sum_{n_2=0}^{n_1-1} \exp(2\mathrm{i}\alpha(n_1 - n_2))(d_2 + d_3 \exp(2\mathrm{i}\alpha n_2))\Delta_{n_2}^{(0)} + \cdots$$

$$+ \sum_{n>n_1>n_2>\cdots>n_{2k-2}} \exp(2\mathrm{i}\alpha[n_1 - n_2 + \cdots + n_{2k-3} - n_{2k-2}])$$

$$(d_{2k-2} + d_{2k-1}\exp(2\mathrm{i}\alpha n_{2k-2})) \prod_{j=0}^{k-2} \Delta_{n_{2j+2}}^{(j)} \Bigg] \tag{232}$$

with constants $d_0, \ldots, d_{2k-1}$. *Thus, we have* $s = \mathscr{I}_n^{(k')}(\alpha, \{\{s_n\}\}, \{\{\omega_n\}\}, \{\Delta_n^{(k)}\})$ *for* $k' \geqslant k$ *for sequences of this form.*

5.3. A two-step approach

In favorable cases, one may use a two-step approach for the construction of sequence transformations:

Step 1: Use asymptotic analysis of the remainder $R_n = s_n - s$ of the given problem to find the adequate model sequence (or hierarchy of model sequences) for large n.

Step 2: Use the methods described in Sections 5.1 or 5.2 to construct the sequence transformation adapted to the problem.

This is, of course, a mathematically promising approach. A good example for the two-step approach is the derivation of the $d^{(m)}$ transformations by Levin and Sidi [54] (cf. also Section 3.2).

But there are two difficulties with this approach.

The first difficulty is a practical one. In many cases, the problems to be treated in applications are simply too complicated to allow to perform Step 1 of the two-step approach.

The second difficulty is a more mathematical one. The optimal system of functions $f_j(n)$ used in the asymptotic expansion

$$s_n - s \sim \sum_{j=0}^{\infty} c_j f_j(n) \tag{233}$$

with $f_{j+1}(n) = \mathrm{o}(f_j(n))$, i.e., the optimal *asymptotic scale* [102, p. 2], is not clear a priori. For instance, as the work of Weniger has shown, sequence transformations like the Levin transformation that are based on expansions in powers of $1/n$, i.e., the asymptotic scale $\phi_j(n) = 1/(n+\beta)^j$, are not always superior to, and even often worse than those based upon factorial series, like Weniger's $\mathscr{S}$

transformation that is based on the asymptotic scale $\psi_j(n)=1/(n+\beta)_j$. To find an optimal asymptotic scale in combination with nonlinear sequence transformations seems to be an open mathematical problem.

Certainly, the proper choice of remainder estimates [50] is also crucial in the context of Levin-type sequence transformations. See also Section 9.

6. Properties of Levin-type transformations

6.1. Basic properties

Directly from the definition in Eqs. (65) and (66), we obtain the following theorem. The proof is left to the interested reader.

Theorem 5. *Any Levin-type sequence transformation* $\mathscr{T}$ *is quasilinear, i.e., we have*

$$\mathscr{T}_n^{(k)}(\{\{As_n+B\}\},\{\{\omega_n\}\}) = A\mathscr{T}_n^{(k)}(\{\{s_n\}\},\{\{\omega_n\}\})+B \tag{234}$$

for arbitrary constants A and B. It is multiplicatively invariant in ω_n, *i.e., we have*

$$\mathscr{T}_n^{(k)}(\{\{s_n\}\},\{\{C\omega_n\}\}) = \mathscr{T}_n^{(k)}(\{\{s_n\}\},\{\{\omega_n\}\}) \tag{235}$$

for arbitrary constants $C\neq 0$.

For a coefficient set Λ define the sets $Y_n^{(k)}[\Lambda]$ by

$$Y_n^{(k)}[\Lambda] = \left\{ (x_0,\ldots,x_k)\in\mathbb{F}^{k+1} \,\middle|\, \sum_{j=0}^{k} \lambda_{n,j}^{(k)}/x_j \neq 0 \right\}. \tag{236}$$

Since $\mathscr{T}_n^{(k)}(\{\{s_n\}\},\{\{\omega_n\}\})$ for given coefficient set Λ depends only on the $2k+2$ numbers $s_n,\ldots,s_{n+k}$ and $\omega_n,\ldots,\omega_{n+k}$, it may be regarded as a mapping

$$U_n^{(k)} : \mathbb{C}^{k+1}\times Y_n^{(k)}[\Lambda] \Rightarrow \mathbb{C}, \quad (x,y)\mapsto U_n^{(k)}(x\,|\,y) \tag{237}$$

such that

$$\mathscr{T}_n^{(k)} = U_n^{(k)}(s_n,\ldots,s_{n+k}\,|\,\omega_n,\ldots,\omega_{n+k}). \tag{238}$$

The following theorem is a generalization of theorems for the $\mathscr{J}$ transformation [36, Theorem 5] and the $\mathscr{I}$ transformation [44, Theorem 5].

Theorem 6. (I−0) *The* $\mathscr{T}^{(k)}$ *transformation can be regarded as continous mapping* $U_n^{(k)}$ *on* $\mathbb{C}^{k+1}\times Y_n^{(k)}[\Lambda]$ *where* $Y_n^{(k)}[\Lambda]$ *is defined in Eq.* (236):

(I − 1) *According to Theorem* 5, $U_n^{(k)}$ *is a homogeneous function of first degree in the first* $(k+1)$ *variables and a homogeneous function of degree zero in the last* $(k+1)$ *variables. Hence, for all vectors* $\boldsymbol{x}\in\mathbb{C}^{k+1}$ *and* $\boldsymbol{y}\in Y_n^{(k)}[\Lambda]$ *and for all complex constants s and* $t\neq 0$ *the equations*

$$U_n^{(k)}(s\boldsymbol{x}\,|\,\boldsymbol{y}) = sU_n^{(k)}(\boldsymbol{x}\,|\,\boldsymbol{y}),$$

$$U_n^{(k)}(\boldsymbol{x}\,|\,t\boldsymbol{y}) = U_n^{(k)}(\boldsymbol{x}\,|\,\boldsymbol{y}) \tag{239}$$

hold.

(I − 2) $U_n^{(k)}$ *is linear in the first* $(k+1)$ *variables. Thus, for all vectors* $\boldsymbol{x} \in \mathbb{C}^{k+1}$, $\boldsymbol{x}' \in \mathbb{C}^{k+1}$, *und* $\boldsymbol{y} \in Y_n^{(k)}[\Lambda]$

$$U_n^{(k)}(\boldsymbol{x}+\boldsymbol{x}' \,|\, \boldsymbol{y}) = U_n^{(k)}(\boldsymbol{x} \,|\, \boldsymbol{y}) + U_n^{(k)}(\boldsymbol{x}' \,|\, \boldsymbol{y}) \tag{240}$$

holds.

(I − 3) *For all constant vectors* $\boldsymbol{c} = (c, c, \ldots, c) \in \mathbb{C}^{k+1}$ *and all vectors* $\boldsymbol{y} \in Y_n^{(k)}[\Lambda]$ *we have*

$$U_n^{(k)}(\boldsymbol{c} \,|\, \boldsymbol{y}) = c. \tag{241}$$

Proof. These are immediate consequences of the definitions. □

6.2. The limiting transformation

We note that if a limiting transformation $\mathring{\mathscr{T}}\,[\mathring{\Lambda}\,]$ exists, it is also of Levin-type, and thus, the above theorems apply to the limiting transformation as well.

Also, we have the following result for the kernel of the limiting transformation:

Theorem 7. *Suppose that for a Levin-type sequence transformation* $\mathscr{T}^{(k)}$ *of order k there exists a limiting transformation* $\mathring{\mathscr{T}}^{(k)}$ *with characteristic polynomial* $\mathring{\Pi} \in \mathbb{P}^k$ *given by*

$$\mathring{\Pi}^{(k)}(z) = \sum_{j=0}^{k} \mathring{\lambda}_j^{(k)} z^j = \prod_{\ell=1}^{M} (z-\zeta_\ell)^{m_\ell}, \tag{242}$$

where the zeroes $\zeta_\ell \neq 0$ *have multiplicities* m_ℓ. *Then the kernel of the limiting transformation consists of all sequences* $\{\{s_n\}\}$ *with elements of the form*

$$\sigma_n = \sigma + \omega_n \sum_{\ell=1}^{M} \zeta_\ell^n P_\ell(n), \tag{243}$$

where $P_\ell \in \mathbb{P}^{m_\ell - 1}$ *are arbitrary polynomials and* $\{\{\omega_n\}\} \in \mathring{\mathbb{Y}}^{(k)}$.

Proof. This follows directly from the observation that for such sequences $(\sigma_n - \sigma)/\omega_n$ is nothing but a finite linear combination of the solutions $\varphi_{n,\ell,j_\ell}^{(k)} = n^{j_\ell} \zeta_\ell^n$ with $\ell = 1, \ldots, M$ and $j_\ell = 0, \ldots, m_\ell - 1$ of the recursion relation

$$\sum_{j=0}^{k} \mathring{\lambda}_j^{(k)} v_{n+j} = 0 \tag{244}$$

and thus, it is annihilated by $\mathscr{P}[\,\mathring{\Pi}^{(k)}]$. □

6.3. Application to power series

Here, we generalize some results of Weniger [88] that regard the application of Levin-type sequence transformations to power series.

We use the definitions in Eq. (27). Like Padé approximants, Levin-type sequence transformations yield rational approximants when applied to the partial sums $f_n(z)$ of a power series $f(z)$ with terms $a_j = c_j z^j$. These approximations offer a practical way for the analytical continuation of power series to regions outside of their circle of convergence. Furthermore, the poles of the rational approximations model the singularities of $f(z)$. They may also be used to approximate further terms beyond the last one used in constructing the rational approximant.

When applying a Levin-type sequence transformation $\mathscr{T}$ to a power series, remainder estimates $\omega_n = m_n z^{\gamma+n}$ will be used. We note that t variants correspond to $m_n = c_n$, $\gamma = 0$, u variants correspond to $m_n = c_n(n+\beta)$, $\gamma = 0$, $\tilde{t}$ variants to $m_n = c_{n+1}$, $\gamma = 1$. Thus, for these variants, m_n is independent of z (Case A). For v variants, we have $m_n = c_{n+1}c_n/(c_n - c_{n+1}z)$, and $\gamma = 1$. In this case, $1/m_n \in \mathbb{P}^{(1)}$ is a linear function of z (Case B).

Application of $\mathscr{T}$ yields after some simplification

$$\mathscr{T}_n^{(k)}(\{\{f_n(z)\}\}, \{\{m_n z^{\gamma+n}\}\}) = \frac{\sum_{\ell=0}^{n+k} z^\ell \sum_{j=\max(0,k-\ell)}^{k} (\lambda_{n,j}^{(k)}/m_{n+j}) c_{\ell-(k-j)}}{\sum_{j=0}^{k} (\lambda_{n,j}^{(k)}/m_{n+j}) z^{k-j}} = \frac{P_n^{(k)}[T](z)}{Q_n^{(k)}[T](z)}, \tag{245}$$

where in Case A, we have $P_n^{(k)}[T] \in \mathbb{P}^{n+k}$, $Q_n^{(k)}[T] \in \mathbb{P}^k$, and in Case B, we have $P_n^{(k)}[T] \in \mathbb{P}^{n+k+1}$, $Q_n^{(k)}[T] \in \mathbb{P}^{k+1}$. One needs the $k+1+\gamma$ partial sums $f_n(z), \ldots, f_{n+k+\gamma}(z)$ to compute these rational approximants. This should be compared to the fact that for the computation of the Padé approximant $[n+k+\gamma/k+\gamma]$ one needs the $2k+2\gamma+1$ partial sums $f_n(z), \ldots, f_{n+2k+2\gamma}(z)$.

We show that Taylor expansion of these rational approximants reproduces all terms of power series that have been used to calculate the rational approximation.

Theorem 8. *We have*

$$\mathscr{T}_n^{(k)}(\{\{f_n(z)\}\}, \{\{m_n z^{\gamma+n}\}\}) - f(z) = \mathrm{O}(z^{n+k+1+\tau}), \tag{246}$$

where $\tau = 0$ for t and u variants corresponding to $m_n = c_n$, $\gamma = 0$, or $m_n = c_n(n+\beta)$, $\gamma = 0$, respectively, while $\tau = 1$ holds for the v variant corresponding to $m_n = c_{n+1}c_n/(c_n - c_{n+1}z)$, $\gamma = 1$, and for the $\tilde{t}$ variants corresponding to $m_n = c_{n+1}$, $\gamma = 1$, one obtains $\tau = 1$ if $\mathscr{T}$ is convex.

Proof. Using the identity

$$\mathscr{T}_n^{(k)}(\{\{f_n(z)\}\}, \{\{m_n z^{\gamma+n}\}\}) = f(z) + \mathscr{T}_n^{(k)}(\{\{f_n(z) - f(z)\}\}, \{\{m_n z^{\gamma+n}\}\}) \tag{247}$$

that follows from Theorem 5, we obtain after some easy algebra

$$\mathscr{T}_n^{(k)}(\{\{f_n(z)\}\}, \{\{m_n z^{\gamma+n}\}\}) - f(z) = z^{n+k+1} \frac{\sum_{\ell=0}^{\infty} z^\ell \sum_{j=0}^{k} (\lambda_{n,j}^{(k)}/m_{n+j}) c_{\ell+n+j+1}}{\sum_{j=0}^{k} (\lambda_{n,j}^{(k)}/m_{n+j}) z^{k-j}}. \tag{248}$$

This shows that the right-hand side is at least $\mathrm{O}(z^{n+k+1})$ since the denominator is $\mathrm{O}(1)$ due to $\lambda_{n,k}^{(k)} \neq 0$. For the $\tilde{t}$ variant, the term corresponding to $\ell = 0$ in the numerator is $\sum_{j=0}^{k} \lambda_{n,j}^{(k)} = \Pi_n^{(k)}(1)$ that vanishes for convex $\mathscr{T}$. For the v variant, that term is $\sum_{j=0}^{k} \lambda_{n,j}^{(k)}(c_{n+j} - c_{n+j+1}z)/c_{n+j}$ that simplifies to $(-z)\sum_{j=0}^{k} \lambda_{n,j}^{(k)} c_{n+j+1}/c_{n+j}$ for convex $\mathscr{T}$. This finishes the proof. □

7. Convergence acceleration results for Levin-type transformations

7.1. General results

We note that Germain-Bonne [23] developed a theory of the regularity and convergence acceleration properties of sequence transformations that was later extended by Weniger [84, Section 12; 88, Section 6] to sequence transformations that depend explicitly on n and on an auxiliary sequence of remainder estimates. The essential results of this theory apply to convergence acceleration of linearly convergent sequences. Of course, this theory can be applied to Levin-type sequence transformations. However, for the latter transformations, many results can be obtained more easily and also, one may obtain results of a general nature that are also applicable to other convergence types like logarithmic convergence. Thus, we are not going to use the Germain–Bonne–Weniger theory in the present article.

Here, we present some general convergence acceleration results for Levin-type sequence transformations that have a limiting transformation. The results, however, do not completely determine which transformation provides the best extrapolation results for a given problem sequence since the results are asymptotic in nature, but in practice, one is interested in obtaining good extrapolation results from as few members of the problem sequence as possible. Thus, it may well be that transformations with the same asymptotic behavior of the results perform rather differently in practice.

Nevertheless, the results presented below provide a first indication which results one may expect for large classes of Levin-type sequence transformations.

First, we present some results that show that the limiting transformation essentially determines for which sequences Levin-type sequence transformations are accelerative. The speed of convergence will be analyzed later.

Theorem 9. *Assume that the following asymptotic relations hold for large n:*

$$\lambda_{n,j}^{(k)} \sim \mathring{\lambda}_j^{(k)}, \quad \mathring{\lambda}_k^{(k)} \neq 0, \tag{249}$$

$$\frac{s_n - s}{\omega_n} \sim \sum_{\nu=1}^{A} c_\nu \zeta_\nu^n, \quad c_\nu \zeta_\nu \neq 0, \quad \mathring{\Pi}^{(k)}(\zeta_\nu) = 0, \tag{250}$$

$$\frac{\omega_{n+1}}{\omega_n} \sim \rho \neq 0, \quad \mathring{\Pi}^{(k)}(1/\rho) \neq 0. \tag{251}$$

Then, $\{\{\mathcal{T}_n^{(k)}\}\}$ accelerates $\{\{s_n\}\}$ to s, i.e., we have

$$\lim_{n\to\infty} \frac{\mathcal{T}_n^{(k)} - s}{s_n - s} = 0. \tag{252}$$

Proof. Rewriting

$$\frac{\mathcal{T}_n^{(k)} - s}{s_n - s} = \frac{\omega_n}{s_n - s} \frac{\sum_{j=0}^{k} \lambda_{n,j}^{(k)} (s_{n+j} - s)/\omega_{n+j}}{\sum_{j=0}^{k} \lambda_{n,j}^{(k)} \omega_n/\omega_{n+j}}. \tag{253}$$

one may perform the limit for $n \to \infty$ upon using the assumptions according to

$$\frac{\mathscr{T}_n^{(k)} - s}{s_n - s} \to \frac{\sum_{j=0}^{k} \mathring{\lambda}_j^{(k)} \sum_\nu c_\nu \zeta_\nu^{n+j}}{\sum_\nu c_\nu \zeta_\nu^n \sum_{j=0}^{k} \mathring{\lambda}_j^{(k)} \rho^{-j}} = \frac{\sum_\nu c_\nu \zeta_\nu^n \mathring{\Pi}^{(k)}(\zeta_\nu)}{\mathring{\Pi}^{(k)}(1/\rho) \sum_\nu c_\nu \zeta_\nu^n} = 0 \tag{254}$$

since $\omega_n/\omega_{n+j} \to \rho^{-j}$. □

Thus, the zeroes ζ_ν of the characteristic polynomial of the limiting transformation are of particular importance.

It should be noted that the above assumptions correspond to a more complicated convergence type than linear or logarithmic convergence if $|\zeta_1| = |\zeta_2| \geqslant |\zeta_3| \geqslant \cdots$. This is the case, for instance, for the $\mathscr{H}^{(k)}$ transformation where the limiting transformation has the characteristic polynomial $P^{(2k)}(\alpha)$ with k-fold zeroes at $\exp(\imath\alpha)$ and $\exp(-\imath\alpha)$. Another example is the $\mathscr{I}^{(k)}$ transformation where the limiting transformation has characteristic polynomials $Q^{(2k)}(\alpha)$ with zeroes at $\exp(\pm\imath\alpha)/\Theta_j$, $j = 0, \ldots, k-1$.

Specializing to $A = 1$ in Theorem 9, we obtain the following corollary:

Corollary 10. *Assume that the following asymptotic relations hold for large n:*

$$\lambda_{n,j}^{(k)} \sim \mathring{\lambda}_j^{(k)}, \quad \mathring{\lambda}_k^{(k)} \neq 0, \tag{255}$$

$$\frac{s_n - s}{\omega_n} \sim cq^n, \quad cq \neq 0, \quad \mathring{\Pi}^{(k)}(q) = 0, \tag{256}$$

$$\frac{\omega_{n+1}}{\omega_n} \sim \rho \neq 0, \quad \mathring{\Pi}^{(k)}(1/\rho) \neq 0. \tag{257}$$

Then, $\{\{\mathscr{T}_n^{(k)}\}\}$ accelerates $\{\{s_n\}\}$ to s, i.e., we have

$$\lim_{n\to\infty} \frac{\mathscr{T}_n^{(k)} - s}{s_n - s} = 0. \tag{258}$$

Note that the assumptions of Corollary 10 imply

$$\frac{s_{n+1} - s}{s_n - s} = \frac{s_{n+1} - s}{\omega_{n+1}} \frac{\omega_n}{s_n - s} \frac{\omega_{n+1}}{\omega_n} \sim \rho \frac{cq^{n+1}}{cq^n} = \rho q \tag{259}$$

and thus, Corollary 10 corresponds to linear convergence for $0 < |\rho q| < 1$ and to logarithmic convergence for $\rho q = 1$.

Many important sequence transformations have convex limiting transformations, i.e., the characteristic polynomials satisfy $\mathring{\Pi}^{(k)}(1) = 0$. In this case, they accelerate linear convergence. More exactly, we have the following corollary:

Corollary 11. *Assume that the following asymptotic relations hold for large n:*

$$\lambda_{n,j}^{(k)} \sim \mathring{\lambda}_j^{(k)}, \quad \mathring{\lambda}_k^{(k)} \neq 0, \tag{260}$$

$$\frac{s_n - s}{\omega_n} \sim c, \quad c \neq 0, \quad \mathring{\Pi}^{(k)}(1) = 0, \tag{261}$$

$$\frac{\omega_{n+1}}{\omega_n} \sim \rho \neq 0, \quad \mathring{\Pi}^{(k)}(1/\rho) \neq 0. \tag{262}$$

Then, $\{\{\mathcal{T}_n^{(k)}\}\}$ accelerates $\{\{s_n\}\}$ *to* s, *i.e., we have*

$$\lim_{n\to\infty} \frac{\mathcal{T}_n^{(k)} - s}{s_n - s} = 0. \tag{263}$$

Hence, any Levin-type sequence transformation with a convex limiting transformation accelerates linearly convergent sequences with

$$\lim_{n\to\infty} \frac{s_{n+1} - s}{s_n - s} = \rho, \quad 0 < |\rho| < 1 \tag{264}$$

such that $\mathring{\Pi}^{(k)}(1/\rho) \neq 0$ *for suitably chosen remainder estimates* ω_n *satisfying* $(s_n - s)/\omega_n \to c \neq 0$.

Proof. Specializing Corollary 10 to $q = 1$, it suffices to prove the last assertion. Here, the proof follows from the observation that $(s_{n+1} - s)/(s_n - s) \sim \rho$ and $(s_n - s)/\omega_n \sim c$ imply $\omega_{n+1}/\omega_n \sim \rho$ for large n in view of the assumptions. □

Note that Corollary 11 applies for instance to suitable variants of the Levin transformation, the ${}_p\mathbf{J}$ transformation and, more generally, of the $\mathcal{J}$ transformation. In particular, it applies to t, $\tilde{t}$, u and v variants, since in the case of linear convergence, one has $\Delta s_n/\Delta s_{n-1} \sim \rho$ which entails $(s_n - s)/\omega_n \sim c$ for all these variants by simple algebra.

Now, some results for the speed of convergence are given. Matos [59] presented convergence theorems for sequence transformations based on annihilation difference operators with characteristic polynomials with constants coefficients that are close in spirit to the theorems given below. However, it should be noted that the theorems presented here apply to large classes of Levin-type transformations that have a limiting transformation (the latter, of course, has a characteristic polynomial with constants coefficients).

Theorem 12. (C-1) *Suppose that for a Levin-type sequence transformation* $\mathcal{T}^{(k)}$ *of order* k *there is a limiting transformation* $\mathring{\mathcal{T}}^{(k)}$ *with characteristic polynomial* $\mathring{\Pi} \in \mathbb{P}^k$ *given by Eq.* (242) *where the multiplicities* m_ℓ *of the zeroes* $\zeta_\ell \neq 0$ *satisfy* $m_1 \leqslant m_2 \leqslant \cdots \leqslant m_M$. *Let*

$$\lambda_{n,j}^{(k)} \frac{n^{m_1-1}}{(n+j)^{m_1-1}} \sim \mathring{\lambda}_j^{(k)} \left(\sum_{t=0}^{\infty} \frac{e_t^{(k)}}{(n+j)^t} \right), \quad e_0^{(k)} = 1 \tag{265}$$

for $n \to \infty$.

(C-2) *Assume that* $\{\{s_n\}\} \in \mathbb{S}^{\mathbb{K}}$ *and* $\{\{\omega_n\}\} \in \mathbb{O}^{\mathbb{K}}$. *Assume further that for* $n \to \infty$ *the asymptotic expansion*

$$\frac{s_n - s}{\omega_n} \sim \sum_{\ell=1}^{M} \zeta_\ell^n \sum_{r=0}^{\infty} c_{\ell,r} n^{-r} \tag{266}$$

holds, and put

$$r_\ell = \min\{r \in \mathbb{N}_0 \mid f_{\ell,r+m_1} \neq 0\}, \tag{267}$$

where

$$f_{\ell,v} = \sum_{r=0}^{v} e_{v-r}^{(k)} c_{\ell,r} \tag{268}$$

and

$$B_\ell = (-1)^{m_\ell} \frac{\mathrm{d}^{m_\ell} \mathring{\Pi}^{(k)}}{\mathrm{d}x^{m_\ell}}(\zeta_\ell) \tag{269}$$

for $\ell = 1,\dots,M$.

(C-3) *Assume that the following limit exists and satisfies*

$$0 \neq \lim_{n\to\infty} \frac{\omega_{n+1}}{\omega_n} = \rho \notin \{\zeta_\ell^{-1} \mid \ell = 1,\dots,M\}. \tag{270}$$

Then we have

$$\frac{\mathscr{T}_n^{(k)}(\{\{s_n\}\},\{\{\omega_n\}\}) - s}{\omega_n} \sim \frac{\sum_{\ell=1}^{M} f_{\ell,r_\ell+m_1} \zeta_\ell^{n+m_\ell} \binom{r_\ell + m_\ell}{r_\ell} B_\ell / n^{r_\ell+m_\ell-m_1}}{\mathring{\Pi}^{(k)}(1/\rho)} \frac{1}{n^{2m_1}}. \tag{271}$$

Thus, $\{\{\mathscr{T}_n^{(k)}(\{\{s_n\}\},\{\{\omega_n\}\})\}\}$ *accelerates* $\{\{s_n\}\}$ *to* s *at least with order* $2m_1$, *i.e.*,

$$\frac{\mathscr{T}_n^{(k)} - s}{s_n - s} = \mathrm{O}(n^{-2m_1-\tau}), \quad \tau \geqslant 0 \tag{272}$$

if $c_{\ell,0} \neq 0$ *for all* ℓ.

Proof. We rewrite $\mathscr{T}_n^{(k)}(\{\{s_n\}\},\{\{\omega_n\}\}) = \mathscr{T}_n^{(k)}$ as defined in Eq. (11) in the form

$$\mathscr{T}_n^{(k)} - s = \omega_n \frac{\sum_{j=0}^{k} \lambda_{n,j}^{(k)}(s_{n+j} - s)/\omega_{n+j}}{\sum_{j=0}^{k} \lambda_{n,j}^{(k)} \omega_n/\omega_{n+j}} \sim \omega_n \frac{\sum_{j=0}^{k} \mathring{\lambda}_j^{(k)} \sum_{t=0}^{\infty} \frac{e_t^{(k)}}{(n+j)^t} \frac{(n+j)^{m_1-1}}{n^{m_1-1}} \frac{s_{n+j}-s}{\omega_{n+j}}}{\sum_{j=0}^{k} \mathring{\lambda}_j^{(k)} \frac{1}{\rho^j}} \tag{273}$$

for large n where we used Eq. (265) in the numerator, and in the denominator the relation $\omega_n/\omega_{n+j} \to \rho^{-j}$ that follows by repeated application of Eq. (270). Insertion of (266) now yields

$$\mathscr{T}_n^{(k)} - s \sim \frac{\omega_n}{n^{m_1-1} \mathring{\Pi}^{(k)}(1/\rho)} \sum_{\ell=1}^{M} \sum_{r=0}^{\infty} f_{\ell,r+m_1} \sum_{j=0}^{k} \mathring{\lambda}_j^{(k)} \frac{\zeta_\ell^{n+j}}{(n+j)^{r+1}}, \tag{274}$$

where Eq. (268) was used. Also the fact was used that $\mathscr{P}[\mathring{\Pi}^{(k)}]$ annihilates any linear combination of the solutions $\varphi_{n,\ell,j_\ell}^{(k)} = n^{j_\ell} \zeta_\ell^n$ with $\ell = 1,\dots,M$ and $j_\ell = 0,\dots,m_1 - 1$ of the recursion relation (244) since each ζ_ℓ is a zero with multiplicity exceeding $m_1 - 1$. Invoking Lemma C.1 given in Appendix C one obtains

$$\mathscr{T}_n^{(k)} - s \sim \frac{\omega_n}{n^{m_1-1} \mathring{\Pi}^{(k)}(1/\rho)} \sum_{\ell=1}^{M} \sum_{r=0}^{\infty} f_{\ell,r+m_1} \zeta_\ell^{n+m_\ell} \binom{r + m_\ell}{r} \frac{(-1)^{m_\ell}}{n^{r+m_\ell+1}} \frac{\mathrm{d}^{m_\ell} \mathring{\Pi}^{(k)}}{\mathrm{d}x^{m_\ell}}(\zeta_\ell). \tag{275}$$

The proof of Eq. (271) is completed taking leading terms in the sums over r. Since $s_n - s \sim \omega_n Z^n \sum_{\ell\in I} (\zeta_\ell/Z)^n c_{\ell,0}$ where $Z = \max\{|\zeta_\ell| \,|\, \ell = 1,\dots,M\}$, and $I = \{\ell = 1,\dots,M \mid Z = |\zeta_\ell|\}$, Eq. (272) is obtained where $\tau = \min\{r_\ell + m_\ell - m_1 \mid \ell \in I\}$. □

If $\omega_{n+1}/\omega_n \sim \rho$, where $\overset{\circ}{\Pi}{}^{(k)}(1/\rho)=0$, i.e., if (C-3) of Theorem 12 does not hold, then the denominators vanish asymptotically. In this case, one has to investigate whether the numerators or the denominators vanish faster.

Theorem 13. *Assume that* (C-1) *and* (C-2) *of Theorem* 12 *hold.*
(C-3′) *Assume that for* $n\to\infty$ *the asymptotic relation*

$$\frac{\omega_{n+1}}{\omega_n} \sim \rho \exp(\varepsilon_n), \quad \rho \neq 0 \tag{276}$$

holds where

$$\frac{1}{\lambda!}\frac{\mathrm{d}^\lambda \overset{\circ}{\Pi}{}^{(k)}}{\mathrm{d}x^\lambda}(1/\rho) = \begin{cases} 0 & \text{for } \lambda = 0,\ldots,\mu-1 \\ C \neq 0 & \text{for } \lambda = \mu \end{cases} \tag{277}$$

and

$$\varepsilon_n \to 0, \quad \frac{\varepsilon_{n+1}}{\varepsilon_n} \to 1 \tag{278}$$

for large n. Define δ_n *via* $\exp(-\varepsilon_n) = 1 + \delta_n\rho$.
Then we have for large n

$$\frac{\mathcal{T}_n^{(k)}(\{\{s_n\}\},\{\{\omega_n\}\}) - s}{\omega_n} \sim \frac{\sum_{\ell=1}^{M} f_{\ell,r_\ell+m_1}\zeta_\ell^{n+m_\ell}\binom{r_\ell+m_\ell}{r_\ell}B_\ell/n^{r_\ell+m_\ell-m_1}}{C(\delta_n)^\mu}\frac{1}{n^{2m_1}}. \tag{279}$$

Proof. The proof proceeds as the proof of Theorem 12 but in the denominator we use

$$\sum_{j=0}^{k}\lambda_{n,j}^{(k)}\frac{\omega_n}{\omega_{n+j}} \sim C(\delta_n)^\mu \tag{280}$$

that follows from Lemma C.2 given in Appendix C. □

Thus, the effect of the sequence transformation in this case essentially depends on the question whether $(\delta_n)^{-\mu}n^{-2m_1}$ goes to 0 for large n or not. In many important cases like the Levin transformation and the ${}_p\mathbf{J}$ transformations, we have $M=1$ and $m_1=k$. We note that Theorem 11 becomes especially important in the case of logarithmic convergence since for instance for $M=1$ one observes that $(s_{n+1}-s)/(s_n-s)\sim 1$ and $(s_n-s)/\omega_n \sim \zeta_1^n c_{1,0} \neq 0$ imply $\omega_{n+1}/\omega_n \sim 1/\zeta_1$ for large n such that the denominators vanish asymptotically. In this case, we have $\mu=m_1$ whence $(\delta_n)^{-\mu}n^{-2m_1} = \mathrm{O}(n^{-m_1})$ if $\delta_n = \mathrm{O}(1/n)$. This reduction of the speed of convergence of the acceleration process from $\mathrm{O}(n^{-2k})$ to $\mathrm{O}(n^{-k})$ in the case of logarithmic convergence is a generic behavior that is reflected in a number of theorems regarding convergence acceleration properties of Levin-type sequence transformations. Examples are Sidi's theorem for the Levin transformation given below (Theorem 15), and for the ${}_p\mathbf{J}$ transformation the Corollaries 18 and 19 given below, cf. also [84, Theorems 13.5, 13.9, 13.11, 13.12, 14.2].

The following theorem was given by Matos [59] where the proof may be found. To formulate it, we define that a sequence $\{\{u_n\}\}$ has *property M* if it satisfies

$$\frac{u_{n+1}}{u_n} \sim 1 + \frac{\alpha}{n} + r_n \quad \text{with } r_n = \mathrm{o}(1/n), \qquad \Delta^\ell r_n = \mathrm{o}(\Delta^\ell(1/n)) \quad \text{for } n\to\infty. \tag{281}$$

Theorem 14 (Matos [59, Theorem 13]). *Let $\{\{s_n\}\}$ be a sequence such that*

$$s_n - s = \omega_n(a_1 g_1^{(1)}(n) + \cdots + a_k g_1^{(k)}(n) + \rho_n) \tag{282}$$

with $g_1^{(j+1)}(n) = \mathrm{o}(g_1^{(j)}(n))$, $\rho_n = \mathrm{o}(g_1^{(k)}(n))$ for $n \to \infty$. Let us consider an operator L of the form (193) for which we know a basis of solutions $\{\{u_n^{(j)}\}\}$, $j = 1,\dots,k$, and each one can be written as

$$u_n^{(j)} \sim \sum_{m=1}^{\infty} \alpha_m^{(j)} g_m^{(j)}(n), \qquad g_{m+1}^{(j)}(n) = \mathrm{o}(g_m^{(j)}(n)) \tag{283}$$

as $n \to \infty$ for all $m \in \mathbb{N}$ and $j = 1,\dots,k$. Suppose that

(a) $g_2^{(j+1)}(n) = \mathrm{o}(g_2^{(j)}(n))$ *for* $n \to \infty$, $j = 1,\dots,k-1$,
(b) $g_2^{(1)}(n) = \mathrm{o}(g_1^{(k)}(n))$, *and* $\rho_n \sim K g_2^{(1)}(n)$ *for* $n \to \infty$,
(c) $\{\{g_m^{(j)}(n)\}\}$ *has property M* for
$m \in \mathbb{N}$, $j = 1,\dots,k$. (284)

Then

1. *If $\{\{\omega_n\}\}$ satisfies $\lim_{n\to\infty} \omega_n/\omega_{n+1} = \lambda \neq 1$, the sequence transformation $\mathscr{T}_n^{(k+1)}$ corresponding to the operator L accelerates the convergence of $\{\{s_n\}\}$. Moreover, the acceleration can be measured by*

$$\frac{\mathscr{T}_n^{(k+1)} - s}{s_n - s} \sim C n^{-k} \frac{g_2^{(1)}(n)}{g_1^{(1)}(n)}, \qquad n \to \infty. \tag{285}$$

2. *If $\{\{1/\omega_n\}\}$ has property M, then the speed of convergence of $\mathscr{T}_n^{(k+1)}$ can be measured by*

$$\frac{\mathscr{T}_n^{(k+1)} - s}{s_n - s} \sim C \frac{g_2^{(1)}(n)}{g_1^{(1)}(n)}, \qquad n \to \infty. \tag{286}$$

7.2. *Results for special cases*

In the case that peculiar properties of a Levin-type sequence transformation are used, more stringent theorems can often be proved as regards convergence acceleration using this particular transformation.

In the case of the Levin transformation, Sidi proved the following theorem:

Theorem 15 (Sidi [76] and Brezwski and Redivo Zaglia [14, Theorem 2.32]). *If $s_n = s + \omega_n f_n$ where $f_n \sim \sum_{j=0}^{\infty} \beta_j / n^j$ with $\beta_0 \neq 0$ and $\omega_n \sim \sum_{j=0}^{\infty} \delta_j / n^{j+a}$ with $a > 0$, $\delta_0 \neq 0$ for $n \to \infty$ then, if $\beta_k \neq 0$*

$$\mathscr{L}_n^{(k)} - s \sim \frac{\delta_0 \beta_k}{\binom{-a}{k}} \cdot n^{-a-k} \quad (n \to \infty). \tag{287}$$

For the W algorithm and the $d^{(1)}$ transformation that may be regarded as direct generalizations of the Levin transformation, Sidi has obtained a large number of results. The interested reader is referred to the literature (see [77,78] and references therein).

Convergence results for the Levin transformation, the Drummond transformation and the Weniger transformations may be found in Section 13 of Weniger's report [84].

Results for the $\mathscr{J}$ transformation and in particular, for the ${}_p\mathbf{J}$ transformation are given in [39,40]. Here, we recall the following theorems:

Theorem 16. *Assume that the following holds:*

(A-0) *The sequence $\{\{s_n\}\}$ has the (anti)limit s.*

(A-1a) *For every n, the elements of the sequence $\{\{\omega_n\}\}$ are strictly alternating in sign and do not vanish.*

(A-1b) *For all n and k, the elements of the sequence $\{\{\delta_n^{(k)}\}\} = \{\{\Delta r_n^{(k)}\}\}$ are of the same sign and do not vanish.*

(A-2) *For all $n \in \mathbb{N}_0$ the ratio $(s_n - s)/\omega_n$ can be expressed as a series of the form*

$$\frac{s_n - s}{\omega_n} = c_0 + \sum_{j=1}^{\infty} c_j \sum_{n>n_1>n_2>\cdots>n_j} \delta_{n_1}^{(0)} \delta_{n_2}^{(1)} \cdots \delta_{n_j}^{(j-1)} \tag{288}$$

with $c_0 \neq 0$.

Then the following holds for $s_n^{(k)} = \mathscr{J}_n^{(k)}(\{\{s_n\}\}, \{\{\omega_n\}\}, \{\{\delta_n^{(k)}\}\})$:

(a) *The error $s_n^{(k)} - s$ satisfies*

$$s_n^{(k)} - s = \frac{b_n^{(k)}}{\nabla_n^{(k-1)} \nabla_n^{(k-2)} \cdots \nabla_n^{(0)} [1/\omega_n]} \tag{289}$$

with

$$b_n^{(k)} = c_k + \sum_{j=k+1}^{\infty} c_j \sum_{n>n_{k+1}>n_{k+2}>\cdots>n_j} \delta_{n_{k+1}}^{(k)} \delta_{n_{k+2}}^{(k+1)} \cdots \delta_{n_j}^{(j-1)}. \tag{290}$$

(b) *The error $s_n^{(k)} - s$ is bounded in magnitude according to*

$$|s_n^{(k)} - s| \leqslant |\omega_n b_n^{(k)} \delta_n^{(0)} \delta_n^{(1)} \cdots \delta_n^{(k-1)}|. \tag{291}$$

(c) *For large n the estimate*

$$\frac{s_n^{(k)} - s}{s_n - s} = \mathrm{O}(\delta_n^{(0)} \delta_n^{(1)} \cdots \delta_n^{(k-1)}) \tag{292}$$

holds if $b_n^{(k)} = \mathrm{O}(1)$ and $(s_n - s)/\omega_n = \mathrm{O}(1)$ as $n \to \infty$.

Theorem 17. *Define $s_n^{(k)} = \mathscr{J}_n^{(k)}(\{\{s_n\}\}, \{\{\omega_n\}\}, \{\{\delta_n^{(k)}\}\})$ and $\omega_n^{(k)} = 1/D_n^{(k)}$ where the $D_n^{(k)}$ are defined as in Eq. (94). Put $e_n^{(k)} = 1 - \omega_{n+1}^{(k)}/\omega_n^{(k)}$ and $b_n^{(k)} = (s_n^{(k)} - s)/\omega_n^{(k)}$. Assume that (A-0) of Theorem 16 holds and that the following conditions are satisfied:*

(B-1) *Assume that*

$$\lim_{n\to\infty} \frac{b_n^{(k)}}{b_n^{(0)}} = B_k \tag{293}$$

exists and is finite.

(B-2) *Assume that*

$$\Omega_k = \lim_{n\to\infty} \frac{\omega_{n+1}^{(k)}}{\omega_n^{(k)}} \neq 0 \tag{294}$$

and

$$F_k = \lim_{n\to\infty} \frac{\delta_{n+1}^{(k)}}{\delta_n^{(k)}} \neq 0 \tag{295}$$

exist for all $k \in \mathbb{N}_0$. *Hence the limits* $\Phi_k = \lim_{n\to\infty} \Phi_n^{(k)}$ (*cf.* Eq. (97)) *exist for all* $k \in \mathbb{N}_0$. *Then, the following holds*:

(a) *If* $\Omega_0 \notin \{\Phi_0 = 1, \Phi_1, \dots, \Phi_{k-1}\}$, *then*

$$\lim_{n\to\infty} \frac{s_n^{(k)} - s}{s_n - s} \left\{ \prod_{l=0}^{k-1} \delta_n^{(l)} \right\}^{-1} = B_k \frac{[\Omega_0]^k}{\prod_{l=0}^{k-1}(\Phi_l - \Omega_0)} \tag{296}$$

and, hence,

$$\frac{s_n^{(k)} - s}{s_n - s} = \mathrm{O}(\delta_n^{(0)}\delta_n^{(1)} \cdots \delta_n^{(k-1)}) \tag{297}$$

holds in the limit $n \to \infty$.

(b) *If* $\Omega_l = 1$ *for* $l \in \{0, 1, 2, \dots, k\}$ *then*

$$\lim_{n\to\infty} \frac{s_n^{(k)} - s}{s_n - s} \left\{ \prod_{l=0}^{k-1} \frac{\delta_n^{(l)}}{e_n^{(l)}} \right\}^{-1} = B_k \tag{298}$$

and, hence,

$$\frac{s_n^{(k)} - s}{s_n - s} = \mathrm{O}\left(\prod_{l=0}^{k-1} \frac{\delta_n^{(l)}}{e_n^{(l)}} \right) \tag{299}$$

holds in the limit $n \to \infty$.

This theorem has the following two corollaries for the ${}_p\mathbf{J}$ transformation [39]:

Corollary 18. *Assume that the following holds*:

(C-1) *Let* $\beta > 0$, $p \geqslant 1$ *and* $\delta_n^{(k)} = \triangle[(n+\beta+(p-1)k)^{-1}]$. *Thus, we deal with the* ${}_p\mathbf{J}$ *transformation and, hence, the equations* $F_k = \lim_{n\to\infty} \delta_{n+1}^{(k)}/\delta_n^{(k)} = 1$ *and* $\Phi_k = 1$ *hold for all* k.

(C-2) *Assumptions* (A-2) *of Theorem* 16 *and* (B-1) *of Theorem* 17 *are satisfied for the particular choice* (C-1) *for* $\delta_n^{(k)}$.

(C-3) *The limit* $\Omega_0 = \lim_{n\to\infty} \omega_{n+1}/\omega_n$ *exists, and it satisfies* $\Omega_0 \notin \{0, 1\}$. *Hence, all the limits* $\Omega_k = \lim_{n\to\infty} \omega_{n+1}^{(k)}/\omega_n^{(k)}$ *exist for* $k \in \mathbb{N}$ *exist and satisfy* $\Omega_k = \Omega_0$.

Then the transformation $s_n^{(k)} = {}_p\mathbf{J}_n^{(k)}(\beta, \{\{s_n\}\}, \{\{\omega_n\}\})$ *satisfies*

$$\lim_{n\to\infty} \frac{s_n^{(k)} - s}{s_n - s} \left\{ \prod_{l=0}^{k-1} \delta_n^{(l)} \right\}^{-1} = B_k \left\{ \frac{\Omega_0}{1 - \Omega_0} \right\}^k \tag{300}$$

and, hence,

$$\frac{s_n^{(k)} - s}{s_n - s} = \mathrm{O}((n+\beta)^{-2k}) \tag{301}$$

holds in the limit $n \to \infty$.

Note that Corollary 18 can be applied in the case of linear convergence because then $0 < |\Omega_0| < 1$ holds.

Corollary 18 allows to conclude that in the case of linear convergence, the ${}_p\mathbf{J}$ transformations should be superior to Wynn's epsilon algorithm [104]. Consider for instance the case that

$$s_n \sim s + \lambda^n n^\theta \sum_{n=0}^{\infty} c_j/n^j, \quad c_0 \neq 0, \quad n \to \infty \tag{302}$$

is an asymptotic expansion of the sequence elements s_n. Assuming $\lambda \neq 1$ and $\theta \notin \{0, 1, \ldots, k-1\}$ it follows that [102, p. 127; 84, p. 333, Eq. (13.4–7)]

$$\frac{\varepsilon_{2k}^{(n)} - s}{s_n - s} = \mathrm{O}(n^{-2k}), \quad n \to \infty. \tag{303}$$

This is the same order of convergence acceleration as in Eq. (301). But it should be noted that for the computation of $\varepsilon_{2k}^{(n)}$ the $2k+1$ sequence elements $\{s_n, \ldots, s_{n+2k}\}$ are required. But for the computation of ${}_p\mathbf{J}_n^{(k)}$ only the $k+1$ sequence elements $\{s_n, \ldots, s_{n+k}\}$ are required in the case of the t and u variants, and additionally s_{n+k+1} in the case of the $\tilde{t}$ variant. Again, this is similar to Levin-type accelerators [84, p. 333].

The following corollary applies to the case of logarithmic convergence:

Corollary 19. *Assume that the following holds*:

(D-1) *Let $\beta > 0$, $p \geqslant 1$ and $\delta_n^{(k)} = \Delta[(n+\beta+(p-1)k)^{-1}]$. Thus, we deal with the ${}_p\mathbf{J}$ transformation and, hence, the equations $F_k = \lim_{n\to\infty} \delta_{n+1}^{(k)}/\delta_n^{(k)} = 1$ and $\Phi_k = 1$ hold for all k.*

(D-2) *Assumptions* (A-2) *of Theorem* 16 *and* (B-1) *of Theorem* 15 *are satisfied for the particular choice* (C-1) *for $\delta_n^{(k)}$.*

(D-3) *Some constants $a_l^{(j)}$, $j = 1, 2$, exist such that*

$$e_n^{(l)} = 1 - \omega_{n+1}^{(l)}/\omega_n^{(l)} = \frac{a_l^{(1)}}{n+\beta} + \frac{a_l^{(2)}}{(n+\beta)^2} + \mathrm{O}((n+\beta)^{-3}) \tag{304}$$

holds for $l=0$. This implies that this equation, and hence, $\Omega_l=1$ holds for $l \in \{0, 1, 2, \ldots, k\}$. Assume further that $a_l^{(1)} \neq 0$ for $l \in \{0, 1, 2, \ldots, k-1\}$.

Then the transformation $s_n^{(k)} = {}_p\mathbf{J}_n^{(k)}(\beta, \{\{s_n\}\}, \{\{\omega_n\}\})$ satisfies

$$\lim_{n\to\infty} \frac{s_n^{(k)} - s}{s_n - s} \left\{ \prod_{l=0}^{k-1} \frac{\delta_n^{(l)}}{e_n^{(l)}} \right\}^{-1} = B_k \tag{305}$$

and, hence,

$$\frac{s_n^{(k)} - s}{s_n - s} = \mathrm{O}((n+\beta)^{-k}) \tag{306}$$

holds in the limit $n \to \infty$.

For convergence acceleration results regarding the $\mathscr{H}$ and $\mathscr{I}$ transformations, see [34,44].

8. Stability results for Levin-type transformations

8.1. General results

We remind the reader of the definition of the *stability indices* $\boldsymbol{\Gamma}_n^{(k)}(\mathscr{T}) \geqslant 1$ as given in Eq. (76). We consider the sequence $\{\{\omega_n\}\} \in \mathbb{O}^{\mathbb{K}}$ as given. We call the transformation $\mathscr{T}$ *stable along the path* $\mathscr{P} = \{(n_\ell, k_\ell) \mid [n_\ell > n_{\ell-1}$ and $k_\ell \geqslant k_{\ell-1}]$ or $[n_\ell \geqslant n_{\ell-1}$ and $k_\ell > k_{\ell-1}]\}$ in the $\mathscr{T}$ table if the limit of its stability index along the path $\mathscr{P}$ exists and is bounded, i.e., if

$$\lim_{\ell\to\infty} \boldsymbol{\Gamma}_{n_\ell}^{(k_\ell)}(\mathscr{T}) = \lim_{\ell\to\infty} \sum_{j=0}^{k_\ell} |\gamma_{n_\ell,j}(k_\ell)(\omega_n)| < \infty, \tag{307}$$

where the $\gamma_{n,j}^{(k)}(\omega_n)$ are defined in Eq. (75). The transformation $\mathscr{T}$ is called *S-stable*, if it is stable along all paths $\mathscr{P}^{(k)} = \{(n,k) \mid n = 0, 1, \ldots\}$ for fixed k, i.e., along all columns in the $\mathscr{T}$ table.

The case of stability along diagonal paths is much more difficult to treat analytically unless Theorem 22 applies. Up to now it seems that such diagonal stability issues have only been analysed by Sidi for the case of the $d^{(1)}$ transformation (see [78] and references therein). We will treat only S-stability in the sequel.

The higher the stability index $\boldsymbol{\Gamma}(\mathscr{T})$ is, the smaller is the numerical stability of the transformation $\mathscr{T}$: If ε_j is the numerical error of s_j,

$$\varepsilon_j = s_j - \mathtt{fl}(s_j), \tag{308}$$

then the difference between the true value $\mathscr{T}_n^{(k)}$ and the numerically computed approximation $\mathtt{fl}(\mathscr{T}_n^{(k)})$ may be bounded according to

$$|\mathscr{T}_n^{(k)} - \mathtt{fl}(\mathscr{T}_n^{(k)})| \leqslant \boldsymbol{\Gamma}_n^{(k)}(\mathscr{T}) \left(\max_{j\in\{0,1,\ldots,k\}} |\varepsilon_{n+j}| \right), \tag{309}$$

cf. also [78].

Theorem 20. *If the Levin-type sequence transformation $\mathscr{T}^{(k)}$ has a limiting transformation $\mathring{\mathscr{T}}^{(k)}$ with characteristic polynomial $\mathring{\Pi}^{(k)} \in \mathbb{P}^{(k)}$ for all $k \in \mathbb{N}$, and if $\{\{\omega_n\}\} \in \mathbb{O}^{\mathbb{K}}$ satisfies $\omega_{n+1}/\omega_n \sim \rho \neq 0$ for large n with $\mathring{\Pi}^{(k)}(1/\rho) \neq 0$ for all $k \in \mathbb{N}$ then the transformation $\mathscr{T}$ is S-stable. If additionally, the coefficients $\mathring{\lambda}_j^{(k)}$ of the characteristic polynomial alternate in sign, i.e., if $\mathring{\lambda}_j^{(k)} = (-1)^j |\mathring{\lambda}_j^{(k)}|/\tau_k$ with $|\tau_k| = 1$, then the limits $\mathring{\Gamma}^{(k)}(\mathscr{T}) = \lim_{n\to\infty} \Gamma_n^{(k)}(\mathscr{T})$ obey*

$$\mathring{\boldsymbol{\Gamma}}^{(k)}(\mathscr{T}) = \tau_k \frac{\mathring{\Pi}^{(k)}(-1/|\rho|)}{|\mathring{\Pi}^{(k)}(1/\rho)|}. \tag{310}$$

Proof. We have for fixed k

$$\gamma_{n,j}^{(k)}(\omega_n) = \lambda_{n,j}^{(k)} \frac{\omega_n}{\omega_{n+j}} \left[\sum_{j'=0}^{k} \lambda_{n,j'}^{(k)} \frac{\omega_n}{\omega_{n+j'}} \right]^{-1} \sim \mathring{\lambda}_j^{(k)} \rho^{-j} \left[\sum_{j'=0}^{k} \mathring{\lambda}_{j'}^{(k)} \rho^{-j'} \right]^{-1} = \frac{\mathring{\lambda}_j^{(k)} \rho^{-j}}{\mathring{\Pi}^{(k)}(1/\rho)}, \tag{311}$$

whence

$$\lim_{n\to\infty} \Gamma_n^{(k)}(\mathscr{T}) = \lim_{n\to\infty} \sum_{j=0}^{k} |\gamma_{n,j}^{(k)}(\omega_n)| = \frac{\sum_{j=0}^{k} |\mathring{\lambda}_j^{(k)}||\rho|^{-j}}{|\mathring{\Pi}^{(k)}(1/\rho)|} < \infty. \tag{312}$$

If the $\mathring{\lambda}_j^{(k)}$ alternate in sign, we obtain for these limits

$$\mathring{\Gamma}^{(k)}(\mathscr{T}) = \tau_k \frac{\sum_{j=0}^{k} \mathring{\lambda}_j^{(k)}(-|\rho|)^{-j}}{|\mathring{\Pi}^{(k)}(1/\rho)|}. \tag{313}$$

This implies Eq. (310). □

Corollary 21. *Assume that the Levin-type sequence transformation $\mathscr{T}^{(k)}$ has a limiting transformation $\mathring{\mathscr{T}}^{(k)}$ with characteristic polynomial $\mathring{\Pi}^{(k)} \in \mathbb{P}^{(k)}$ and the coefficients $\mathring{\lambda}_j^{(k)}$ of the characteristic polynomial alternate in sign, i.e., if $\mathring{\lambda}_j^{(k)} = (-1)^j |\mathring{\lambda}_j^{(k)}|/\tau_k$ with $|\tau_k| = 1$ for all $k \in \mathbb{N}$. The sequence $\{\{\omega_n\}\} \in \mathbb{O}^{\mathbb{K}}$ is assumed to be alternating and to satisfy $\omega_{n+1}/\omega_n \sim \rho < 0$ for large n. Then the transformation $\mathscr{T}$ is S-stable. Additionally the limits are $\mathring{\Gamma}^{(k)}(\mathscr{T}) = 1$.*

Proof. Since

$$\begin{aligned}\sum_{j'=0}^{k} \frac{\lambda_{n,j'}^{(k)}}{\tau_k^{-1}} \frac{\omega_n}{\omega_{n+j'}} \sim \frac{\mathring{\Pi}^{(k)}(1/\rho)}{\tau_k^{-1}} &= \sum_{j'=0}^{k} \frac{\mathring{\lambda}_{j'}^{(k)}}{\tau_k^{-1}} (-1)^{j'} |\rho|^{-j'} \\ &= \sum_{j'=0}^{k} |\mathring{\lambda}_{j'}^{(k)}||\rho|^{-j'} \geqslant |\mathring{\lambda}_k^{(k)}|/|\rho|^k > 0.\end{aligned} \tag{314}$$

$1/\rho$ cannot be a zero of $\mathring{\Pi}^{(k)}$. Then, Theorem 20 entails that $\mathscr{T}$ is S-stable. Furthermore, Eq. (310) is applicable and yields $\mathring{\Gamma}^{(k)}(\mathscr{T}) = 1$. □

This result can be improved if all the coefficients $\lambda_{n,j}^{(k)}$ are alternating:

Theorem 22. *Assume that the Levin-type sequence transformation $\mathscr{T}^{(k)}$ has a characteristic polynomials $\Pi_n^{(k)} \in \mathbb{P}^{(k)}$ with alternating coefficients $\lambda_{n,j}^{(k)}$ i.e., $\lambda_{n,j}^{(k)} = (-1)^j |\lambda_{n,j}^{(k)}|/\tau_k$ with $|\tau_k| = 1$ for all $n \in \mathbb{N}_0$ and $k \in \mathbb{N}$. The sequence $\{\{\omega_n\}\} \in \mathbb{O}^{\mathbb{K}}$ is assumed to be alternating and to satisfy $\omega_{n+1}/\omega_n < 0$ for all $n \in \mathbb{N}_0$. Then we have $\Gamma_n^{(k)}(\mathscr{T}) = 1$. Hence, the transformation $\mathscr{T}$ is stable along all paths for such remainder estimates.*

Proof. We have for fixed n and k

$$\gamma_{n,j}^{(k)}(\omega_n) = \frac{\lambda_{n,j}^{(k)}\omega_n/\omega_{n+j}}{\sum_{j'=0}^{k} \lambda_{n,j'}^{(k)}\omega_n/\omega_{n+j'}} = \frac{\lambda_{n,j}^{(k)}\tau_k(-1)^j|\omega_n/\omega_{n+j}|}{\sum_{j'=0}^{k} \lambda_{n,j'}^{(k)}\tau_k(-1)^{j'}\,|\omega_n/\omega_{n+j'}|} = \frac{|\lambda_{n,j}^{(k)}||\omega_n/\omega_{n+j}|}{\sum_{j'=0}^{k} |\lambda_{n,j'}^{(k)}|\,|\omega_n/\omega_{n+j'}|} \geqslant 0. \tag{315}$$

Note that the denominators cannot vanish and are bounded from below by $|\lambda_{n,k}^{(k)}\omega_n/\omega_{n+k}| > 0$. Hence, we have $\gamma_{n,j}^{(k)}(\boldsymbol{\omega}_n) = |\gamma_{n,j}^{(k)}(\boldsymbol{\omega}_n)|$ and consequently, $\boldsymbol{\Gamma}_n^{(k)}(\mathscr{T}) = 1$ since $\sum_{j=0}^{k} \gamma_{n,j}^{(k)}(\boldsymbol{\omega}_n) = 1$ according to Eq. (75). □

8.2. *Results for special cases*

Here, we collect some special results on the stability of various Levin-type sequence transformations that have been reported in [46] and generalize some results of Sidi on the S-stability of the $d^{(1)}$ transformation.

Theorem 23. *If the sequence ω_{n+1}/ω_n possesses a limit according to*

$$\lim_{n\to\infty} \omega_{n+1}/\omega_n = \rho \neq 0 \tag{316}$$

and if $\rho \notin \{1, \Phi_1, \ldots, \Phi_k, \ldots\}$ such that the limiting transformation exists, the $\mathscr{J}$ transformation is S-stable with the same limiting stability indices as the transformation $\mathring{\mathscr{J}}$, i.e., we have

$$\lim_{n\to\infty} \boldsymbol{\Gamma}_n^{(k)} = \frac{\sum_{j=0}^{k} |\lambda_j^{(k)} \rho^{k-j}|}{\prod_{j'=0}^{k-1} |\Phi_{j'} - \rho|} < \infty. \tag{317}$$

If all Φ_k are positive then

$$\lim_{n\to\infty} \boldsymbol{\Gamma}_n^{(k)} = \prod_{j=0}^{k-1} \frac{\Phi_j + |\rho|}{|\Phi_j - \rho|} < \infty \tag{318}$$

holds.

As corollaries, we get the following results

Corollary 24. *If the sequence ω_{n+1}/ω_n possesses a limit according to*

$$\lim_{n\to\infty} \omega_{n+1}/\omega_n = \rho \notin \{0, 1\}, \tag{319}$$

the ${}_p\mathbf{J}$ transformation for $p > 1$ and $\beta > 0$ is S-stable and we have

$$\lim_{n\to\infty} \boldsymbol{\Gamma}_n^{(k)} = \frac{\sum_{j=0}^{k} \binom{k}{j} |\rho^{k-j}|}{|1-\rho|^k} = \frac{(1+|\rho|)^k}{|1-\rho|^k} < \infty. \tag{320}$$

Corollary 25. *If the sequence ω_{n+1}/ω_n possesses a limit according to*

$$\lim_{n\to\infty} \omega_{n+1}/\omega_n = \rho \notin \{0, 1\}, \tag{321}$$

the Weniger $\mathscr{S}$ transformation [84, Section 8] *for $\beta > 0$ is S-stable and we have*

$$\lim_{n\to\infty} \boldsymbol{\Gamma}_n^{(k)}(\mathscr{S}) = \frac{\sum_{j=0}^{k} \binom{k}{j} |\rho^{k-j}|}{|1-\rho|^k} = \frac{(1+|\rho|)^k}{|1-\rho|^k} < \infty. \tag{322}$$

Corollary 26. *If the sequence* ω_{n+1}/ω_n *possesses a limit according to*

$$\lim_{n\to\infty} \omega_{n+1}/\omega_n = \rho \notin \{0,1\}, \tag{323}$$

the Levin $\mathscr{L}$ *transformation* [53,84] *is S-stable and we have*

$$\lim_{n\to\infty} \Gamma_n^{(k)}(\mathscr{L}) = \frac{\sum_{j=0}^{k} \binom{k}{j} |\rho^{k-j}|}{|1-\rho|^k} = \frac{(1+|\rho|)^k}{|1-\rho|^k} < \infty. \tag{324}$$

Corollary 27. *Assume that the elements of the sequence* $\{t_n\}_{n\in\mathbb{N}}$ *satisfy* $t_n \neq 0$ *for all* n *and* $t_n \neq t_{n'}$ *for all* $n \neq n'$. *If the sequence* t_{n+1}/t_n *possesses a limit*

$$\lim_{n\to\infty} t_{n+1}/t_n = \tau \quad \text{with } 0 < \tau < 1 \tag{325}$$

and if the sequence ω_{n+1}/ω_n *possesses a limit according to*

$$\lim_{n\to\infty} \omega_{n+1}/\omega_n = \rho \notin \{0,1,\tau^{-1},\ldots,\tau^{-k},\ldots\}, \tag{326}$$

then the generalized Richardson extrapolation process $\mathscr{R}$ *introduced by Sidi* [73] *that is identical to the* $\mathscr{J}$ *transformation with* $\delta_n^{(k)} = t_n - t_{n+k+1}$ *as shown in* [36], *i.e., the W algorithm is S-stable and we have*

$$\lim_{n\to\infty} \Gamma_n^{(k)}(\mathscr{R}) = \frac{\sum_{j=0}^{k} |\tilde{\lambda}_j^{(k)} \rho^{k-j}|}{\prod_{j'=0}^{k-1} |\tau^{-j'} - \rho|} = \prod_{j'=0}^{k-1} \frac{1+\tau^{j'}|\rho|}{|1-\tau^{j'}\rho|} < \infty. \tag{327}$$

Here

$$\tilde{\lambda}_j^{(k)} = (-1)^{k-j} \sum_{\substack{j_0+j_1+\ldots+j_{k-1}=j,\\ j_0\in\{0,1\},\ldots,j_{k-1}\in\{0,1\}}} \prod_{m=0}^{k-1} (\tau)^{-m\,j_m}, \tag{328}$$

such that

$$\sum_{j=0}^{k} \tilde{\lambda}_j^{(k)} \rho^{k-j} = \prod_{j=0}^{k-1} (\tau^{-j} - \rho) = \tau^{-k(k-1)/2} \prod_{j=0}^{k-1} (1 - \tau^j \rho). \tag{329}$$

Note that the preceding corollary is essentially the same as a result of Sidi [78, Theorem 2.2] that now appears as a special case of the more general Theorem 23 that applies to a much wider class of sequence transformations. As noted above, Sidi has also derived conditions under which the $d^{(1)}$ transformation is stable along the paths $\mathscr{P}_n = \{(n,k)|k = 0,1,\ldots\}$ for fixed n. For details and more references see [78]. Analogous work for the $\mathscr{J}$ transformation is in progress.

An efficient algorithm for the computation of the stability index of the $\mathscr{J}$ transformation can be given in the case $\delta_n^{(k)} > 0$. Since the $\mathscr{J}$ transformation is invariant under $\delta_n^{(k)} \to \alpha^{(k)}\delta_n^{(k)}$ for any $\alpha^{(k)} \neq 0$ according to Homeier [36, Theorem 4], $\delta_n^{(k)} > 0$ can always be achieved if for given k, all $\delta_n^{(k)}$ have the same sign. This is the case, for instance, for the ${}_p\mathbf{J}$ transformation [36,39].

Theorem 28. *Define*

$$F_n^{(0)} = (-1)^n |D_n^{(0)}|, \qquad F_n^{(k+1)} = (F_{n+1}^{(k)} - F_n^{(k)})/\delta_n^{(k)} \tag{330}$$

and $\hat{F}_n^{(0)} = F_n^{(0)}$, $\hat{F}_n^{(k)} = (\delta_n^{(0)} \cdots \delta_n^{(k-1)}) F_n^{(k)}$. *If all* $\delta_n^{(k)} > 0$ *then*

1. $F_n^{(k)} = (-1)^{n+k} |F_n^{(k)}|$,
2. $\lambda_{n,j}^{(k)} = (-1)^{j+k} |\lambda_{n,j}^{(k)}|$, *and*
3. $$\Gamma_n^{(k)} = \frac{|\hat{F}_n^{(k)}|}{|\hat{D}_n^{(k)}|} = \frac{|F_n^{(k)}|}{|D_n^{(k)}|}. \tag{331}$$

This generalizes Sidi's method for the computation of stability indices [78] to a larger class of sequence transformations.

9. Application of Levin-type sequence transformations

9.1. Practical guidelines

Here, we address shortly the following questions:

When should one try to use sequence transformations? One can only hope for good convergence acceleration, extrapolation, or summation results if (a) the s_n have some asymptotic structure for large n and are not erratic or random, (b) a sufficiently large number of decimal digits is available. Many problems can be successfully tackled if 13–15 digits are available but some require a much larger number of digits in order to overcome some inevitable rounding errors, especially for the acceleration of logarithmically convergent sequences. The asymptotic information that is required for a successful extrapolation is often hidden in the last digits of the problem data.

How should the transformations be applied? The recommended mode of application is that one computes the highest possible order k of the transformation from the data. In the case of triangular recursive schemes like that of the $\mathscr{J}$ transformation and the Levin transformation, this means that one computes as transformed sequence $\{\mathscr{T}_0^{(n)}\}$. For L-shaped recursive schemes as in the case of the $\mathscr{H}$, $\mathscr{I}$, and $\mathscr{K}$ transformations, one usually computes as transformed sequence $\{\{\mathscr{T}_{n-2[\![n/2]\!]}^{[\![n/2]\!]}\}\}$. The error ε of the current estimate can usually be approximated a posteriori using sums of magnitudes of differences of a few entries of the $\mathscr{T}$ table, e.g.,

$$\varepsilon \approx |\mathscr{T}_1^{(n)} - \mathscr{T}_0^{(n)}| + |\mathscr{T}_0^{(n-1)} - \mathscr{T}_0^{(n)}| \tag{332}$$

for transformations with triangular recursive schemes. Such a simple approach works surprisingly well in practice. The loss of decimal digits can be estimated computing stability indices. An example is given below.

What happens if one of the denominator vanishes? The occurrence of zeroes in the D table for specific combinations of n and k is usually no problem since the recurrences for numerators and denominators still work in this case. Thus, no special devices are required to jump over such singular points in the $\mathscr{T}$ table.

Which transformation and which variant should be chosen? This depends on the type of convergence of the problem sequence. For linearly convergent sequences, t, $\tilde{t}$, u and v variants of the Levin transformation, or the ${}_p\mathbf{J}$ transformation, especially the ${}_2\mathbf{J}$ transformation are usually a good choice [39] as long as one is not too close to a singularity or to a logarithmically convergent problem. Especially well behaved is usually the application to alternating series since then, the stability is very good as discussed above. For the summation of alternating divergent sequences and series, usually the t and the $\tilde{t}$ variants of the Levin transformation, the ${}_2\mathbf{J}$ and the Weniger $\mathcal{S}$ and $\mathcal{M}$ transformations provide often surprisingly accurate results. In the case of logarithmic convergence, t and $\tilde{t}$ variants become useless, and the order of acceleration is dropping from $2k$ to k when the transformation is used columnwise. If a Kummer-related series is available (cf. Section 2.2.1), then K and lu variants leading to linear sequence transformations can be efficient [50]. Similarly, linear variants can be based on some good asymptotic estimates ${}^{\mathrm{asy}}\omega_n$, that have to be obtained via a separate analysis [50]. In the case of logarithmcic convergence, it pays to consider special devices like using subsequences $\{\{s_{\xi_n}\}\}$ where the ξ_n grow exponentially like $\xi_n = [\![\sigma\xi_{n-1}]\!] + 1$ like in the d transformations. This choice can be also used in combination with the $\mathcal{F}$ transformation. Alternatively, one can use some other transformations like the condensation transformation [51,65] or interpolation to generate a linearly convergent sequence [48], before applying an usually nonlinear sequence transformation. A somewhat different approach is possible if one can obtain a few terms a_n with large n easily [47].

What to do near a singularity? When extrapolating power series or, more generally, sequences depending on certain parameters, quite often extrapolation becomes difficult near the singularities of the limit function. In the case of linear convergence, one can often transform to a problem with a larger distance to the singularity: If Eq. (28) holds, then the subsequence $\{\{s_{\tau n}\}\}$ satisfies

$$\lim_{n\to\infty} (s_{\tau(n+1)} - s)/(s_{\tau n} - s) = \rho^{\tau}. \tag{333}$$

This is a method of Sidi that has can, however, be applied to large classes of sequence transformations [46].

What to do for more complicated convergence type? Here, one should try to rewrite the problem sequence as a sum of sequences with more simple convergence behavior. Then, nonlinear sequence transformations are used to extrapolate each of these simpler series, and to sum the extrapolation results to obtain an estimate for the original problem. This is for instance often possible for (generalized) Fourier series where it leads to complex series that may asymptotically be regarded as power series. For details, the reader is referred to the literature [14,35,40–45,77]. If this approach is not possible one is forced to use more complicated sequence transformations like the $d^{(m)}$ transformations or the (generalized) $\mathcal{H}$ transformation. These more complicated sequence transformations, however, do require more numerical effort to achieve a desired accuracy.

9.2. Numerical examples

In Table 3, we present results of the application of certain variants of the $\mathcal{F}$ transformation and the W algorithm to the series

$$S(z,a) = 1 + \sum_{j=1}^{\infty} z^j \prod_{\ell=0}^{j-1} \frac{1}{\ln(a+\ell)} \tag{334}$$

Table 3
Comparison of the $\mathscr{F}$ transformation and the W algorithm for series (334)[a]

n	A_n	B_n	C_n	D_n
14	13.16	13.65	7.65	11.13
16	15.46	15.51	9.43	12.77
18	18.01	17.84	11.25	14.43
20	21.18	20.39	13.10	16.12
22	23.06	23.19	14.98	17.81
24	25.31	26.35	16.89	19.53
26	27.87	28.17	18.83	21.26
28	30.83	30.59	20.78	23.00
30	33.31	33.19	22.76	24.76
n	E_n	F_n	G_n	H_n
14	14.07	13.18	9.75	10.47
16	15.67	15.49	11.59	12.05
18	17.94	18.02	13.46	13.66
20	20.48	20.85	15.37	15.29
22	23.51	23.61	17.30	16.95
24	25.66	25.63	19.25	18.62
26	27.89	28.06	21.23	20.31
28	30.46	30.67	23.22	22.02
29	31.82	32.20	24.23	22.89
30	33.43	33.45	25.24	23.75

[a]Plotted is the negative decadic logarithm of the relative error.
A_n: $\mathscr{F}_0^{(n)}(\{\{S_n(z,a)\}\},\{\{(2+\ln(n+a))\,\triangle\, S_n(z,a)\}\},\{\{1+\ln(n+a)\}\})$,
B_n: $W_0^{(n)}(\{\{S_n(z,a)\}\},\{\{(2+\ln(n+a))\,\triangle\, S_n(z,a)\}\},\{\{1/(1+\ln(n+a))\}\})$,
C_n: $\mathscr{F}_0^{(n)}(\{\{S_n(z,a)\}\},\{\{(n+1)\,\triangle\, S_n(z,a)\}\},\{\{1+n+a\}\})$,
D_n: $W_0^{(n)}(\{\{S_n(z,a)\}\},\{\{(n+1)\,\triangle\, S_n(z,a)\}\},\{\{1/(1+n+a)\}\})$,
E_n: $\mathscr{F}_0^{(n)}(\{\{S_n(z,a)\}\},\{\{\triangle S_n(z,a)\}\},\{\{1+\ln(n+a)\}\})$,
F_n: $W_0^{(n)}(\{\{S_n(z,a)\}\},\{\{\triangle S_n(z,a)\}\},\{\{1/(1+\ln(n+a))\}\})$,
G_n: $\mathscr{F}_0^{(n)}(\{\{S_n(z,a)\}\},\{\{\triangle S_n(z,a)\}\},\{\{1+n+a\}\})$,
H_n: $W_0^{(n)}(\{\{S_n(z,a)\}\},\{\{\triangle S_n(z,a)\}\},\{\{1/(1+n+a)\}\})$.

with partial sums

$$S_n(z,a) = 1 + \sum_{j=1}^{n} z^j \prod_{\ell=0}^{j-1} \frac{1}{\ln(a+\ell)} \tag{335}$$

for $z = 1.2$ and $a = 1.01$. Since the terms a_j satisfy $a_{j+1}/a_j = z/\ln(a+j)$, the ratio test reveals that $S(z,a)$ converges for all z and, hence, represents an analytic function. Nevertheless, only for $j \geqslant -a+\exp(|z|)$, the ratio of the terms becomes less than unity in absolute value. Hence, for larger z the series converges rather slowly.

It should be noted that for cases C_n and G_n, the $\mathscr{F}$ transformation is identical to the Weniger transformation $\mathscr{S}$, i.e., to the ${}_3\mathbf{J}$ transformation, and for cases C_n and H_n the W algorithm is identical to the Levin transformation. In the upper part of the table, we use u-type remainder estimates while

Table 4
Acceleration of $\zeta(-1/10+10\mathrm{i},1,95/100)$ with the $\mathcal{J}$ transformation[a]

n	A_n	B_n	C_n	D_n	E_n	F_n	G_n	H_n
10	2.59e − 05	3.46e + 01	2.11e − 05	4.67e + 01	1.84e − 05	4.14e + 01	2.63e − 05	3.90e + 01
20	1.72e − 05	6.45e + 05	2.53e − 05	5.53e + 07	1.38e − 04	3.40e + 09	1.94e − 05	2.47e + 06
30	2.88e − 05	3.52e + 10	8.70e − 06	2.31e + 14	8.85e − 05	1.22e + 17	2.02e − 05	6.03e + 11
40	4.68e − 06	1.85e + 15	8.43e − 08	1.27e + 20	4.06e − 06	2.78e + 23	1.50e − 06	9.27e + 16
42	2.59e − 06	1.46e + 16	2.61e − 08	1.51e + 21	2.01e − 06	4.70e + 24	6.64e − 07	8.37e + 17
44	1.33e − 06	1.10e + 17	7.62e − 09	1.76e + 22	1.73e − 06	7.85e + 25	2.76e − 07	7.24e + 18
46	6.46e − 07	8.00e + 17	1.80e − 09	2.02e + 23	1.31e − 05	1.30e + 27	1.09e − 07	6.08e + 19
48	2.97e − 07	5.62e + 18	1.07e − 08	2.29e + 24	1.52e − 04	2.12e + 28	4.16e − 08	5.00e + 20
50	1.31e − 07	3.86e + 19	1.51e − 07	2.56e + 25	1.66e − 03	3.43e + 29	1.54e − 08	4.05e + 21

[a] A_n: relative error of ${}_1\mathbf{J}_0^{(n)}(1,\{\{s_n\}\},\{\{(n+1)(s_n-s_{n-1})\}\})$, B_n: stability index of ${}_1\mathbf{J}_0^{(n)}(1,\{\{s_n\}\},\{\{(n+1)(s_n-s_{n-1})\}\})$, C_n: relative error of ${}_2\mathbf{J}_0^{(n)}(1,\{\{s_n\}\},\{\{(n+1)(s_n-s_{n-1})\}\})$, D_n: stability index of ${}_2\mathbf{J}_0^{(n)}(1,\{\{s_n\}\},\{\{(n+1)(s_n-s_{n-1})\}\})$, E_n: relative error of ${}_3\mathbf{J}_0^{(n)}(1,\{\{s_n\}\},\{\{(n+1)(s_n-s_{n-1})\}\})$, F_n: stability index of ${}_3\mathbf{J}_0^{(n)}(1,\{\{s_n\}\},\{\{(n+1)(s_n-s_{n-1})\}\})$, G_n: relative error of $\mathcal{J}_0^{(n)}(\{\{s_n\}\},\{\{(n+1)(s_n-s_{n-1})\}\},\{1/(n+1)-1/(n+k+2)\})$, H_n: Stability index of $\mathcal{J}_0^{(n)}(\{\{s_n\}\},\{\{(n+1)(s_n-s_{n-1})\}\},\{1/(n+1)-1/(n+k+2)\})$.

in the lower part, we use $\tilde{t}$ variants. It is seen that the choices $x_n = 1+\ln(a+n)$ for the $\mathcal{F}$ transformation and $t_n = 1/(1+\ln(a+n))$ for the W algorithm perform for both variants nearly identical (columns A_n, B_n, E_n and F_n) and are superior to the choices $x_n = 1+n+a$ and $t_n = 1/(1+n+a)$, respectively, that correspond to the Weniger and the Levin transformation as noted above. For the latter two transformations, the Weniger ${}^{\tilde{t}}\mathcal{S}$ transformation is slightly superior the ${}^{\tilde{t}}\mathcal{L}$ transformation for this particular example (columns G_n vs. H_n) while the situation is reversed for the u-type variants displayed in colums C_n and D_n.

The next example is taken from [46], namely the "inflated Riemann ζ function", i.e., the series

$$\zeta(\varepsilon,1,q) = \sum_{j=0}^{\infty} \frac{q^j}{(j+1)^{\varepsilon}}, \tag{336}$$

that is a special case of the Lerch zeta function $\zeta(s,b,z)$ (cf. [30, p. 142, Eq. (6.9.7); 20, Section 1.11]). The partial sums are defined as

$$s_n = \sum_{j=0}^{n} \frac{q^j}{(j+1)^{\varepsilon}}. \tag{337}$$

The series converges linearly for $0<|q|<1$ for any complex ε. In fact, we have in this case $\rho = \lim_{n\to\infty}(s_{n+1}-s)/(s_n-s) = q$. We choose $q=0.95$ and $\varepsilon = -0.1+10\mathrm{i}$. Note that for this value of ε, there is a singularity of $\zeta(\varepsilon,1,q)$ at $q=1$ where the defining series diverges since $\Re(\varepsilon)<1$.

The results of applying u variants of the ${}_p\mathbf{J}$ transformation with $p=1,2,3$ and of the Levin transformation to the sequence of partial sums is displayed in Table 4. For each of these four variants of the $\mathcal{J}$ transformation, we give the relative error and the stability index. The true value of the series (that is used to compute the errors) was computed using a more accurate method described below. It is seen that the ${}_2\mathbf{J}$ transformation achieves the best results. The attainable accuracy for this transformation is limited to about 9 decimal digits by the fact that the stability index displayed in the column D_n of Table 4 grows relatively fast. Note that for $n=46$, the number of digits (as given by

Table 5
Acceleration of $\zeta(-1/10+10\mathrm{i},1,95/100)$ with the $\mathcal{J}$ transformation ($\tau=10$)[a]

n	A_n	B_n	C_n	D_n	E_n	F_n	G_n	H_n
10	2.10e − 05	2.08e + 01	8.17e − 06	3.89e + 01	1.85e − 05	5.10e + 01	1.39e − 05	2.52e + 01
12	2.49e − 06	8.69e + 01	1.43e − 07	3.03e + 02	9.47e − 06	8.98e + 02	1.29e − 06	1.26e + 02
14	1.93e − 07	3.11e + 02	5.98e − 09	1.46e + 03	8.24e − 07	4.24e + 03	6.86e − 08	5.08e + 02
16	1.11e − 08	9.82e + 02	2.02e − 11	6.02e + 03	6.34e − 08	2.09e + 04	2.57e − 09	1.77e + 03
18	5.33e − 10	2.87e + 03	1.57e − 12	2.29e + 04	4.08e − 09	9.52e + 04	7.81e − 11	5.66e + 03
20	2.24e − 11	7.96e + 03	4.15e − 14	8.26e + 04	2.31e − 10	4.12e + 05	2.07e − 12	1.73e + 04
22	8.60e − 13	2.14e + 04	8.13e − 16	2.89e + 05	1.16e − 11	1.73e + 06	4.95e − 14	5.08e + 04
24	3.07e − 14	5.61e + 04	1.67e − 17	9.87e + 05	5.17e − 13	7.07e + 06	1.10e − 15	1.46e + 05
26	1.04e − 15	1.45e + 05	3.38e − 19	3.31e + 06	1.87e − 14	2.84e + 07	2.33e − 17	4.14e + 05
28	3.36e − 17	3.69e + 05	6.40e − 21	1.10e + 07	3.81e − 16	1.13e + 08	4.71e − 19	1.16e + 06
30	1.05e − 18	9.30e + 05	1.15e − 22	3.59e + 07	1.91e − 17	4.43e + 08	9.19e − 21	3.19e + 06

[a] A_n: relative error of ${}_1\mathbf{J}_0^{(n)}(1,\{\{s_{10\,n}\}\},\{\{(10\,n+1)(s_{10\,n}-s_{10\,n-1})\}\})$, B_n: stability index of ${}_1\mathbf{J}_0^{(n)}(1,\{\{s_{10\,n}\}\},\{\{(10\,n+1)(s_{10\,n}-s_{10\,n-1})\}\})$, C_n: relative error of ${}_2\mathbf{J}_0^{(n)}(1,\{\{s_{10\,n}\}\},\{\{(10\,n+1)(s_{10\,n}-s_{10\,n-1})\}\})$, D_n: stability index of ${}_2\mathbf{J}_0^{(n)}(1,\{\{s_{10\,n}\}\},\{\{(10\,n+1)(s_{10\,n}-s_{10\,n-1})\}\})$, E_n: relative error of ${}_3\mathbf{J}_0^{(n)}(1,\{\{s_{10\,n}\}\},\{\{(10\,n+1)(s_{10\,n}-s_{10\,n-1})\}\})$, F_n: stability index of ${}_3\mathbf{J}_0^{(n)}(1,\{\{s_{10\,n}\}\},\{\{(10\,n+1)(s_{10\,n}-s_{10\,n-1})\}\})$, G_n: relative error of $\mathcal{J}_0^{(n)}(\{\{s_{10\,n}\}\},\{\{(10\,n+1)(s_{10\,n}-s_{10\,n-1})\}\},\{1/(10\,n+10)-1/(10\,n+10\,k+10)\})$ H_n: Stability index of H_n: $\mathcal{J}_0^{(n)}(\{\{s_{10\,n}\}\},\{\{(10\,n+1)(s_{10\,n}-s_{10\,n-1})\}\},\{1/(n+1)-1/(n+k+2)\})$.

the negative decadic logarithm of the relative error) and the decadic logarithm of the stability index sum up to approximately 32 which corresponds to the maximal number of decimal digits that could be achieved in the run. Since the stability index increases with n, indicating decreasing stability, it is clear that for higher values of n the accuracy will be lower.

The magnitude of the stability index is largely controlled by the value of ρ, compare Corollary 24. If one can treat a related sequence with a smaller value of ρ, the stability index will be smaller and thus, the stability of the extrapolation will be greater.

Such a related sequence is given by putting $\check{s}_\ell = s_{\xi_\ell}$ for $\ell \in \mathbb{N}_0$, where the sequence ξ_ℓ is a monotonously increasing sequence of nonnegative integers. In the case of linear convergent sequences, the choice $\xi_\ell = \tau\ell$ with $\tau \in \mathbb{N}$ can be used as in the case of the $d^{(1)}$ transformation. It is easily seen that the new sequence also converges linearly with $\rho=\lim_{n\to\infty}(\check{s}_{n+1}-s)/(\check{s}_n-s)=q^\tau$. For $\tau>1$, both the effectiveness and the stability of the various transformations are increased as shown in Table 5 for the case $\tau=10$. Note that this value was chosen to display basic features relevant to the stability analysis, and is not necessarily the optimal value. As in Table 4, the relative errors and the stability indices of some variants of the $\mathcal{J}$ transformation are displayed. These are nothing but the ${}_p\mathbf{J}$ transformation for $p=1,2,3$ and the Levin transformation as applied to the sequence $\{\{\check{s}_n\}\}$ with remainder estimates $\omega_n=(n\tau+\beta)(s_{n\tau}-s_{n\tau-1})$ for $\beta=1$. Since constant factors in the remainder estimates are irrelevant since the $\mathcal{J}$ transformation is invariant under any scaling $\omega_n \to \alpha\omega_n$ for $\alpha \neq 0$, the same results would have been obtained for $\omega_n=(n+\beta/\tau)(s_{n\tau}-s_{n\tau-1})$.

If the Levin transformation is applied to the series with partial sums $\check{s}_n=s_{\tau n}$, and if the remainder estimates $\omega_n=(n+\beta/\tau)(s_{\tau n}-s_{(\tau n)-1})$ are used, then one obtains nothing but the $d^{(1)}$ transformation with $\xi_\ell=\tau\ell$ for $\tau\in\mathbb{N}$ [46,77].

Table 6
Stability indices for the ${}_2\mathbf{J}$ transformation ($\tau = 10$)

n	$\Gamma_n^{(1)}$	$\Gamma_n^{(2)}$	$\Gamma_n^{(3)}$	$\Gamma_n^{(4)}$	$\Gamma_n^{(5)}$	$\Gamma_n^{(6)}$	$\Gamma_n^{(7)}$
20	3.07	9.26	$2.70\,10^1$	$7.55\,10^1$	$2.02\,10^2$	$5.20\,10^2$	$1.29\,10^3$
30	3.54	$1.19\,10^1$	$3.81\,10^1$	$1.16\,10^2$	$3.36\,10^2$	$9.36\,10^2$	$2.51\,10^3$
40	3.75	$1.33\,10^1$	$4.49\,10^1$	$1.44\,10^2$	$4.42\,10^2$	$1.30\,10^3$	$3.71\,10^3$
41	3.77	$1.34\,10^1$	$4.54\,10^1$	$1.46\,10^2$	$4.51\,10^2$	$1.34\,10^3$	$3.82\,10^3$
42	3.78	$1.35\,10^1$	$4.59\,10^1$	$1.49\,10^2$	$4.60\,10^2$	$1.37\,10^3$	$3.93\,10^3$
43	3.79	$1.36\,10^1$	$4.64\,10^1$	$1.51\,10^2$	$4.69\,10^2$	$1.40\,10^3$	$4.05\,10^3$
44	3.80	$1.37\,10^1$	$4.68\,10^1$	$1.53\,10^2$	$4.77\,10^2$	$1.43\,10^3$	
45	3.81	$1.38\,10^1$	$4.73\,10^1$	$1.55\,10^2$	$4.85\,10^2$		
46	3.82	$1.39\,10^1$	$4.77\,10^1$	$1.57\,10^2$			
47	3.83	$1.39\,10^1$	$4.81\,10^1$				
48	3.84	$1.40\,10^1$					
49	3.85						
Extr.	4.01	$1.59\,10^1$	$6.32\,10^1$	$2.52\,10^2$	$1.00\,10^3$	$4.00\,10^3$	$1.59\,10^4$
Corollary 24	3.98	$1.59\,10^1$	$6.32\,10^1$	$2.52\,10^2$	$1.00\,10^3$	$4.00\,10^3$	$1.59\,10^4$

It is seen from Table 5 that again the best accuracy is obtained for the ${}_2\mathbf{J}$ transformation. The $d^{(1)}$ transformation is worse, but better than the ${}_p\mathbf{J}$ transformations for $p = 1$ and 3. Note that the stability indices are now much smaller and do not limit the achievable accuracy for any of the transformations up to $n = 30$. The true value of the series was computed numerically by applying the ${}_2\mathbf{J}$ transformation to the further sequence $\{\{s_{40n}\}\}$ and using 64 decimal digits in the calculation. In this way, a sufficiently accurate approximation was obtained that was used to compute the relative errors in Tables 4 and 5. A comparison value was computed using the representation [20, p. 29, Eq. (8)]

$$\zeta(s,1,q) = \frac{\Gamma(1-s)}{z}(\log 1/q)^{s-1} + z^{-1}\sum_{j=0}^{\infty}\zeta(s-j)\frac{(\log q)^j}{j!} \tag{338}$$

that holds for $|\log q| < 2\pi$ and $s \notin \mathbb{N}$. Here, $\zeta(z)$ denotes the Riemann zeta function. Both values agreed to all relevant decimal digits.

In Table 6, we display stability indices corresponding to the acceleration of $\check{s}_n$ with the ${}_2\mathbf{J}$ transformation columnwise, as obtainable by using the sequence elements up to $\check{s}_{50} = s_{500}$. In the row labelled Corollary 24, we display the limits of the $\Gamma_n^{(k)}$ for large n, i.e., the quantities

$$\lim_{n\to\infty} \Gamma_n^{(k)} = \left(\frac{1+q^\tau}{1-q^\tau}\right)^k, \tag{339}$$

that are the limits according to Corollary 24. It is seen that the values for finite n are still relatively far off the limits. In order to check numerically the validity of the corollary, we extrapolated the values of all $\Gamma_n^{(k)}$ for fixed k with n up to the maximal n for which there is an entry in the corresponding column of Table 6 using the u variant of the ${}_1\mathbf{J}$ transformation. The results of the extrapolation are displayed in the row labelled *Extr* in Table 6 and coincide nearly perfectly with the values expected according to Corollary 24.

Table 7
Extrapolation of series representation (340) of the $F_m(z)$ function using the ${}_2\mathbf{J}$ transformation ($z = 8$, $m = 0$)

n	s_n	${}^u\omega_n$	${}^t\omega_n$	${}^K\omega_n$
5	−13.3	0.3120747	0.3143352	0.3132981
6	−14.7	0.3132882	0.3131147	0.3133070
7	−13.1	0.3132779	0.3133356	0.3133087
8	11.4	0.3133089	0.3133054	0.3133087
9	−8.0	0.3133083	0.3133090	0.3133087

As a final example, we consider the evaluation of the $F_m(z)$ functions that are used in quantum chemistry calculations via the series representation

$$F_m(z) = \sum_{j=0}^{\infty} (-z)^j / j!(2m + 2j + 1) \tag{340}$$

with partial sums

$$s_n = \sum_{j=0}^{n} (-z)^j / j!(2m + 2j + 1). \tag{341}$$

In this case, for larger z, the convergence is rather slow although the convergence finally is hyperlinear. As a K variant, one may use

$${}^k\omega_n = \left(\sum_{j=0}^{n} (-z)^j / (j+1)! - (1 - \mathrm{e}^{-z})/z \right). \tag{342}$$

since $(1 - \mathrm{e}^{-z})/z$ is a Kummer related series. The results for several variants in Table 7 show that the K variant is superior to u and t variants in this case.

Many further numerical examples are given in the literature [39,41–44,50,84].

Appendix A. Stieltjes series and functions

A Stieltjes series is a formal expansion

$$f(z) = \sum_{j=0}^{\infty} (-1)^j \mu_j z^j \tag{A.1}$$

with partial sums

$$f_n(z) = \sum_{j=0}^{n} (-1)^j \mu_j z^j. \tag{A.2}$$

The coefficients μ_n are the moments of an uniquely given positive measure $\psi(t)$ that has infinitely many different values on $0 \leqslant t < \infty$ [4, p. 159]:

$$\mu_n = \int_0^{\infty} t^n \, \mathrm{d}\psi(t), \quad n \in \mathbb{N}_0. \tag{A.3}$$

Formally, the Stieltjes series can be identified with a Stieltjes integral

$$f(z) = \int_0^\infty \frac{\mathrm{d}\psi(t)}{1+zt}, \quad |\arg(z)| < \pi. \tag{A.4}$$

If such an integral exists for a function f then the function is called a Stieltjes function. For every Stieltjes function there exist a unique asymptotical Stieltjes series (A.1), uniformly in every sector $|\arg(z)| < \theta$ for all $\theta < \pi$. For any Stieltjes series, however, several different corresponding Stieltjes functions may exist. To ensure uniqueness, additional criteria are necessary [88, Section 4.3].

In the context of convergence acceleration and summation of divergent series, it is important that for given z the tails $f(z) - f_n(z)$ of a Stieltjes series are bounded in absolute value by the next term of the series,

$$|f(z) - f_n(z)| \leqslant \mu_{n+1} z^{n+1} \quad z \geqslant 0. \tag{A.5}$$

Hence, for Stieltjes series the remainder estimates may be chosen as

$$\omega_n = (-1)^{n+1} \mu_{n+1} z^{n+1}. \tag{A.6}$$

This corresponds to $\omega_n = \Delta f_n(z)$, i.e., to a $\tilde{t}$ variant.

Appendix B. Derivation of the recursive scheme (148)

We show that for the divided difference operator $\square_n^{(k)} = \square_n^{(k)}[\{\{x_n\}\}]$ the identity

$$\square_n^{(k+1)}((x)_{\ell+1} g(x)) = \frac{(x_{n+k+1} + \ell)\square_{n+1}^{(k)}((x)_\ell g(x)) - (x_n + \ell)\square_n^{(k)}((x)_\ell g(x))}{x_{n+k+1} - x_n} \tag{B.1}$$

holds. The proof is based on the Leibniz formula for divided differences (see, e.g., [69, p. 50]) that yields upon use of $(x)_{\ell+1} = (x+\ell)(x)_\ell$ and $\square_n^{(k)}(x) = x_n \delta_{k,0} + \delta_{k,1}$

$$\begin{aligned} \square_n^{(k+1)}((x)_{\ell+1} g(x)) &= \ell \square_n^{(k+1)}((x)_\ell g(x)) + \sum_{j=0}^{k+1} \square_n^{(j)}(x) \square_{n+j}^{(k+1-j)}((x)_\ell g(x)) \\ &= (x_n + \ell)\square_n^{(k+1)}((x)_\ell g(x)) + \square_{n+1}^{(k)}((x)_\ell g(x)). \end{aligned} \tag{B.2}$$

Using the recursion relation of the divided differences, one obtains

$$\square_n^{(k+1)}((x)_{\ell+1} g(x)) = (x_n + \ell)\frac{\square_{n+1}^{(k)}((x)_\ell g(x)) - \square_n^{(k)}((x)_\ell g(x))}{x_{n+k+1} - x_n} + \square_{n+1}^{(k)}((x)_\ell g(x)). \tag{B.3}$$

Simple algebra then yields Eq. (B.1).

Comparison with Eq. (140) shows that using the interpolation conditions $g_n = g(x_n) = s_n/\omega_n$ and $\ell = k-1$ in Eq. (B.1) yields the recursion for the numerators in Eq. (148), while the recursion for the denominators in Eq. (148) follows for $\ell = k-1$ and using the interpolation conditions $g_n = g(x_n) = 1/\omega_n$. In each case, the initial conditions follow directly from Eq. (140) in combination with the definition of the divided difference operator: For $k=0$, we use $(a)_{-1} = 1/(a-1)$ and obtain $\square_n^{(k)}(x_n)_{k-1} g_n = (x_n)_{-1} g_n = g_n/(x_n - 1)$.

Appendix C. Two lemmata

Lemma C.1. *Define*

$$A = \sum_{j=0}^{k} \overset{\circ}{\lambda}{}_j^{(k)} \frac{\zeta^{n+j}}{(n+j)^{r+1}}, \tag{C.1}$$

where ζ *is a zero of multiplicity* m *of* $\overset{\circ}{\Pi}{}^{(k)}(z) = \sum_{j=0}^{k} \overset{\circ}{\lambda}{}_j^{(k)} z^j$. *Then*

$$A \sim \zeta^{n+m} \binom{r+m}{r} \frac{(-1)^m}{n^{r+m+1}} \frac{\mathrm{d}^m \overset{\circ}{\Pi}{}^{(k)}}{\mathrm{d}x^m}(\zeta) \quad (n \to \infty). \tag{C.2}$$

Proof. Use

$$\frac{1}{a^{r+1}} = \frac{1}{r!} \int_0^\infty \exp(-at) t^r \, \mathrm{d}t, \quad a > 0 \tag{C.3}$$

to obtain

$$A = \frac{1}{r!} \int_0^\infty \sum_{j=0}^{k} \overset{\circ}{\lambda}{}_j^{(k)} \zeta^{n+j} \exp(-(n+j)t) t^r \, \mathrm{d}t = \frac{\zeta^n}{r!} \int_0^\infty \exp(-nt)\, \overset{\circ}{\Pi}{}^{(k)}(\zeta \exp(-t)) t^r \, \mathrm{d}t. \tag{C.4}$$

Taylor expansion of the polynomial yields due to the zero at ζ

$$\overset{\circ}{\Pi}{}^{(k)}(\zeta \exp(-t)) = \frac{(-\zeta)^m}{m!} \left. \frac{\mathrm{d}^m \overset{\circ}{\Pi}{}^{(k)}(x)}{\mathrm{d}x^m} \right|_{x=\zeta} t^m (1 + \mathrm{O}(t)). \tag{C.5}$$

Invoking Watson's lemma [6, p. 263ff] completes the proof. □

Lemma C.2. *Assume that assumption* (C-3′) *of Theorem* 13 *holds. Further assume* $\lambda_{n,j}^{(k)} \to \overset{\circ}{\lambda}{}_j^{(k)}$ *for* $n \to \infty$. *Then, Eq.* (280) *holds.*

Proof. We have

$$\frac{\omega_{n+j}}{\omega_n} \sim \rho^j \exp\left(\varepsilon_n \sum_{t=0}^{j-1} \frac{\varepsilon_{n+t}}{\varepsilon_n} \right) \sim \rho^j \exp(j\varepsilon_n) \tag{C.6}$$

for large n. Hence,

$$\sum_{j=0}^{k} \lambda_{n,j}^{(k)} \frac{\omega_n}{\omega_{n+j}} \sim \sum_{j=0}^{k} \overset{\circ}{\lambda}{}_j^{(k)} (\rho \exp(\varepsilon_n))^{-j} = \overset{\circ}{\Pi}{}^{(k)}(1/\rho + \delta_n) \tag{C.7}$$

Since the characteristic polynomial $\overset{\circ}{\Pi}{}^{(k)}(z)$ has a zero of order μ at $z = 1/\rho$ according to the assumptions, Eq. (280) follows using Taylor expansion. □

References

[1] M. Abramowitz, I. Stegun, Handbook of Mathematical Functions, Dover Publications, New York, 1970.

[2] A.C. Aitken, On Bernoulli's numerical solution of algebraic equations, Proc. Roy. Soc. Edinburgh 46 (1926) 289–305.

[3] G.A. Baker Jr., Essentials of Padé Approximants, Academic Press, New York, 1975.

[4] G.A. Baker Jr., P. Graves-Morris, Padé Approximants, Part I: Basic Theory, Addison-Wesley, Reading, MA, 1981.

[5] G.A. Baker Jr., P. Graves-Morris, Padé Approximants, 2nd Edition, Cambridge Unievrsity Press, Cambridge, GB, 1996.

[6] C.M. Bender, S.A. Orszag, Advanced Mathematical Methods for Scientists and Engineers, McGraw-Hill, Singapore, 1987.

[7] C. Brezinski, Accélération de la Convergence en Analyse Numérique, Springer, Berlin, 1977.

[8] C. Brezinski, Algorithmes d'Accélération de la Convergence – Étude Numérique, Éditions Technip, Paris, 1978.

[9] C. Brezinski, A general extrapolation algorithm, Numer. Math. 35 (1980) 175–180.

[10] C. Brezinski, Padé-Type Approximation and General Orthogonal Polynomials, Birkhäuser, Basel, 1980.

[11] C. Brezinski, A Bibliography on Continued Fractions, Padé Approximation, Extrapolation and Related Subjects, Prensas Universitarias de Zaragoza, Zaragoza, 1991.

[12] C. Brezinski (Ed.), Continued Fractions and Padé Approximants, North-Holland, Amsterdam, 1991.

[13] C. Brezinski, A.C. Matos, A derivation of extrapolation algorithms based on error estimates, J. Comput. Appl. Math. 66 (1–2) (1996) 5–26.

[14] C. Brezinski, M. Redivo Zaglia, Extrapolation Methods, Theory and Practice, North-Holland, Amsterdam, 1991.

[15] C. Brezinski, M. Redivo Zaglia, A general extrapolation algorithm revisited, Adv. Comput. Math. 2 (1994) 461–477.

[16] J. Cioslowski, E.J. Weniger, Bulk properties from finite cluster calculations, VIII Benchmark calculations on the efficiency of extrapolation methods for the HF and MP2 energies of polyacenes, J. Comput. Chem. 14 (1993) 1468–1481.

[17] J. Čížek, F. Vinette, E.J. Weniger, Examples on the use of symbolic computation in physics and chemistry: applications of the inner projection technique and of a new summation method for divergent series, Int. J. Quantum Chem. Symp. 25 (1991) 209–223.

[18] J. Čížek, F. Vinette, E.J. Weniger, On the use of the symbolic language Maple in physics and chemistry: several examples, in: de Groot R.A, Nadrchal J. (Eds.), Proceedings of the Fourth International Conference on Computational Physics PHYSICS COMPUTING '92, World Scientific, Singapore, 1993, pp. 31–44.

[19] J.E. Drummond, A formula for accelerating the convergence of a general series, Bull. Austral. Math. Soc. 6 (1972) 69–74.

[20] A. Erdélyi, W. Magnus, F. Oberhettinger, F.G. Tricomi, Higher Transcendental Functions, Vol. I, McGraw-Hill, New York, 1953.

[21] T. Fessler, W.F. Ford, D.A. Smith, HURRY: an acceleration algorithm for scalar sequences and series, ACM Trans. Math. Software 9 (1983) 346–354.

[22] W.F. Ford, A. Sidi, An algorithm for a generalization of the Richardson extrapolation process, SIAM J. Numer. Anal. 24 (5) (1987) 1212–1232.

[23] B. Germain-Bonne, Transformations de suites, Rev. Française Automat. Inform. Rech. Oper. 7 (R-1) (1973) 84–90.

[24] P.R. Graves-Morris (Ed.), Padé Approximants, The Institute of Physics, London, 1972.

[25] P.R. Graves-Morris (Ed.), Padé Approximants and their Applications, Academic Press, London, 1973.

[26] P.R. Graves-Morris, E.B. Saff, R.S. Varga (Ed.), Rational Approximation and Interpolation, Springer, Berlin, 1984.

[27] J. Grotendorst, A Maple package for transforming series, sequences and functions, Comput. Phys. Comm. 67 (1991) 325–342.

[28] J. Grotendorst, E.O. Steinborn, Use of nonlinear convergence accelerators for the efficient evaluation of GTO molecular integrals, J. Chem. Phys. 84 (1986) 5617–5623.

[29] J. Grotendorst, E.J. Weniger, E.O. Steinborn, Efficient evaluation of infinite-series representations for overlap, two-center nuclear attraction and Coulomb integrals using nonlinear convergence accelerators, Phys. Rev. A 33 (1986) 3706–3726.

[30] E.R. Hansen, A Table of Series and Products, Prentice-Hall, Englewood-Cliffs, NJ, 1975.

[31] T. Hasegawa, A. Sidi, An automatic integration procedure for infinite range integrals involving oscillatory kernels, Numer. Algorithms 13 (1996) 1–19.
[32] T. Håvie, Generalized Neville type extrapolation schemes, BIT 19 (1979) 204–213.
[33] H.H.H. Homeier, Integraltransformationsmethoden und Quadraturverfahren für Molekülintegrale mit B-Funktionen, Theorie und Forschung, Vol. 121, S. Roderer Verlag, Regensburg, 1990, also Doctoral dissertation, Universität Regensburg.
[34] H.H.H. Homeier, A Levin-type algorithm for accelerating the convergence of Fourier series, Numer. Algorithms 3 (1992) 245–254.
[35] H.H.H. Homeier, Some applications of nonlinear convergence accelerators, Int. J. Quantum Chem. 45 (1993) 545–562.
[36] H.H.H. Homeier, A hierarchically consistent, iterative sequence transformation, Numer. Algorithms 8 (1994) 47–81.
[37] H.H.H. Homeier, Nonlinear convergence acceleration for orthogonal series, in: R. Gruber, M. Tomassini (Eds.), Proceedings of the sixth Joint EPS–APS International Conference on Physics Computing, Physics Computing '94, European Physical Society, Boite Postale 69, CH-1213 Petit-Lancy, Genf, Schweiz, 1994, pp. 47–50.
[38] H.H.H. Homeier, Determinantal representations for the $\mathcal{J}$ transformation, Numer. Math. 71 (3) (1995) 275–288.
[39] H.H.H. Homeier, Analytical and numerical studies of the convergence behavior of the $\mathcal{J}$ transformation, J. Comput. Appl. Math. 69 (1996) 81–112.
[40] H.H.H. Homeier, Extrapolationsverfahren für Zahlen-, Vektor- und Matrizenfolgen und ihre Anwendung in der Theoretischen und Physikalischen Chemie, Habilitation Thesis, Universität Regensburg, 1996.
[41] H.H.H. Homeier, Extended complex series methods for the convergence acceleration of Fourier series, Technical Report TC-NA-97-3, Institut für Physikalische und Theoretische Chemie, Universität Regensburg, D-93040 Regensburg, 1997.
[42] H.H.H. Homeier, On an extension of the complex series method for the convergence acceleration of orthogonal expansions. Technical Report TC-NA-97-4, Institut für Physikalische und Theoretische Chemie, Universität Regensburg, D-93040 Regensburg, 1997.
[43] H.H.H. Homeier., On propertiesand the application of Levin-type sequence transformations for the convergence acceleration of Fourier series, Technical Report TC-NA-97-1, Institut für Physikalische und Theoretische Chemie, Universität Regensburg, D-93040 Regensburg, 1997.
[44] H.H.H. Homeier, An asymptotically hierarchy-consistent iterative sequence transformation for convergence acceleration of Fourier series, Numer. Algorithms 18 (1998) 1–30.
[45] H.H.H. Homeier, On convergence acceleration of multipolar and orthogonal expansions, Internet J. Chem. 1 (28) (1998), online computer file: URL: http://www.ijc.com/articles/1998v1/28/, Proceedings of the Fourth Electronic Computational Chemistry Conference.
[46] H.H.H. Homeier, On the stability of the $\mathcal{J}$ transformation, Numer. Algorithms 17 (1998) 223–239.
[47] H.H.H. Homeier, Convergence acceleration of logarithmically convergent series avoiding summation, Appl. Math. Lett. 12 (1999) 29–32.
[48] H.H.H. Homeier, Transforming logarithmic to linear convergence by interpolation, Appl. Math. Lett. 12 (1999) 13–17.
[49] H.H.H. Homeier, B. Dick, Zur Berechnung der Linienform spektraler Löcher (Engl.: on the computation of the line shape of spectral holes), Technical Report TC-PC-95-1, Institut für Physikalische und Theoretische Chemie, Universität Regensburg, D-93040 Regensburg, 1995, Poster CP 6.15, 59. Physikertagung Berlin 1995, Abstract: Verhandlungen der Deutschen Physikalischen Gesellschaft, Reihe VI, Band 30, 1815 (Physik-Verlag GmbH, D-69469 Weinheim, 1995).
[50] H.H.H. Homeier, E.J. Weniger, On remainder estimates for Levin-type sequence transformations, Comput. Phys. Comm. 92 (1995) 1–10.
[51] U. Jentschura, P.J. Mohr, G. Soff, E.J. Weniger, Convergence acceleration via combined nonlinear-condensation transformations, Comput. Phys. Comm. 116 (1999) 28–54.
[52] A.N. Khovanskii, The Application of Continued Fractions and their Generalizations to Problems in Approximation Theory, Noordhoff, Groningen, 1963.
[53] D. Levin, Development of non-linear transformations for improving convergence of sequences, Int. J. Comput. Math. B 3 (1973) 371–388.

[54] D. Levin, A. Sidi, Two new classes of nonlinear transformations for accelerating the convergence of infinite integrals and series, Appl. Math. Comput. 9 (1981) 175–215.
[55] I.M. Longman, Difficulties of convergence acceleration, in: M.G. de Bruin, H. van Rossum (Eds.), Padé Approximation and its Applications Amsterdam 1980, Springer, Berlin, 1981, pp. 273–289.
[56] L. Lorentzen, H. Waadeland, Continued Fractions with Applications, North-Holland, Amsterdam, 1992.
[57] S.K. Lucas, H.A. Stone, Evaluating infinite integrals involving Bessel functions of arbitrary order, J. Comput. Appl. Math. 64 (1995) 217–231.
[58] W. Magnus, F. Oberhettinger, R.P. Soni, Formulas and Theorems for the Special Functions of Mathematical Physics, Springer, New York, 1966.
[59] A.C. Matos, Linear difference operators and acceleration methods, Publication ANO-370, Laboratoire d'Analyse Numérique et d'Optimisation, Université des Sciences et Technologies de Lille, France, 1997, IMA J. Numer. Anal. (2000), to appear.
[60] K.A. Michalski, Extrapolation methods for Sommerfeld integral tails, IEEE Trans. Antennas Propagation 46 (10) (1998) 1405–1418.
[61] J.R. Mosig, Integral equation technique, in: T. Itoh (Ed.), Numerical Techniques for Microwave and Millimeter-Wave Passive Structures, Wiley, New York, 1989, pp. 133–213.
[62] E.M. Nikishin, V.N. Sorokin, Rational Approximations and Orthogonality, American Mathematical Society, Providence, RI, 1991.
[63] C. Oleksy, A convergence acceleration method of Fourier series, Comput. Phys. Comm. 96 (1996) 17–26.
[64] K.J. Overholt, Extended Aitken acceleration, BIT 5 (1965) 122–132.
[65] P.J. Pelzl, F.W. King, Convergence accelerator approach for the high-precision evaluation of three-electron correlated integrals, Phys. Rev. E 57 (6) (1998) 7268–7273.
[66] P.P. Petrushev, V.A. Popov, Rational Approximation of Real Functions, Cambridge University Press, Cambridge, 1987.
[67] B. Ross, Methods of Summation, Descartes Press, Koriyama, 1987.
[68] E.B. Saff, R.S. Varga (Eds.), Padé and Rational Approximation, Academic Press, New York, 1977.
[69] L. Schumaker, Spline Functions: Basic Theory, Wiley, New York, 1981.
[70] D. Shanks, Non-linear transformations of divergent and slowly convergent sequences, J. Math. Phys. (Cambridge, MA) 34 (1955) 1–42.
[71] A. Sidi, Convergence properties of some nonlinear sequence transformations, Math. Comp. 33 (1979) 315–326.
[72] A. Sidi, Some properties of a generalization of the Richardson extrapolation process, J. Inst. Math. Appl. 24 (1979) 327–346.
[73] A. Sidi, An algorithm for a special case of a generalization of the Richardson extrapolation process, Numer. Math. 38 (1982) 299–307.
[74] A. Sidi, Generalization of Richardson extrapolation with application to numerical integration, in: H. Brass, G. Hämmerlin (Eds.), Numerical Integration, Vol. III, Birkhäuser, Basel, 1988, pp. 237–250.
[75] A. Sidi, A user-friendly extrapolation method for oscillatory infinite integrals, Math. Comp. 51 (1988) 249–266.
[76] A. Sidi, On a generalization of the Richardson extrapolation process, Numer. Math. 47 (1990) 365–377.
[77] A. Sidi, Acceleration of convergence of (generalized) Fourier series by the d-transformation, Ann. Numer. Math. 2 (1995) 381–406.
[78] A. Sidi, Convergence analysis for a generalized Richardson extrapolation process with an application to the $d^{(1)}$ transformation on convergent and divergent logarithmic sequences, Math. Comp. 64 (212) (1995) 1627–1657.
[79] D.A. Smith, W.F. Ford, Acceleration of linear and logarithmic convergence, SIAM J. Numer. Anal. 16 (1979) 223–240.
[80] D.A. Smith, W.F. Ford, Numerical comparisons of nonlinear convergence accelerators, Math. Comp. 38 (158) (1982) 481–499.
[81] E.O. Steinborn, H.H.H. Homeier, J. Fernández Rico, I. Ema, R. López, G. Ramírez, An improved program for molecular calculations with B functions, J. Mol. Struct. (Theochem) 490 (1999) 201–217.
[82] E.O. Steinborn, E.J. Weniger, Sequence transformations for the efficient evaluation of infinite series representations of some molecular integrals with exponentially decaying basis functions, J. Mol. Struct. (Theochem) 210 (1990) 71–78.
[83] H.S. Wall, Analytic Theory of Continued Fractions, Chelsea, New York, 1973.

[84] E.J. Weniger, Nonlinear sequence transformations for the acceleration of convergence and the summation of divergent series, Comput. Phys. Rep. 10 (1989) 189–371.
[85] E.J. Weniger, On the summation of some divergent hypergeometric series and related perturbation expansions, J. Comput. Appl. Math. 32 (1990) 291–300.
[86] E.J. Weniger, On the derivation of iterated sequence transformations for the acceleration of convergence and the summation of divergent series, Comput. Phys. Comm. 64 (1991) 19–45.
[87] E.J. Weniger, Interpolation between sequence transformations, Numer. Algorithms 3 (1992) 477–486.
[88] E.J. Weniger, Verallgemeinerte Summationsprozesse als numerische Hilfsmittel für quantenmechanische und quantenchemische Rechnungen, Habilitationsschrift, Universität Regensburg, 1994.
[89] E.J. Weniger, Computation of the Whittaker function of the second kind by summing its divergent asymptotic series with the help of nonlinear sequence transformations, Comput. Phys. 10 (5) (1996) 496–503.
[90] E.J. Weniger, Construction of the strong coupling expansion for the ground state energy of the quartic, sextic, and octic anharmonic oscillator via a renormalized strong coupling expansion, Phys. Rev. Lett. 77 (14) (1996) 2859–2862.
[91] E.J. Weniger, A convergent renormalized strong coupling perturbation expansion for the ground state energy of the quartic, sextic, and octic anharmonic oscillator, Ann. Phys. 246 (1) (1996) 133–165.
[92] E.J. Weniger, Erratum: nonlinear sequence transformations: a computational tool for quantum mechanical and quantum chemical calculations, Int. J. Quantum Chem. 58 (1996) 319–321.
[93] E.J. Weniger, Nonlinear sequence transformations: a computational tool for quantum mechanical and quantum chemical calculations, Int. J. Quantum Chem. 57 (1996) 265–280.
[94] E.J. Weniger, J. Čížek, Rational approximations for the modified Bessel function of the second kind, Comput. Phys. Comm. 59 (1990) 471–493.
[95] E.J. Weniger, J. Čížek, F. Vinette, Very accurate summation for the infinite coupling limit of the perturbation series expansions of anharmonic oscillators, Phys. Lett. A 156 (1991) 169–174.
[96] E.J. Weniger, J. Čížek, F. Vinette, The summation of the ordinary and renormalized perturbation series for the ground state energy of the quartic, sextic and octic anharmonic oscillators using nonlinear sequence transformations, J. Math. Phys. 34 (1993) 571–609.
[97] E.J. Weniger, C.-M. Liegener, Extrapolation of finite cluster and crystal-orbital calculations on trans-polyacetylene, Int. J. Quantum Chem. 38 (1990) 55–74.
[98] E.J. Weniger, E.O. Steinborn, Comment on "molecular overlap integrals with exponential-type integrals", J. Chem. Phys. 87 (1987) 3709–3711.
[99] E.J. Weniger, E.O. Steinborn, Overlap integrals of B functions, A numerical study of infinite series representations and integral representations, Theor. Chim. Acta 73 (1988) 323–336.
[100] E.J. Weniger, E.O. Steinborn, Nonlinear sequence transformations for the efficient evaluation of auxiliary functions for GTO molecular integrals, in: M. Defranceschi, J. Delhalle (Eds.), Numerical Determination of the Electronic Structure of Atoms, Diatomic and Polyatomic Molecules, Dordrecht, Kluwer, 1989, pp. 341–346.
[101] H. Werner, H.J. Bünger (Eds.), Padé Approximations and its Applications, Bad Honnef 1983, Springer, Berlin, 1984.
[102] J. Wimp, Sequence Transformations and their Applications, Academic Press, New York, 1981.
[103] L. Wuytack (Ed.), Padé Approximations and its Applications, Springer, Berlin, 1979.
[104] P. Wynn, On a device for computing the $e_m(S_n)$ transformation, Math. Tables Aids Comput. 10 (1956) 91–96.

ELSEVIER

Journal of Computational and Applied Mathematics 122 (2000) 149–165

JOURNAL OF COMPUTATIONAL AND APPLIED MATHEMATICS

www.elsevier.nl/locate/cam

Vector extrapolation methods. Applications and numerical comparison

K. Jbilou *, H. Sadok

Université du Littoral, Zone Universitaire de la Mi-voix, Batiment H. Poincaré, 50 rue F. Buisson, BP 699, F-62228 Calais Cedex, France

Received 24 November 1999; received in revised form 15 February 2000

Abstract

The present paper is a survey of the most popular vector extrapolation methods such as the reduced rank extrapolation (RRE), the minimal polynomial extrapolation (MPE), the modified minimal polynomial extrapolation (MMPE), the vector ε-algorithm (VEA) and the topological ε-algorithm (TEA). Using projectors, we derive a different interpretation of these methods and give some theoretical results. The second aim of this work is to give a numerical comparison of the vector extrapolation methods above when they are used for practical large problems such as linear and nonlinear systems of equations. © 2000 Elsevier Science B.V. All rights reserved.

Keywords: Linear systems; Nonlinear systems; Extrapolation; Projection; Vector sequences; Minimal polynomial; Epsilon-algorithm

1. Introduction

In the last decade, many iterative methods for solving large and sparse nonsymmetric linear systems of equations have been developed. The extensions of these methods to nonlinear systems have been considered. As the classical iteration processes may converge slowly, extrapolation methods are required. The aim of vector extrapolation methods is to transform a sequence of vectors generated by some process to a new one with the goal to converge faster than the initial sequence. The most popular vector extrapolation methods can be classified into two categories: the polynomial methods and the ε-algorithms. The first family contains the minimal polynomial extrapolation (MPE) method of Cabay and Jackson [8], the reduced rank extrapolation (RRE) method of Eddy [9] and Mesina [24] and the modified minimal polynomial extrapolation (MMPE) method of Sidi et al. [35], Brezinski [3] and Pugachev [25]. The second class includes the topological ε-algorithm (TEA) of

* Corresponding author.
E-mail addresses: jbilou@lma.univ-littoral.fr (K. Jbilou), sadok@lma.univ-littoral.fr (H. Sadok).

0377-0427/00/$ - see front matter
PII: S0377-0427(00)00357-5

Brezinski [3] and the scalar and vector ε-algorithms (SEA and VEA) of Wynn [39,40]. Some convergence results and properties of these methods were given in [3,16,18,28,30,33–36].

Different recursive algorithms for implementing these methods were also proposed in [5,15,10,39,40]. However, in practice and for large problems, these algorithms become very unstable and are not recommended. When solving large linear and nonlinear systems, Sidi [32] gives a more stable implementation of the RRE and MPE methods using a QR decomposition while Jbilou and Sadok [19] developed an LU-implementation of the MMPE method. These techniques require low storage and work and are more stable numerically.

When applied to linearly generated vector sequences, the MPE, the RRE and the TEA methods are mathematically related to some known Krylov subspace methods. It was shown in [34] that these methods are equivalent to the method of Arnoldi [26], the generalized minimal residual method (GMRES) [27] and the method of Lanczos [21], respectively. The MMPE method is mathematically equivalent to Hessenberg method [30] and [38]. For linear problems, some numerical comparisons have been given in [11].

We note also that, when the considered sequence is not generated linearly, these extrapolation methods are still projection methods but not necessarily Krylov subspace methods [20].

An important property of the vector extrapolation methods above is that they could be applied directly to the solution of linear and nonlinear systems. This comes out from the fact that the definitions of these methods do not require an explicit knowledge of how the sequence is generated. Hence, these vector extrapolation methods are more effective for nonlinear problems [29].

For nonlinear problems, these methods do not need the use of the Jacobian of the function and have the property of quadratic convergence under some assumptions [17]. Note that for some nonlinear problems, vector extrapolation methods such as nonlinear Newton–Krylov methods fail to converge if the initial guess is "away" from a solution. In this case, some techniques such as the linear search backtracing procedure could be added to the basic algorithms; see [2].

The paper is organized as follows. In Section 2, we introduce the polynomial extrapolation methods (RRE, MPE and MMPE) by using the generalized residual. We will also see how these methods could be applied for solving linear and nonlinear systems of equations. In this case some theoretical results are given. Section 3 is devoted to the epsilon-algorithm's family (SEA, VEA and TEA). In Section 4, we give the computational steps and storage required for these methods. Some numerical experiments are given in Section 5 and a comparison with the vector extrapolation methods cited above.

In this paper, we denote by $(.,.)$ the Euclidean inner product in $\mathbb{R}^N$ and by $\|.\|$ the corresponding norm. For an $N \times N$ matrix A and a vector v of $\mathbb{R}^N$ the Krylov subspace $K_k(A, v)$ is the subspace generated by the vectors $v, Av, \ldots, A^{k-1}v$. I_N is the unit matrix and the Kronecker product $\otimes$ is defined by $C \otimes B = [c_{i,j}\, B]$ where B and C are two matrices.

2. The polynomial methods

2.1. Definitions of the RRE, MPE and MMPE methods

Let (s_n) be a sequence of vectors of $\mathbb{R}^N$ and consider the transformation T_k defined by

$$T_k : \mathbb{R}^N \to \mathbb{R}^N,$$
$$s_n \to t_k^{(n)}$$

with

$$t_k^{(n)} = s_n + \sum_{i=1}^{k} a_i^{(n)} g_i(n), \quad n \geqslant 0, \tag{2.1}$$

where the auxiliary vector sequences $(g_i(n))_n$; $i=1,\ldots,k$, are given. The coefficients $a_i^{(n)}$ are scalars.

Let $\tilde{T}_k$ denote the new transformation obtained from T_k by

$$\tilde{t}_k^{(n)} = s_{n+1} + \sum_{i=1}^{k} a_i^{(n)} g_i(n+1), \quad n \geqslant 0. \tag{2.2}$$

For these extrapolation methods, the auxiliary sequences are such that $g_i(n) = \Delta s_{n+i-1}$, $i = 1,\ldots,k$; $n \geqslant 0$, and the coefficients $a_i^{(n)}$ are the same in the two expressions (2.1) and (2.2).

We define the generalized residual of $t_k^{(n)}$ by

$$\begin{aligned} \tilde{r}(t_k^{(n)}) &= \tilde{t}_k^{(n)} - t_k^{(n)} \\ &= \Delta s_n + \sum_{i=1}^{k} a_i^{(n)} \Delta g_i(n). \end{aligned} \tag{2.3}$$

The forward difference operator Δ acts on the index n, i.e., $\Delta g_i(n) = g_i(n+1) - g_i(n)$, $i=1,\ldots,k$.

We will see later that, when solving linear systems of equations, the sequence $(s_n)_n$ is generated by a linear process and then the generalized residual coincides with the classical residual.

The coefficients $a_i^{(n)}$ involved in expression (2.1) are obtained from the orthogonality relation

$$\tilde{r}(t_k^{(n)}) \perp \operatorname{span}\{y_1^{(n)},\ldots,y_k^{(n)}\}, \tag{2.4}$$

where $y_i^{(n)} = \Delta s_{n+i-1}$ for the MPE; $y_i^{(n)} = \Delta^2 s_{n+i-1}$ for the RRE and $y_i^{(n)} = y_i$ for the MMPE where $y_1,\ldots,y_k$ are arbitrary linearly independent vectors of $\mathbb{R}^N$.

Now, if $\tilde{W}_{k,n}$ and $\tilde{L}_{k,n}$ denote the subspaces $\tilde{W}_{k,n} = \operatorname{span}\{\Delta^2 s_n,\ldots,\Delta^2 s_{n+k-1}\}$ and $\tilde{L}_{k,n} = \operatorname{span}\{y_1^{(n)},\ldots,y_k^{(n)}\}$, then from (2.3) and (2.4), the generalized residuals satisfies

$$\tilde{r}(t_k^{(n)}) - \Delta s_n \in \tilde{W}_{k,n} \tag{2.5}$$

and

$$\tilde{r}(t_k^{(n)}) \perp \tilde{L}_{k,n}. \tag{2.6}$$

Conditions (2.5) and (2.6) show that the generalized residual $\tilde{r}(t_k^{(n)})$ is obtained by projecting, the vector Δs_n onto the subspace $\tilde{W}_{k,n}$, orthogonally to $\tilde{L}_{k,n}$.

In a matrix form, $\tilde{r}(t_k^{(n)})$ can be written as

$$\tilde{r}(t_k^{(n)}) = \Delta s_n - \Delta^2 S_{k,n}(L_{k,n}^{\mathrm{T}} \Delta^2 S_{k,n})^{-1} L_{k,n}^{\mathrm{T}} \Delta s_n, \tag{2.7}$$

where $L_{k,n}$, $\Delta S_{k,n}$ and $\Delta^2 S_{k,n}$ are the $k \times k$ matrices whose columns are $y_1^{(n)},\ldots,y_k^{(n)}$, $\Delta s_n,\ldots,\Delta s_{n+k-1}$ and $\Delta^2 s_n,\ldots,\Delta^2 s_{n+k-1}$ respectively. Note that $\tilde{r}(t_k^{(n)})$ is well defined if and only if the $k \times k$ matrix

$L_{k,n}^{\mathrm{T}}\Delta^2 S_{k,n}$ is nonsingular; a necessary condition for this is that the matrices $L_{k,n}$ and $\Delta^2 S_{k,n}$ are full rank. In this case, $t_k^{(n)}$ exists and is uniquely given by

$$t_k^{(n)} = s_n - \Delta S_{k,n}(L_{k,n}^{\mathrm{T}}\Delta^2 S_{k,n})^{-1}L_{k,n}^{\mathrm{T}}\Delta s_n. \tag{2.8}$$

The approximation $t_k^{(n)}$ can also be expressed as

$$t_k^{(n)} = \sum_{j=0}^{k} \beta_j^{(n)} s_{n+j}$$

with

$$\sum_{i=0}^{k} \beta_j^{(n)} = 1$$

and

$$\sum_{j=0}^{k} \alpha_{i,j}^{(n)} \beta_j^{[n)} = 0, \quad j = 0,\ldots,k-1,$$

where the coefficients $\alpha_{i,j}^{(n)}$ are defined by

$$\alpha_{i,j}^{(n)} = (\Delta s_{n+i}, \Delta s_{n+j}) \quad \text{for the MPE method,}$$
$$\alpha_{i,j}^{(n)} = (\Delta^2 s_{n+i}, \Delta s_{n+j}) \quad \text{for the RRE method,}$$
$$\alpha_{i,j}^{(n)} = (y_{i+1}, \Delta s_{n+j}) \quad \text{for the MPE method, } i = 0,\ldots,k-1 \text{ and } j = 0,\ldots,k.$$

From these relations it is not difficult to see that $t_k^{(n)}$ can also be written as a ratio of two determinants as follows:

$$t_k^{(n)} = \left.\begin{vmatrix} s_n & s_{n+1} & \cdots & s_{n+k} \\ \alpha_{0,0}^{(n)} & \alpha_{0,1}^{(n)} & \cdots & \alpha_{0,k}^{(n)} \\ \vdots & \vdots & & \vdots \\ \alpha_{k-1,0}^{(n)} & \alpha_{k-1,1}^{(n)} & \cdots & \alpha_{k-1,k}^{(n)} \end{vmatrix} \middle/ \begin{vmatrix} 1 & 1 & \cdots & 1 \\ \alpha_{0,0}^{(n)} & \alpha_{0,1}^{(n)} & \cdots & \alpha_{0,k}^{(n)} \\ \vdots & \vdots & & \vdots \\ \alpha_{k-1,0}^{(n)} & \alpha_{k-1,1}^{(n)} & \cdots & \alpha_{k-1,k}^{(n)} \end{vmatrix}\right. . \tag{2.9}$$

The determinant in the numerator of (2.9) is the vector obtained by expanding this determinant with respect to its first row by the classical rule.

Note that the determinant in the denominator of (2.9) is equal to $\det(L_{k,n}^{\mathrm{T}}\Delta^2 S_{k,n})$ which is assumed to be nonzero. The computation of the approximation $t_k^{(n)}$ needs the values of the terms $s_n, s_{n+1},\ldots,s_{n+k+1}$.

2.2. Application to linear systems

Consider the system of linear equations

$$Cx = f, \tag{2.10}$$

where C is a real nonsingular $N \times N$ matrix, f is a vector of $\mathbb{R}^N$ and x^* denotes the unique solution.

Instead of applying the extrapolation methods for solving (2.10), we will use them for the preconditioned linear system

$$M^{-1}Cx = M^{-1}f, \tag{2.11}$$

where M is a nonsingular matrix.

Starting from an initial vector s_0, we construct the sequence $(s_j)_j$ by

$$s_{j+1} = Bs_j + b; \quad j = 0, 1, \ldots \tag{2.12}$$

with $B = I - A$; $A = M^{-1}C$ and $b = M^{-1}f$.

Note that if the sequence (s_j) is convergent, its limit $s = x^*$ is the solution of the linear system (2.10).

From (2.12) we have $\Delta s_j = b - As_j = r(s_j)$, the residual of the vector s_j. Therefore using (2.3) and (2.12), it follows that the generalized residual of the approximation $t_k^{(n)}$ is the true residual

$$\tilde{r}(t_k^{(n)}) = r(t_k^{(n)}) = b - At_k^{(n)}. \tag{2.13}$$

Note also that, since $\Delta^2 s_n = -A\Delta s_n$, we have $\Delta^2 S_{k,n} = -A\Delta S_{k,n}$.

For simplicity and unless specified otherwise, we set $n = 0$, we denote $t_k^{(0)} = t_k$ and we drop the index n in our notations. Let d be the degree of the minimal polynomial $\mathscr{P}_d$ of B for the vector $s_0 - x^*$ and, as $A = I - B$ is nonsingular, P_d is also the minimal polynomial of B for $r_0 = \Delta s_0$. Therefore, the matrices $\Delta S_k = [\Delta s_0, \ldots, \Delta s_{k-1}]$ and $\Delta^2 S_k = [\Delta^2 s_0, \ldots, \Delta^2 s_{k-1}]$ have full rank for $k \leqslant d$. We also note that the approximation t_d exits and is equal to the solution of the linear system (2.10).

The three extrapolation methods make use implicitly of the polynomial $\mathscr{P}_d$ and since this polynomial is not known in practice, the aim of these methods is to approximate it.

When applied to the sequence generated by (2.12), the vector extrapolation methods above produce approximations t_k such that the corresponding residuals $r_k = b - At_k$ satisfy the relations

$$r_k \in \tilde{W}_k = A\tilde{V}_k \tag{2.14}$$

and

$$r_k \perp \tilde{L}_k, \tag{2.15}$$

where $\tilde{V}_k = \text{span}\{\Delta s_0, \ldots, \Delta s_{k-1}\}$ and $\tilde{L}_k \equiv \tilde{W}_k$ for RRE, $\tilde{L}_k \equiv \tilde{V}_k$ for MPE and $\tilde{L}_k \equiv \tilde{Y}_k = \text{span}\{y_1, \ldots, y_k\}$ for MMPE where $y_1, \ldots, y_k$ are linearly independent vectors.

Note that, since $\tilde{W}_k \equiv K_k(A, Ar_0)$, the extrapolation methods above are Krylov subspace methods. RRE is an orthogonal projection and is theoretically equivalent to GMRES while MPE and MMPE are oblique projection methods and are equivalent to the method of Arnoldi and to the Hessenberg method [38], respectively. From this observation, we conclude that for $k \leqslant d$, the approximation t_k exists and is unique, unconditionally for RRE, and this is not always the case for MPE and MMPE. In fact, for the last two methods the approximation t_k $(k < d)$ exists if and only if $\det(\Delta S_k^{\mathrm{T}} \Delta^2 S_k) \neq 0$ for MPE and $\det(Y_k^{\mathrm{T}} \Delta^2 S_k) \neq 0$ for MMPE where $Y_k = [y_1, \ldots, y_k]$.

Let P_k be the orthogonal projector onto $\tilde{W}_k$. Then from (2.14) and (2.15), the residual generated by RRE can be expressed as

$$r_k^{\text{rre}} = r_0 - P_k r_0. \tag{2.16}$$

We also consider the oblique projectors Q_k and R_k onto $\tilde{W}_k$ and orthogonally to $\tilde{V}_k$ and $\tilde{Y}_k$ respectively. It follows that the residuals produced by MPE and MMPE can be written as

$$r_k^{\text{mpe}} = r_0 - Q_k r_0 \tag{2.17}$$

and

$$r_k^{\text{mmpe}} = r_0 - R_k r_0. \tag{2.18}$$

The acute angle θ_k between r_0 and the subspace $\tilde{W}_k$ is defined by

$$\cos\theta_k = \max_{z \in \tilde{W}_k - \{0\}} \left(\frac{|(r_0, z)|}{||r_0|| \, ||z||} \right). \tag{2.19}$$

Note that θ_k is the acute angle between the vector r_0 and $P_k r_0$.

In the sequel we give some relations satisfied by the residual norms of the three extrapolation methods.

Theorem 1. *Let ϕ_k be the acute angle between r_0 and $Q_k r_0$ and let ψ_k denote the acute angle between r_0 and $R_k r_0$. Then we have the following relations*:

(1) $||r_k^{\text{rre}}||^2 = (\sin^2\theta_k)||r_0||^2$.
(2) $||r_k^{\text{mpe}}||^2 = (\tan^2\phi_k)||r_0||^2$.
(3) $||r_k^{\text{rre}}|| \leqslant (\cos\phi_k)||r_k^{\text{mpe}}||$.

Moreover if for MMPE $y_j = r_0$ *for some* $j = 1, \ldots, k$, *then we also have*

(4) $||r_k^{\text{mmpe}}||^2 = (\tan^2\psi_k)||r_0||^2$.
(5) $||r_k^{\text{rre}}|| \leqslant (\cos\psi_k)||r_k^{\text{mmpe}}||$.

Proof. Parts (1)–(3) have been proved in [18]

(4) From (2.18), we get

$$(r_k^{\text{mmpe}}, r_k^{\text{mmpe}}) = (r_k^{\text{mmpe}}, r_0 - R_k r_0).$$

Since $(r_k^{\text{mmpe}}, r_0) = 0$, it follows that

$$\begin{aligned}(r_k^{\text{mmpe}}, r_k^{\text{mmpe}}) &= (r_k^{\text{mmpe}}, -R_k r_0) \\ &= -||r_k^{\text{mmpe}}|| \, ||R_k r_0|| \cos(r_k^{\text{mmpe}}, R_k r_0) \\ &= ||r_k^{\text{mmpe}}|| \, ||R_k r_0|| \sin\psi_k.\end{aligned}$$

On the other hand,

$$||r_0|| = ||R_k r_0|| \cos\psi_k,$$

hence

$$||r_k^{\text{mmpe}}|| = ||r_0|| \tan\psi_k.$$

(5) Using statements (1) and (4), we get

$$\frac{||r_k^{\text{mmpe}}||^2}{||r_k^{\text{rre}}||^2} = \frac{1 - \cos^2\psi_k}{1 - \cos^2\theta_k} (\cos^2\psi_k)^{-1}.$$

But $\cos\psi_k \leqslant \cos\theta_k$, therefore

$$||r_k^{\mathrm{rre}}|| \leqslant ||r_k^{\mathrm{mmpe}}|| \cos\psi_k. \qquad \square$$

Remark.

- From relations (1), (2) and (4) of Theorem 1, we see that the residuals of the RRE are always defined while those produced by MPE and MMPE may not exist.
- We also observe that if a stagnation occurs in RRE ($||r_k^{\mathrm{rre}}|| = ||r_0||$ for some $k < d$), then $\cos\theta_k = 0$ and, from (2.19), this implies that $\cos\phi_k = \cos\psi_k = 0$ and hence the approximations produced by MPE and MMPE are not defined.

When the linear process (2.12) is convergent, it is more useful in practice to apply the extrapolation methods after a fixed number p of basic iterations. We note also that, when these methods are used in their complete form, the required work and storage grow linearly with the iteration step. To overcome this drawback we use them in a cycling mode and this means that we have to restart the algorithms after a chosen number m of iterations.

The algorithm is summarized as follows:

1. $k = 0$, choose x_0 and the numbers p and m.
2. Basic iteration
   set $t_0 = x_0$
   $z_0 = t_0$
   $z_{j+1} = B z_j + b$, $j = 0, \ldots, p-1$.
3. Extrapolation scheme
   $s_0 = z_p$
   $s_{j+1} = B s_j + b$, $j = 0, \ldots, m$,
   compute the approximation t_m by RRE, MPE or MMPE.
4. Set $x_0 = t_m$, $k = k + 1$ and go to 2.

Stable schemes for the computation of the approximation t_k are given in [32, 19]. In [32], Sidi gave an efficient implementation of the MPE and RRE methods which is based on the QR decomposition of the matrix ΔS_k. In [19], we used an LU decomposition of ΔS_k with a pivoting strategy. These implementations require low work and storage and are more stable numerically.

2.3. Application to nonlinear systems

Consider the system of nonlinear equations

$$G(x) = x, \tag{2.20}$$

where $G : \mathbb{R}^N \Rightarrow \mathbb{R}^N$ and let x^* be a solution of (2.20).

For any arbitrary vector x, the residual is defined by

$$r(x) = G(x) - x.$$

Let $(s_j)_j$ be the sequence of vectors generated from an initial guess s_0 as follows:

$$s_{j+1} = G(s_j), \quad j = 0, 1, \ldots . \tag{2.21}$$

Note that

$$r(s_j)=\tilde{r}(s_j)=\Delta s_j, \quad j=,1,\dots .$$

As for linear problems, it is more useful to run some basic iterations before the application of an extrapolation method for solving (2.20). Note also that the storage and the evaluation of the function G increase with the iteration step k. So, in practice, it is recommended to restart the algorithms after a fixed number of iterations. Another important remark is the fact that the extrapolation methods are more efficient if they are applied to a preconditioned nonlinear system

$$\tilde{G}(x)=x, \tag{2.22}$$

where the function $\tilde{G}$ is obtained from G by some preconditioning nonlinear technique.

An extrapolation algorithm for solving the nonlinear problem (2.22) is summarized as follows:

1. $k=0$, choose x_0 and the integers p and m.
2. Basic iteration
   set $t_0=x_0$
   $w_0=t_0$
   $w_{j+1}=\tilde{G}(w_j)$, $j=0,\dots,p-1$.
3. Extrapolation phase
   $s_0=w_p$;
   if $\|s_1-s_0\|<\varepsilon$ stop;
   otherwise generate $s_{j+1}=\tilde{G}(s_j)$, $j=0,\dots,m$,
   compute the approximation t_m by RRE, MPE or MMPE;
4. set $x_0=t_m$, $k=k+1$ and go to 2.

As for systems of linear equations, efficient computation of the approximation t_m produced by RRE, MPE and MMPE have been derived in [32,19]. These implementations give as an estimation of the residual norm at each iteration and it allows to stop the algorithms without having to compute the true residual which requires an extra evaluation of the function $\tilde{G}$.

Important properties of vector extrapolation methods is the fact that they do not use the knowledge of the Jacobian of the function $\tilde{G}$ and have a quadratic convergence (when they are used in their complete form).

We also note that the results of Theorem 1 are still valid for nonlinear problems by replacing in the relations of this theorem the residual r_k by the generalized residual $\tilde{r}_k$.

Vector extrapolation methods such as MMPE can also be used for computing eigenelements of a matrix [16].

3. The ε-algorithms

3.1. The scalar ε-algorithm

Let (x_n) be a scalar sequence and consider the Hankel determinant

$$H_k(x_n) = \begin{vmatrix} x_n & \dots & x_{n+k-1} \\ \vdots & \vdots & \vdots \\ x_{n+k-1} & \dots & x_{n+2k-2} \end{vmatrix}, \quad \text{with } H_0(x_n) = 0, \ \forall n.$$

Shanks's transformation [31] e_k is defined by

$$e_k(x_n) = \frac{H_{k+1}(x_n)}{H_k(\Delta^2 x_n)}. \tag{3.1}$$

For the kernel of the transformation e_k, it was proved (see [6]) that

$$\forall n, \ e_k(x_n) = x \Leftrightarrow \exists a_0, \dots, a_k \text{ with } a_k \neq 0 \text{ and } a_0 + \dots + a_k \neq 0 \text{ such that } \forall n,$$

$$\sum_{i=0}^{k} a_i(x_{n+i} - x) = 0.$$

To implement Shank's transformation without computing determinants, Wynn [39] discovered a simple recursion called the scalar epsilon algorithm (SEA) defined by

$$\varepsilon_{-1}^{(n)} = 0, \quad \varepsilon_0^{(n)} = x_n, \quad n = 0, 1, \dots,$$

$$\varepsilon_{k+1}^{(n)} = \varepsilon_{k-1}^{(n+1)} + \frac{1}{\varepsilon_k^{(n+1)} - \varepsilon_k^{(n)}} \ k, \quad n = 0, 1, \dots .$$

The scalar ε-algorithm is related to Shanks's transformation by

$$\varepsilon_{2k}^{(n)} = e_k(x_n) \quad \text{and} \quad \varepsilon_{2k+1}^{(n)} = \frac{1}{e_k(\Delta x_n)}.$$

For more details and properties of SEA, see [6] and the references therein. For vector sequences (s_n), one can apply the scalar ε-algorithm to each component of s_n. However, one disadvantage of this technique is that it ignores the connexions between the components. Another problem is the fact that some transformed components fail to exist or may be very large numerically. These drawbacks limit the application of SEA to vector sequences.

3.2. The vector ε-algorithm

In order to generalize the scalar ε-algorithm to the vector case, we have to define the inverse of a vector. One possibility that was considered by Wynn [40] is to use the inverse defined by

$$z^{-1} = \frac{z}{||z||^2}, \quad z \in \mathbb{R}^N.$$

Therefore, for vector sequences (s_n) the vector ε-algorithm of Wynn is defined by

$$\varepsilon_{-1}^{(n)} = 0, \quad \varepsilon_0^{(n)} = s_n, \quad n = 0, 1, \dots,$$

$$\varepsilon_{k+1}^{(n)} = \varepsilon_{k-1}^{(n+1)} + [\varepsilon_k^{(n+1)} - \varepsilon_k^{(n)}]^{-1}, \quad k, n = 0, 1, \dots .$$

For the real case, it was proved by McLeod [23] that if $\forall n \geqslant N_0$, $\sum_{i=0}^{k} a_i(s_{n+i} - s) = 0$, with $a_k \neq 0$ and $a_0 + \dots + a_k \neq 0$, then $\varepsilon_{2k}^{(n)} = s$; $\forall n \geqslant N_0$. This result has been proved by Graves-Morris [13] in the complex case.

When applied to the vector sequence generated by (2.12), the scalar and the vector ε-algorithms give the solution of the linear system (2.10) that is $\forall n$, $\varepsilon_{2N}^{(n)} = x^*$, see [6]. As will be seen in the last section, the intermediate quantities $\varepsilon_{2k}^{(n)}$, $k < N$, are approximations of the solution x^*.

We note also that the vector ε-algorithm has been used for solving nonlinear problems by applying it to the nonlinear sequence defined by (2.21); see [7,12].

However, the vector ε-algorithm requires higher work and storage as compared to the vector polynomial methods described in Section 2. In fact, computing the approximation $\varepsilon_{2k}^{(n)}$ needs the terms $s_n, \ldots, s_{n+2k}$ which requires a storage of $2k+1$ vectors of $\mathbb{R}^N$ while the three methods (RRE, MPE and MMPE) require only $k+2$ terms $s_n, \ldots, s_{n+k+1}$. Computational work and storage requirements are given in Section 4.

3.3. The topological ε-algorithm

In [3], Brezinski proposed another generalization of the scalar ε-algorithm for vector sequences which is quite different from the vector ε-algorithm and was called the topological ε-algorithm (TEA).

This approach consists in computing approximations $e_k(s_n) = t_k^{(n)}$ of the limit or the anti-limit of the sequence (s_n) such that

$$t_k^{(n)} = s_n + \sum_{i=1}^{k} a_i^{(n)} \Delta s_{n+i-1}, \quad n \geqslant 0. \tag{3.2}$$

We consider the new transformations $\tilde{t}_{k,j}$, $j = 1, \ldots, k$ defined by

$$\tilde{t}_{k,j}^{(n)} = s_{n+j} + \sum_{i=1}^{k} a_i^{(n)} \Delta s_{n+i+j-1}, \quad j = 1, \ldots, k.$$

We set $\tilde{t}_{k,0}^{(n)} = t_k^{(n)}$ and define the jth generalized residual as follows:

$$\begin{aligned}\tilde{r}_j(t_k^{(n)}) &= \tilde{t}_{k,j}^{(n)} - \tilde{t}_{k,j-1}^{(n)}\\ &= \Delta s_{n+j-1} + \sum_{i=1}^{k} a_i^{(n)} \Delta^2 s_{n+i+j-2}, \quad j = 1, \ldots, k.\end{aligned}$$

Therefore, the coefficients involved in expression (3.2) of $t_k^{(n)}$ are computed such that each jth generalized residual is orthogonal to some chosen vector $y \in \mathbb{R}^N$, that is

$$(y, \tilde{r}_j(t_k^{(n)})) = 0, \quad j = 1, \ldots, k. \tag{3.3}$$

Hence the vector $a_n = (a_1^{(n)}, \ldots, a_k^{(n)})^{\mathrm{T}}$ is the solution of the $k \times k$ linear system (3.3) which is written as

$$T_{k,n}\, a_n = \Delta S_{k,n}^{\mathrm{T}} y, \tag{3.4}$$

where $T_{k,n}$ is the matrix whose columns are $\Delta^2 S_{k,n}^{\mathrm{T}}\, y, \ldots, \Delta^2 S_{k,n+k-1}^{\mathrm{T}}\, y$ (assumed to be nonsingular) and $\Delta^j S_{k,n}$, $j = 1,2$ are the $N \times k$ matrices whose columns are $\Delta^j s_n, \ldots, \Delta^j s_{n+k-1}$, $j = 1,2$.

Note that the $k \times k$ matrix $T_{k,n}$ is also given by the formula

$$T_{k,n} = \mathscr{S}_{k,n}\,(I_N \otimes y),$$

where $\mathscr{S}_{k,n}$ is the $k \times Nk$ matrix whose block columns are $\Delta^2 S_{k,n}^{\mathrm{T}}, \ldots, \Delta^2 S_{k,n+k-1}^{\mathrm{T}}$.

Invoking (3.2) and (3.4), $t_k^{(n)}$ can be expressed in a matrix form as

$$t_k^{(n)} = s_n - \Delta S_{k,n}\, T_{k,n}^{-1}\, \Delta S_{k,n}^{\mathrm{T}} y. \tag{3.5}$$

Using Schur's formula, $t_k^{(n)}$ can be expressed as a ratio of two determinants

$$t_k^{(n)} = \left| \begin{matrix} s_n & \Delta S_{k,n} \\ \Delta S_{k,n}^{\mathrm{T}} y & T_{k,n} \end{matrix} \right| \Big/ \det(T_{k,n}).$$

For the kernel of the topological ε-algorithm it is easy to see that if $\forall n$, $\exists a_0, \ldots, a_k$ with $a_k \neq 0$ and $a_0 + \cdots + a_k \neq 0$ such that $\sum_{i=0}^{k} a_i(s_{n+i} - s) = 0$, then $\forall n$, $t_k^{(n)} = s$.

The vectors $e_k(s_n) = t_k^{(n)}$ can be recursively computed by the topological ε-algorithm discovered by Brezinski [3]

$$\varepsilon_{-1}^{(n)} = 0; \quad \varepsilon_0^{(n)} = s_n, \quad n = 0, 1, \ldots,$$

$$\varepsilon_{2k+1}^{(n)} = \varepsilon_{2k-1}^{(n+1)} + \frac{y}{(y, \Delta\varepsilon_{2k}^{(n)})},$$

$$\varepsilon_{2k+2}^{(n)} = \varepsilon_{2k}^{(n+1)} + \frac{\Delta\varepsilon_{2k}^{(n)}}{(\Delta\varepsilon_{2k+1}^{(n)}, \Delta\varepsilon_{2k}^{(n)})} n, \quad k = 0, 1, \ldots .$$

The forward difference operator Δ acts on the superscript n and we have

$$\varepsilon_{2k}^{(n)} = e_k(s_n) = t_k^{(n)}, \quad \text{and} \quad \varepsilon_{2k+1}^{(n)} = \frac{y}{(y, e_k(\Delta s_n))}, \quad n, k = 0, 1, \ldots .$$

We notice that, for the complex case, we can use the product $(y, z) = \sum_{i=1}^{N} y_i \bar{z}_i$, hence (y,z) is not equal to (z, y). The order of vectors in the scalar product is important, and similar methods have been studied in detail by Tan [37].

3.4. Application of VEA and TEA to linear and nonlinear systems

Consider again the system of linear equations (2.10) and let (s_n) be the sequence of vectors generated by the linear process (2.12).

Using the fact that $\Delta^2 s_{n+i} = B\Delta^2 s_{n+i-1}$, the matrix $T_{k,n}$ has now the following expression:

$$T_{k,n} = -L_k^{\mathrm{T}} A \Delta S_{k,n}, \tag{3.6}$$

where L_k is the $N \times k$ matrix whose columns are $y, B^{\mathrm{T}} y, \ldots, B^{\mathrm{T}^{k-1}} y$. As n will be a fixed integer, we set $n = 0$ for simplicity and denote $T_{k,0}$ by T_k and $\Delta S_{k,0}$ by ΔS_k.

On the other hand, it is not difficult to see that

$$\Delta S_k^{\mathrm{T}} y = L_k^{\mathrm{T}} r_0. \tag{3.7}$$

Therefore, using (3.6), (3.7) with (3.5), the kth residual produced by TEA is given by

$$r_k^{\text{tea}} = r_0 - A\Delta S_k\, (L_k^{\mathrm{T}} A \Delta S_k)^{-1} L_k^{\mathrm{T}} r_0. \tag{3.8}$$

Let E_k denotes the oblique projector onto the Krylov subspace $K_k(A, Ar_0)$ and orthogonally to the Krylov subspace $K_k(B^{\mathrm{T}}, y) = K_k(A^{\mathrm{T}}, y)$. Then from (3.8) the residual generated by TEA can be written as follows:

$$r_k^{\text{tea}} = r_0 - E_k r_0. \tag{3.9}$$

This shows that the topological ε-algorithm is mathematically equivalent to the method of Lanczos [4]. Note that the kth approximation defined by TEA exists if and only if the $k \times k$ matrix $L_k^T A \Delta S_k$ is nonsingular.

The following result gives us some relations satisfied by the residual norms in the case where $y = r_0$.

Theorem 2. *Let φ_k be the acute angle between r_0 and $E_k r_0$ and let $y = r_0$. Then we have the following relations*:
(1) $||r_k^{\text{tea}}|| = |\tan \varphi_k| \, ||r_0||; \; k > 1$.
(2) $||r_k^{\text{rre}}|| \leqslant (\cos \varphi_k) \, ||r_k^{\text{tea}}||$.

Proof. (1) Follows from (3.9) and the fact that $r_0 = y$ is orthogonal to r_k^{tea}.

(2) From (2.19) we have $\cos \varphi_k \leqslant \cos \theta_k$, then using relations (1) of Theorem 1 and (1) of Theorem 2 the result follows. □

Remark.
- Relation (1) of Theorem 2 shows that the residuals of the TEA are defined if and only if cos $\varphi_k \neq 0$.
- We also observe that, if a stagnation occurs in RRE ($||r_k^{\text{rre}}|| = ||r_0||$ for some k, then $\cos \theta_k = 0$ and this implies that $\cos \varphi_k = 0$, which shows that the TEA-approximation is not defined.

The topological ε-algorithm can also be applied for solving nonlinear systems of equations. For this, TEA is applied to the sequence (s_n) generated by the nonlinear process (2.22). We note that TEA does not need the knowledge of the Jacobian of the function $\tilde{G}$ and has the property of quadratic convergence [22].

When applied for the solution of linear and nonlinear problems, work and storage required by VEA and TEA grow with the iteration step. So, in practice and for large problems, the algorithms must be restarted. It is also useful to run some basic iterations before the extrapolation phase.

The application of VEA or TEA for linear and nonlinear systems leads to the following algorithm where $\tilde{G}(x)$ is to be replaced by $Bx + b$ for linear problems:

1. $k = 0$, choose x_0 and the integers p and m.
2. Basic iteration
   set $t_0 = x_0$
   $w_0 = t_0$
   $w_{j+1} = \tilde{G}(w_j)$, $j = 0, \ldots, p-1$.
3. Extrapolation phase
   $s_0 = w_p$;
   if $||s_1 - s_0|| < \varepsilon$ stop;
   otherwise generate $s_{j+1} = \tilde{G}(s_j)$, $j = 0, \ldots, 2m-1$,
   compute the approximation $t_m = \varepsilon_{2m}^{(0)}$ by VEA or TEA;
4. set $x_0 = t_m$, $k = k + 1$ and go to 2.

Table 1
Memory requirements and computational costs (multiplications and additions) for RRE, MPE, MMPE, VEA and TEA

Method	RRE	MPE	MMPE	VEA	TEA
Multiplications and additions	$2Nk^2$	$2Nk^2$	Nk^2	$10Nk^2$	$10Nk^2$
Mat–Vec with A or evaluation of $\tilde{G}$	$k+1$	$k+1$	$k+1$	$2k$	$2k$
Memory locations	$(k+1)N$	$(k+1)N$	$(k+1)N$	$(2k+1)N$	$(2k+1)N$

4. Operation count and storage

Table 1 lists the operation count (multiplications and additions) and the storage requirements to compute the approximation $t_k^{(0)}$ with RRE, MPE and MMPE and the approximation $\varepsilon_{2k}^{(0)}$ with VEA and TEA. In practice, the dimension N of vectors is very large and k is small, so we listed only the main computational effort.

For RRE and MPE, we used the QR-implementation given in [32], whereas the LU-implementation developed in [19] was used for MMPE.

To compute $t_k^{[0]}$ with the three polynomial vector extrapolation methods, the vectors $s_0, s_1, \ldots, s_{k+1}$ are required while the terms $s_0, \ldots, s_{2k}$ are needed for the computation of $\varepsilon_{2k}^{(0)}$ with VEA and TEA. So, when solving linear systems of equations, $k+1$ matrix–vector (Mat–Vec) products are required with RRE, MPE and MMPE and $2k$ matrix–vector products are needed with VEA and TEA. For nonlinear problems the comparison is still valid by replacing "Mat–Vec" with "evaluation of the function $\tilde{G}$".

All these operations are listed in Table 1.

As seen in Table 1, the vector and topological ε-algorithms are more expensive in terms of work and storage as compared to the polynomial vector extrapolation methods, namely RRE, MPE and MMPE.

The implementations given in [32,19] for RRE, MPE and MMPE allow us to compute exactly the norm of the residual at each iteration for linear systems or to estimate it for nonlinear problems without actually computing the residuals. This reduce the cost of implementation and is used to stop the algorithms when the accuracy is achieved.

5. Numerical examples

We report in this section a few numerical examples comparing the performances of RRE, MPE, MMPE, VEA and TEA. For RRE and MPE, we used the program given in [32] and for MMPE we used the implementation developed in [19]. The programs used for VEA and TEA were taken out from [6].

The tests were run in double precision on SUN Entreprise 450 SERVER using the standard F77 compiler. We have first considered one example for linear systems and one example for nonlinear systems. In these examples the starting point was chosen $x_0 = \text{rand}(N, 1)$ where the function rand creates an N vector with coefficients uniformly distributed in $[0, 1]$.

For the TEA the vector y was also $y = \text{rand}(N, 1)$.

Table 2

Method	MMPE	MPE	RRE	VEA	TEA
Number of restarts	28	25	26	30	30
Residual norms	2.16d-09	2.d-09	1.d-09	9d-04	3d-01
CPU time	40	80	83	230	206

5.1. Example 1

In the first example, we derived the matrix test problem by discretizing the boundary value problem [1]

$$-u_{xx}(x,y) - u_{yy}(x,y) + 2p_1u_x(x,y) + 2p_2u_y(x,y) - p_3u(x,y) = \phi(x,y) \quad \text{on } \Omega,$$

$$u(x,y) = 1 + xy \quad \text{on } \partial\Omega,$$

by finite differences, where Ω is the unit square $\{(x,y) \in R^2,\ 0 \leqslant x, y \leqslant 1\}$ and p_1, p_2, p_3 are positive constants. The right-hand-side function $\phi(x,y)$ was chosen so that the true solution is $u(x,y) = 1 + xy$ in Ω. We used centred differences to discretize this problem on a uniform $(n+2) \times (n+2)$ grid (including grid points on the boundary). We get a matrix of size $N = n^2$.

We applied the extrapolation methods to the sequence (s_j) defined as in [14] by

$$s_{j+1} = B_\omega s_k + c_\omega, \tag{5.1}$$

where

$$c_\omega = \omega(2-\omega)(D-\omega U)^{-1}D(D-\omega L)^{-1}b, \tag{5.2}$$

$$B_\omega = (D-\omega U)^{-1}(\omega L + (1-\omega)D)(D-\omega L)^{-1}(\omega U + (1-\omega)D) \tag{5.3}$$

and $A = D - L - U$, the classical splitting decomposition.

When (s_j) converges, the fixed point of iteration (5.1) is the solution of the SSOR preconditioned system $(I - B_\omega)x = c_\omega$.

The stopping criterion was $\|(I - B_\omega)x_k - c_\omega\| < 10^{-8}$ for this linear problem.

We let $n = 70$ and choose $p_1 = 1$, $p_2 = 1$ and $p_3 = 10$.

For this experiment, the system has dimension 4900×4900. The width of extrapolation is $m = 20$ and $\omega = 0.5$.

In Table 2, we give the l_2-norm of the residuals obtained at the end of each cycle and the CPU time for the five methods (MMPE, MPE, RRE, VEA and TEA).

A maximum of 30 cycles was allowed to all the algorithms. Remark that for this experiment, TEA failed to converge and for the VEA we obtained only a residual norm of $9 \cdot 10^{-4}$.

5.2. Example 2

We consider now the following nonlinear partial differential equation:

$$-u_{xx}(x,y) - u_{yy}(x,y) + 2p_1u_x(x,y) + 2p_2u_y(x,y) - p_3u(x,y) + 5e^{u(x,y)} = \phi(x,y) \quad \text{on } \Omega,$$

$$u(x,y) = 1 + xy \quad \text{on } \partial\Omega,$$

Table 3

Method	MMPE	MPE	RRE	VEA	TEA
Number of restarts	20	18	19	22	30
Residual norms	2.9d-09	9.2d-08	2.8d-08	9.6d-09	2.9d-05
CPU time	13.59	13.90	14.72	51.24	65.90

over the unit square of $\mathbb{R}^2$ with Dirichlet boundary condition. This problem is discretized by a standard five-point central difference formula with uniform grid of size $h = 1/(n+1)$. We get the following nonlinear system of dimension $N \times N$, where $N = n^2$:

$$AX + 5e^X - b = 0. \tag{5.4}$$

The right-hand-side function $\phi(x, y)$ was chosen so that the true solution is $u(x, y) = 1 + xy$ in Ω.

The sequence (s_j) is generated by using the nonlinear SSOR method. Hence we have $s_{j+1} = G(s_j)$, where

$$G(X) = B_\omega X + \omega(2-\omega)(D-\omega U)^{-1}D(D-\omega L)^{-1}(b - 5e^X),$$

the matrix B_ω is given in (5.3).

In the following tests, we compare the five extrapolation methods using the SSOR preconditioning. The stopping criterion was $||x_k - G(x_k)|| < 10^{-8}$.

In our tests, we choose $n = 72$ and hence the system has dimension $N = 4900$. With $m = 20$ and $\omega = 0.5$, we obtain the results of Table 3.

The convergence of the five extrapolation methods above is relatively sensitive to the choice of the parameter ω. We note that for this experiment, the TEA algorithm stagnates after 30 restarts. The VEA algorithm requires more CPU time as compared to the three polynomial extrapolation methods.

6. Conclusion

We have proposed a review of the most known vector extrapolation methods namely the polynomial ones (MMPE, RRE and MPE) and the ε-algorithms (TEA and VEA). We also give some numerical comparison of these methods. The numerical tests presented in this paper show the advantage of the vector polynomial methods. We note also that VEA is numerically more stable than TEA. However, the last two algorithms require more storage and operation counts as compared to the polynomial methods. The advantage of vector extrapolation methods when compared to the classical Krylov subspace methods is that they generalize in a straightforward manner from linear to nonlinear problems.

Acknowledgements

We would like to thank M. Redivo-Zaglia and C. Brezinski for providing us their programs for the ε-algorithms and A. Sidi for the RRE and MPE programs. We also wish to thank the referees for their valuable comments and suggestions.

References

[1] Z. Bai, D. Hu, L. Reichel, A Newton basis GMRES implementation, IMA J. Numer. Anal. 14 (1994) 563–581.

[2] P. Brown, Y. Saad, Hybrid Krylov methods for nonlinear systems of equations, SIAM J. Sci. Statist. Comput. 11 (1990) 450–481.

[3] C. Brezinski, Généralisation de la transformation de Shanks, de la table de la Table de Padé et de l'epsilon-algorithm, Calcolo 12 (1975) 317–360.

[4] C. Brezinski, Padé-type Approximation and General Orthogonal Polynomials, International Series of Numerical Methods, Vol. 50, Birkhäuser, Basel, 1980.

[5] C. Brezinski, Recursive interpolation, extrapolation and projection, J. Comput. Appl. Math. 9 (1983) 369–376.

[6] C. Brezinski, M. Redivo Zaglia, Extrapolation Methods, Theory and Practice, North-Holland, Amsterdam, 1991.

[7] C. Brezinski, A.C. Rieu, The solution of systems of equations using the vector ε-algorithm and an application to boundary value problems, Math. Comp. 28 (1974) 731–741.

[8] S. Cabay, L.W. Jackson, A polynomial extrapolation method for finding limits and antilimits for vector sequences, SIAM J. Numer. Anal. 13 (1976) 734–752.

[9] R.P. Eddy, Extrapolation to the limit of a vector sequence, in: P.C.C Wang (Ed.), Information Linkage Between Applied Mathematics and Industry, Academic Press, New-York, 1979, pp. 387–396.

[10] W.D. Ford, A. Sidi, Recursive algorithms for vector extrapolation methods, Appl. Numer. Math. 4 (6) (1988) 477–489.

[11] W. Gander, G.H. Golub, D. Gruntz, Solving linear equations by extrapolation, in: J.S. Kovalic (Ed.), Supercomputing, Nato ASI Series, Springer, Berlin, 1990.

[12] E. Gekeler, On the solution of systems of equations by the epsilon algorithm of Wynn, Math. Comp. 26 (1972) 427–436.

[13] P.R. Graves-Morris, Vector valued rational interpolants I, Numer. Math. 42 (1983) 331–348.

[14] L. Hageman, D. Young, Applied Iterative Methods, Academic Press, New York, 1981.

[15] K. Jbilou, A general projection algorithm for solving systems of linear equations, Numer. Algorithms 4 (1993) 361–377.

[16] K. Jbilou, On some vector extrapolation methods, Technical Report, ANO(305), Université de Lille1, France, 1993.

[17] K. Jbilou, H. Sadok, Some results about vector extrapolation methods and related fixed point iterations, J. Comput. Appl. Math. 36 (1991) 385–398.

[18] K. Jbilou, H. Sadok, Analysis of some vector extrapolation methods for linear systems, Numer. Math. 70 (1995) 73–89.

[19] K. Jbilou, H. Sadok, LU-implementation of the modified minimal polynomial extrapolation method, IMA J. Numer. Anal. 19 (1999) 549–561.

[20] K. Jbilou, H. Sadok, Hybrid vector sequence transformations, J. Comput. Appl. Math. 81 (1997) 257–267.

[21] C. Lanczos, Solution of systems of linear equations by minimized iterations, J. Res. Natl. Bur. Stand. 49 (1952) 33–53.

[22] H. Le Ferrand, The quadratic convergence of the topological ε-algorithm for systems of nonlinear equations, Numer. Algorithms 3 (1992) 273–284.

[23] J.B. McLeod, A note on the ε-algorithm, Computing 7 (1971) 17–24.

[24] M. Mešina, Convergence acceleration for the iterative solution of $x = Ax + f$, Comput. Methods Appl. Mech. Eng. 10 (2) (1977) 165–173.

[25] B.P. Pugatchev, Acceleration of the convergence of iterative processes and a method for solving systems of nonlinear equations, USSR. Comput. Math. Math. Phys. 17 (1978) 199–207.

[26] Y. Saad, Krylov subspace methods for solving large unsymmetric linear systems, Math. Comp. 37 (1981) 105–126.

[27] Y. Saad, M.H. Schultz, GMRES: A generalized minimal residual algorithm for solving nonsymmetric linear systems, SIAM J. Sci. Statist. Comput. 7 (1986) 856–869.

[28] H. Sadok, Quasilinear vector extrapolation methods, Linear Algebra Appl. 190 (1993) 71–85.

[29] H. Sadok, About Henrici's transformation for accelerating vector sequences, J. Comput. Appl. Math. 29 (1990) 101–110.

[30] H. Sadok, Méthodes de projection pour les systèmes linéaires et non linéaires, Habilitation Thesis, Université de Lille1, France, 1994.

[31] D. Shanks, Nonlinear transformations of divergent and slowly convergent sequences, J. Math. Phys. 34 (1955) 1–42.
[32] A. Sidi, Efficient implementation of minimal polynomial and reduced rank extrapolation methods, J. Comput. Appl. Math. 36 (1991) 305–337.
[33] A. Sidi, Convergence and stability of minimal polynomial and reduced rank extrapolation algorithms, SIAM J. Numer. Anal. 23 (1986) 197–209.
[34] A. Sidi, Extrapolation vs. projection methods for linear systems of equations, J. Comput. Appl. Math. 22 (1) (1988) 71–88.
[35] A. Sidi, W.F. Ford, D.A. Smith, Acceleration of convergence of vector sequences, SIAM J. Numer. Anal. 23 (1986) 178–196.
[36] D.A. Smith, W.F. Ford, A. Sidi, Extrapolation methods for vector sequences, SIAM Rev. 29 (1987) 199–233; Correction, SIAM Rev. 30 (1988) 623–624.
[37] R.C.E. Tan, Implementation of the topological ε-algorithm, SIAM J. Sci. Statist. Comput. 9 (1988) 839–848.
[38] J.H. Wilkinson, The Algebraic Eigenvalue Problem, Clarendon Press, Oxford, England, 1965.
[39] P. Wynn, On a device for computing the $e_m(s_n)$ transformation, MTAC 10 (1956) 91–96.
[40] P. Wynn, Acceleration technique for iterated vector and matrix problems, Math. Comp. 16 (1962) 301–322.

ELSEVIER

Journal of Computational and Applied Mathematics 122 (2000) 167–201

JOURNAL OF COMPUTATIONAL AND APPLIED MATHEMATICS

www.elsevier.nl/locate/cam

Multivariate Hermite interpolation by algebraic polynomials: A survey

R.A. Lorentz

SCAI - Institute for Algorithms and Scientific Computing, GMD - German National Research Center for Information Technology, Schloß Birlinghoven, D - 53754 Sankt Augustin, Germany

Received 10 June 1999; received in revised form 15 February 2000

Abstract

This is a survey of that theory of multivariate Lagrange and Hermite interpolation by algebraic polynomials, which has been developed in the past 20 years. Its purpose is not to be encyclopedic, but to present the basic concepts and techniques which have been developed in that period of time and to illustrate them with examples. It takes "classical" Hermite interpolation as a starting point, but then successively broadens the assumptions so that, finally, interpolation of arbitrary functionals and the theory of singularities from algebraic geometry is discussed. © 2000 Elsevier Science B.V. All rights reserved.

Keywords: Multivariate Hermite interpolation; Algebraic polynomials; Lagrange; Least interpolation; Lifting schemes

1. Introduction

1.1. Motivation

This is a survey of interpolation by multivariate algebraic polynomials covering roughly the last 20 years.

Why should one study multivariate polynomials? Firstly of all, they are a building block of surprisingly many numerical methods, most often locally. For example, finite elements and splines, both univariate and multivariate, are piecewise polynomials. Secondly, theorems on the quality of approximation of functions or on the quality of a numerical scheme almost invariably reduce to local interpolation by polynomials, even when the approximating functions, respectively the basis of the numerical scheme is of another type. Take any modern textbook on numerical analysis. Except for the part on linear algebra, polynomials are probably mentioned on every third or fourth page. Finally, despite their fundamental importance for numerical methods, the theory of multivariate polynomial

0377-0427/00/$ - see front matter © 2000 Elsevier Science B.V. All rights reserved.
PII: S0377-0427(00)00367-8

interpolation is underdeveloped and there is hardly any awareness of the issues involved. Indeed, there is hardly any awareness that basic questions are still unresolved.

The three reasons just given to motivate the study of polynomial interpolation are practically oriented. But there is another reason to study polynomials: they are beautiful! Why is number theory so appealing? Because it is beautiful. The objects studied, integers, are the simplest in all of mathematics. I see polynomials as the next step up in the scale of complexity. They are also simple and, again, they are beautiful!

This survey neither intends to be encyclopedic in that all of the newest results are mentioned nor does it intend to be historic, in that it reports on the first occurence of any particular idea. Instead, the main constructions and the basic ideas behind them will be presented. In addition, many examples will be given. It is my hope that the reader will then *understand* what has been done and be in a position to apply the methods. No proofs will be given, but often the ideas behind the proofs. In addition, the references are mainly chosen according to how well they explain an idea or survey a group of ideas and as to how useful their list of references is.

Most of the results obtained in multivariate interpolation have been obtained in the 30 years surveyed here. The past 10 years have seen a new wave of constructive methods which show how to construct the interpolants and which have been used to find new interpolation schemes. These results are surveyed in detail by Gasca and Sauer in [15], which is also a survey of multivariate interpolation in the same time period. For this reason, these techniques will not be discussed in very much detail here. Rather links to [15] will be made wherever appropriate.

1.2. Interpolation

What is interpolation? In the most general case we will be considering here, we are given a normed linear space Y, a finite linear subspace V of Y, a finite set of bounded functionals $\mathscr{F}=\{F_q\}_{q=1}^m$ and real numbers $\{c_q\}_{q=1}^m$. The *interpolation problem* is to find a $P\in V$ such that

$$F_qP=c_q,\quad q=1,\ldots,m. \tag{1}$$

We will often abbreviate this formulation by saying that we interpolate the functionals $\mathscr{F}$ from V. The interpolating element is called the *interpolant*.

The interpolation problem is called *regular* if the above equation has a unique solution for each choice of values $\{c_q\}_{q=1}^m$. Otherwise, the interpolation is *singular*. In order that an interpolation be regular, it is necessary that

$$\dim V=m=\text{ the number of functionals.} \tag{2}$$

Often the values are given by applying the functionals to an element f of Y:

$$c_q=F_qf,\quad q=1,\ldots,m. \tag{3}$$

If so, the interpolation problem can be formulated in a somewhat different way, which we will have cause to use later. Let $\mathscr{G}=\operatorname{span}\{F_q\}_{q=1}^m$. Then $\mathscr{G}$ is an m-dimensional subspace of Y^*, the dual of Y. The interpolation problem (3) is equivalent to: given $f\in Y$, find a $P\in V$ such that

$$FP=Ff\quad\text{for any } F\in\mathscr{G}. \tag{4}$$

An example of all of the above is what I call "classical" Hermite interpolation. To describe it, we first need some polynomial spaces:

$$\Pi_n^d = \left\{ \sum_{|j| \leqslant n} a_j z^j \mid j \in \mathbb{N}_0^d, z \in \mathbb{R}^d, a_j \in \mathbb{R} \right\}. \tag{5}$$

Here and in the following, we use multivariate notation: $z = (x_1, \ldots, x_d)$, $j = (j_1, \ldots, j_d)$, $z^j = x_i^{j_1} x_2^{j_2} \cdots x_d^{j_d}$ for $z \in \mathbb{R}^d$ and $j \in \mathbb{N}_0^d$. Moreover $|j| = j_1 + \cdots + j_d$. The above space is called the space of polynomials of *total degree n* and will be our interpolation space V.

The functionals we interpolate are partial derivatives: given a set of distinct points $\{z_q\}_{q=1}^m$ in $\mathbb{R}^d$ and nonnegative integers $k_1, \ldots, k_m$, our functionals are

$$F_{q,\alpha} f = D^\alpha f(z_q), \quad 0 \leqslant |\alpha| \leqslant k_q, 1 \leqslant q \leqslant m, \tag{6}$$

where

$$D^\alpha = \frac{\partial^{|\alpha|}}{\partial x_1^{\alpha_1} \cdots x_d^{\alpha_d}}.$$

Since

$$\dim \Pi_n^d = \binom{d+n}{d}$$

and the number of partial derivatives (including the function value) to be interpolated at z_q is

$$\binom{d+k_p}{d},$$

we also require that

$$\binom{d+n}{d} = \sum_{q=1}^m \binom{d+k_p}{d}. \tag{7}$$

"Classical" Hermite interpolation, or *Hermite interpolation of total degree* is thus the problem of finding a $P \in \Pi_n^d$ satisfying

$$D^\alpha P(z_q) = c_{q,\alpha}, \quad 0 \leqslant |\alpha| \leqslant k_q, \ 1 \leqslant q \leqslant m, \tag{8}$$

for some given values c_q. If all the $k_q = 0$, then we have *Lagrange* interpolation: find $P \in \Pi_n^d$ such that

$$P(z_q) = c_q, \quad 1 \leqslant q \leqslant m. \tag{9}$$

Here, of course, $m = \dim \Pi_n^d$.

What determines which kind of of multivariate interpolation is the "right" or "most natural" one? The simplest answer is to just look at the univariate case and find the most natural generalization. That is how one arrives at the multivariate interpolation just described. But, as we shall see, there are other perfectly good multivariate interpolations. For, while in the univariate case, the polynomial interpolation space, namely Π_n^1, is canonical, in the multivariate case, we have many polynomial spaces of the same dimension. The derivatives we have chosen are in the direction of the coordinate axes. See Section 3 for Hermite interpolations involving directional derivatives. Moreover, we will see (in Section 4) that interpolating not point values of functions or their derivatives, but mean values of them over line segments, triangles, etc., leads to interpolation schemes retaining many of the properties of univariate interpolation.

Other criteria for the choice of the interpolation derive from the use to which the interpolant is to be put. If, for example, the interpolant interpolates the values resulting from the application of the functionals to a function, as in (3), then one would like to know how well the interpolant approximates the function (in the norm of Y). and tailor the interpolation accordingly. For exactly this reason, there is a host of papers concerned with developing Newton like interpolations. Since this aspect of polynomial interpolation will not be emphasized here, the reader can find them in the survey paper by Gasca and Sauer [15], and in the papers by Sauer and Xu [37,38].

Often such interpolations allow more precise statements about the error of approximation. In addition, these methods are numerically quite stable.

For finite elements, other properties play an important role. Two of them are that the dimension of interpolation space, V, be as small as possible to attain the desired global continuity and that the interpolation spaces be affinely invariant. Consequently, V may not be a full space Π_n^d for some n, d, but may lie between two such spaces $\Pi_n^d \subsetneq V \subsetneq \Pi_{n+1}^d$. Many examples of such interpolations can be found in [7].

As a last application which would require special properties from the interpolant, let me mention cubature formulas. As in the univariate case, one method for constructing cubature formulas is to base them on Lagrange interpolation at the zeros of orthogonal polynomials. Here the nodes are prescribed and a polynomial space must be constructed so that interpolation at these nodes is regular. Again, these spaces will, in general, not coincide with a Π_n^d.

Of course, many nonpolynomial multivariate interpolation methods have benn developed. See [40] for a survey on scattered data interpolation.

2. The issues involved

2.1. Univariate interpolation

Let us first look at univariate interpolation, since everything works well there. Given m distinct points $\{z_q\}_{q=1}^m$ in $\mathbb{R}^1$ and m real values $\{c_q\}_{q=1}^m$, there is one and only one polynomial $P \in \Pi_{m-1}^1$ with

$$P(z_q) = c_q, \quad q = 1,\ldots,m, \tag{10}$$

i.e., the univariate Lagrange interpolation problem is regular for any set of nodes. The same is true of univariate Hermite interpolation: Given a nodal set $Z = \{z_q\}_{q=1}^m$, integers k_q and values $c_{q,\alpha}$ for $0 \leqslant \alpha \leqslant k_q$ and $q = 1,\ldots,m$, there is one and only one polynomial $P \in \Pi_n^1$, where

$$n = \sum_{q=1}^{m} (k_q + 1) - 1, \tag{11}$$

such that

$$D^\alpha P(z_q) = c_{q,\alpha} \quad 0 \leqslant \alpha \leqslant k_q, \ q = 1,\ldots,m, \tag{12}$$

i.e., the univariate Hermite interpolation problem is regular for any nodal set and for any choice of derivatives to be interpolated. A word of caution here. There is another type of univariate interpolation, called Birkhoff interpolation, which interpolates derivatives but differs from Hermite interpolation in that gaps in the derivatives are allowed. In this theory, for example, one could interpolate $f(z_q)$ and $f''(z_q)$ but not $f'(z_q)$. Such interpolations are not necessarily regular. See [27] for the univariate theory and [28] for the multivariate theory of Birkhoff interpolation.

How does one prove regularity? There are several ways. One method is constructive in that one finds a basis λ_q, $q=1,\ldots,m$ of Π_n^1 (or of V in general) dual to the functionals we are interpolating in the sense that $F_q\lambda_r = \delta_{qr}$. The elements of such a basis are sometimes called the *fundamental* functions of the interpolation. The Lagrange interpolant, for example, could then be written as

$$P(z) = \sum_{q=1}^{m} f(z_q)\lambda_q(z), \tag{13}$$

if the data to be interpolated are given by $c_q = f(z_q)$. The fundamental functions then satisfy $\lambda_r(z_q) = \delta_{qr}$. For Lagrange interpolation, it is easy to find the dual basis

$$\lambda_q(z) = \frac{\prod_{1\leqslant r\leqslant m,\ r\neq q}(z-z_r)}{\prod_{1\leqslant r\leqslant m,\ r\neq q}(z_q-z_r)}. \tag{14}$$

It is also not hard to find the dual basis for univariate Hermite interpolation.

Another, but nonconstructive approach to proving regularity starts by choosing a basis for Π_n^1 (or of V in general). Then Eq. (10) or (12) (or (1) in general) become a linear system of equations for the coefficients of the representation of the interpolant in the chosen basis. We will call the matrix M of this linear system the *Vandermonde matrix* and the determinant D of M the *Vandermonde determinant*. We sometimes write $M(F,V)$ or $M(Z)$ to make the dependency of M or D on the functionals and on the details of the interpolation more explicit.

The method is based on the fact that our interpolation is regular if and only if $D \neq 0$. The interpolation is singular if $D=0$. For Lagrange interpolation, we have the famous formula for the (original) Vandermonde determinant

$$D(Z) = \prod_{1\leqslant q<r\leqslant m} (z_q - z_r)$$

if the monomial basis $\{x^\alpha\}_{\alpha=1}^m$ is taken for Π_{m-1}^1. We can immediately read off of this formula that Lagrange interpolation is regular if and only if the nodes z_q are distinct. A similar formula can be found for the Vandermonde determinant of Hermite interpolation.

A third method, another constructive method, is known as the Newton method. The idea is to start by interpolating one functional, then increase the number of functionals interpolated stepwise (either one at a time, or in packets) until the required set of functionals is interpolated. At each step, one adds a polynomial to the previous interpolant, which interpolates zero values for the functionals already interpolated. In this way, the work done in the previous steps is not spoiled.

Let us carry this out for Lagrange interpolation of the function values c_q at the nodes z_q, $q=1,\ldots,m$. We take $P_1(z)\equiv c_1$. Then P_1 interpolates the first functional (functional evaluation at z_1) and $P_1 \in \Pi_0^1$. Let $Q_2(z) = z - z_1$. Then $Q_2(z_1)=0$ and $R_2(z) = c_2Q_2(z)/Q_2(z_2)$ vanishes at z_1 and takes the value c_2 at z_2. So $P_2 = P_1 + R_2$ interpolates the first two functionals. After finding a polynomial $P_j \in \Pi^1_{j-1}$ which interpolates the first j values, We take

$$Q_{j+1}(z) = \prod_{q=1}^{j}(z - z_q).$$

Then $R_{j+1}(z) = c_{j+1}Q_{j+1}(z)/Q_{j+1}(z_{j+1})$ vanishes at the first j nodes while taking the value c_{j+1} at z_{j+1}. Let $P_{j+1} = P_j + R_{j+1}$. Then $P_{j+1} \in \Pi^1_j$ interpolates the first $j+1$ values.

The formula for the Newton interpolant to a continuous function f is

$$P(f)(x) = \sum_{q=1}^{m} f[z_1,\dots,z_q]\prod_{r=1}^{q-1}(z - z_r), \tag{15}$$

where the divided difference $f[z_r,\dots,z_q]$, $r \leqslant q$, is defined by

$$f[z_r] = f(x_r), \tag{16}$$

$$f[z_r,\dots,z_q] = \frac{f[z_r,\dots,z_{q-1}] - f[z_{r+1},\dots,z_q]}{z_r - z_q}. \tag{17}$$

The Newton form of the interpolant is particularly suitable for obtaining error estimates.

We have gone through these well-known univariate methods in such detail because, as we will see, many multivariate proofs and constructions are based on these principles, although the details may be much more involved.

2.2. *Multivariate Lagrange interpolation*

I claim that univariate and multivariate Lagrange interpolation are two very different animals. To see what I mean, let us look at the first nontrivial bivariate Lagrange interpolation: the interpolation of two values at two nodes z_1, z_2 in $\mathbb{R}^2$. But which space of polynomials should we use? The space of linear polynomials Π^2_1 is spanned by 1, x and y and thus has dimension 3, while the space of constants, Π^2_0, has dimension only one! Our interpolation falls into the gap. There is no "natural" space of polynomials which fits our problem, or at least no obvious natural space.

Now let us try Lagrange interpolation on three nodes $Z = \{z_1, z_2, z_3\}$. We choose $V = \Pi^2_1$, so that (2) is satisfied. Choosing the monomial basis for Π^2_1, the Vandermonde determinant of the interpolation is

$$D(Z) = (x_2 - x_1)(y_3 - y_1) - (x_3 - x_1)(y_2 - y_1),$$

where $z_q = (x_q, y_q)$. If the three nodes are the vertices of a non-degenerate triangle, then $D(Z) \neq 0$. But if the nodes are collinear, say $z_1 = (0,0)$, $z_2 = (1,0)$ and $z_3 = (2,0)$, one can check that $D(Z)$ vanishes and the interpolation is singular.

Thus, we have seen two differences between univariate and multivariate Langrange interpolation. In the multivariate case, it is not clear which interpolation spaces we should choose and, even when there is an easy choice, the interpolation is regular for some knot sets and singular for others. We can be more precise about the latter statement.

Theorem 1. *Let $Z = \{z_q\}_{q=1}^m \subset \mathbb{R}^d$ and $n \in \mathbb{Z}_0$ such that $m = \dim \Pi^d_n$. Then Lagrange interpolation is regular for almost all choices of Z in the Lebesgue measure of $\mathbb{R}^{md}$.*

In fact, the Vandermonde determinant $D(Z)$ of the system of equations for the coefficients of the interpolant is a polynomial in the coordinates of the nodes and therefore either vanishes identically or is nonzero almost everywhere.

The most general statement one can make about Z for which Lagrange interpolation is regular is that Z does not lie on an algebraic curve of degree not exceeding n. However, this statement is almost a tautology and is not constructive.

Many people have worked on finding explicit formulas for arrays for which Lagrange interpolation is regular. Some of them are quite ingenious and beautiful, but all of them are only *necessary* conditions for regularity. We will see some of them in Section 5.

2.3. Multivariate Hermite interpolation

Everything we have said about Lagrange interpolation also holds for multivariate Hermite interpolation including Theorem 1. However, yet another complication rears its head. Consider the simplest non-trivial bivariate Hermite interpolation. This is interpolating function values and both first derivatives at each of two nodes z_1 and z_2 in $\mathbb{R}^2$. We interpolate using quadratic polynomials, i.e., polynomials from Π_2^2. Note that the condition (2) is satisfied. There are 6 functionals to be interpolated and $\dim \Pi_2^2 = 6$. We will show that this interpolation is *singular* for *all* choices of nodes z_1, z_2. We will be seeing this example quite often. For this reason, I call it *my favorite singular Hermite interpolation.*

A common method used to demonstrate the singularity of this interpolation is based on the observation that an interpolation satisfying the condition (2) is singular if and only if there is a nonzero polynomial which satisfies the homogeneous interpolation conditions. That is, there is a $P \in V$ satisfying (9) with all $c_q = 0$.

Now returning to our Hermite interpolation, let $\ell(z)=0$ be the equation of the straight line joining z_1 and z_2. ℓ is a linear polynomial, i.e., in Π_1^2. So $\ell^2 \in \Pi_2^2$. We see that ℓ^2 and both of its first derivatives vanish at both z_1 and z_2. Thus the interpolation is singular.

Sitting back and sifting through the rubble, we are forced to distinguish between three possibilities:

(a) the interpolation is regular for any choice of node set,
(b) the interpolation is regular for almost any choice of node set, but not for all,
(c) the interpolation is singular for any choice of node set.

In the situation where the functionals to be interpolated depends on the choice of the nodes, as for Lagrange or Hermite interpolation, we will say that the interpolation is *regular* if (a) holds, that it is *a.e. regular* if (b) holds, and that it is singular if (c) holds.

We have just shown that both (b) and (c) occur. What about (a)? In the next section, we will show that for Hermite interpolation, (a) can occur if and only if the interpolation is on one node ($m = 1$). This is called *Taylor* interpolation. In that section we also run through the known cases of almost everywhere regular and of singular Hermite interpolations.

In Section 4, we will look at two alternatives to “classical” Hermite interpolation. The first one “lifts” univariate Hermite interpolation to higher dimensions. The second, called the “least interpolation”, constructs the polynomial space to match the functionals and can be used in a much more general context.

The above proof of singularity brings us quite close to a classical question of algebraic geometry.

Given the functionals (6) of Hermite interpolation, what is the lowest degree of a nonzero polynomial Q satisfying the homogeneous equation? One then says that Q has singularities of order k_q+1 at z_q. If (7) is satisfied, then the interpolation is singular if and only the lowest degree is the n of (7). But one can still formulate the question even if (7) does not hold. This question is considered in Section 6.

3. Multivariate Hermite interpolation

3.1. Types of Hermite interpolation

Although we only discussed classical Hermite interpolation in the introduction, a type which we will denote by *total degree* from now on, there are other types. They can be subsumed in the following definition of Hermite interpolation from [38], which replaces derivatives in the coordinate directions by directional derivatives.

Let $z\in\mathbb{R}^d$ and $m\in\mathbb{N}_0$. Given an index set

$$E=(1|\varepsilon_1^1,\ldots,\varepsilon_{r_1^d}^1|\ldots|\varepsilon_1^m,\ldots,\varepsilon_{r_m^d}^m), \tag{18}$$

where $\varepsilon_i^k=1$ or 1, and a diagram $T_z=T_{z,E}$ defined by

$$T_z=(z|y_1^1,\ldots,y_{r_1^d}^1|\ldots|y_1^m,\ldots,y_{r_m^d}^m), \tag{19}$$

where $y_i^k\in\mathbb{R}^d$ and $y_i^k=0$ if $\varepsilon_i^k=0$, one says that E is of tree structure if for each $\varepsilon_i^k=1$, $k>1$, there exists a unique $j=j(i)$ such that $\varepsilon_j^{k-1}=1$. ε_j^{k-1} is called the predecessor of ε_i^k. Moreover, the edges of the tree connect only at a vertex and its predecessor. Note that this definition of a tree is more restrictive than the usual definition of a tree in graph theory.

The trees used here will be specified by their maximal chains. A sequence $\mu=(i_1,\ldots,i_k)$ is called a chain in the tree structure E, if $\varepsilon_{i_1}^1=\cdots=\varepsilon_{i_k}^k=1$, where for each j, $1\leqslant j\leqslant k-1$, ε_i^k is the predecessor of ε_{i+1}^{k+1}. It is called a maximal chain if its last vertex $\varepsilon_{i_k}^k$ is not the predecessor of another element in E.

Let μ be a chain of T_z, $(\mu\in T_z)$. Define

$$\mathbf{y}^\mu=y_{i_1}^1\cdots y_{i_k}^k;\quad \sigma(\mu)=k \tag{20}$$

and the differential operator

$$D_{\mathbf{y}^\mu}^{\sigma(\mu)}=D_{y_{i_1}^1}\cdots D_{y_{i_k}^k}. \tag{21}$$

The chain and the diagram T_z define a product of directional derivatives.

To define Hermite interpolation at a point z, we choose a tree T_z with the additional property that every vertex of T_z is connected to the root z by exactly one chain. Then Hermite interpolation at the point z is defined to be the interpolation of all of the functionals

$$D_{y_{i_1}^1}\cdots D_{y_{i_k}^k}f(z) \tag{22}$$

associated to the vertices of the tree via the unique chain connecting that vertex to the root. One of the important characteristics of this definition is that there are no gaps in the chains of increasing orders of derivatives. Hermite interpolation at nodes $z_1,\ldots,z_m$ would be interpolating the functionals of Eq. (22) for trees $T_{z_1},\dots,T_{z_m}$ by polynomials from Π_n^d. It is assumed that the number of functionals interpolated equals the dimension of the interpolating space.

By restricting the directions y_j^k to the coordinate directions and choosing the tree appropriately, one may obtain Hermite interpolation of total degree.

Another special case is denoted by *Hermite interpolation of coordinate degree*. To motivate this second type, let us look at Lagrange interpolation on a node set which forms a rectangular grid in $\mathbb{R}^2$. Let $x_1 < x_2 < \cdots < x_{n_1+1}$ and $y_1 < y_2 < \cdots < y_{n_2+1}$ be points in $\mathbb{R}$. It seems natural to measure the values of a function on the rectangular grid

$$Z = \{(x_i, y_j) \mid 1 \leqslant i \leqslant n_1 + 1;\ 1 \leqslant j \leqslant n_2 + 1\}. \tag{23}$$

Now let us interpolate these values. Which space of polynomials fits this grid? Surely not some Π_n^2! Instead, we introduce the space of polynomials $\Pi^d_{(n_1,\dots,n_d)}$ of *coordinate degree* $(n_1,\dots,n_d)$.

It is convenient to first introduce more general polynomial spaces of which the space of polynomials of total degree will be a special case. Let $A \subset \mathbb{N}_0^d$. Then the polynomial space Π_A is defined by

$$\Pi_A = \left\{ \sum_{j\in A} a_j z^j \right\}. \tag{24}$$

For example, we recover the space Π_n^d in the form $\Pi_n^d = \Pi_A$ with

$$A = \{ j \in \mathbb{N}_0^d \mid |j| \leqslant n\}.$$

Now the space of polynomials $\Pi^d_{(n_1,\dots,n_d)}$ of *coordinate degree* $(n_1,\dots,n_d)$ is Π_A with

$$A = \{ j \in \mathbb{N}_0^d \mid 0 \leqslant j_i \leqslant n_i,\ i = 1,\dots,d\}. \tag{25}$$

For Lagrange interpolation on the grid (23), at least, $\Pi^d_{(n_1,\dots,n_d)}$ is the right space. Lagrange interpolation on the grid (23) with polynomials from $\Pi^d_{(n_1,\dots,n_d)}$ is regular.

This motivates the definition of *Hermite interpolation of coordinate degree*: Let a positive integer m, d-tuples of non-negative integers $(n_1,\dots,n_d)$, $k_q = (k_{q,1},\dots,k_{q,d})$ for $q = 1,\dots,m$, a node set $Z = \{z_1,\dots,z_m\}$, $z_q \in \mathbb{R}^d$ and a set of values $c_{q,\alpha}$ be given. Find a $P \in \Pi^d_{(n_1,\dots,n_d)}$ with

$$D^\alpha P(z_q) = c_{q,\alpha} \quad 0 \leqslant \alpha_i \leqslant k_{q,i},\ 1 \leqslant i \leqslant d,\ 1 \leqslant q \leqslant m. \tag{26}$$

The numbers n and k_q are assumed to satisfy

$$\prod_{i=1}^d (n_i + 1) = \sum_{q=1}^m \prod_{i=1}^d (k_{q,i} + 1). \tag{27}$$

If we interpolate the same derivatives at each node, then we have *uniform* Hermite interpolation of type either total or coordinate degree. Of course, (7) and (27) are to be satisfied with all k_q equal.

3.2. Everywhere regular schemes

In this subsection, we will consider those interpolation schemes which are regular for any location of the nodes. For Hermite interpolation of type either total or coordinate degree, this is only the case if there is only one node in the interpolation, $Z = \{z\}$. Such interpolations are called Taylor interpolations. Then there is a one-to-one correspondence between the partial derivatives to be interpolated (at z) and the monomial basis of Π_n^d, respectively $\Pi_{(n_1,\dots,n_d)}^d$: $D^\alpha \leftrightarrow z^\alpha$. The Vandermonde matrix based on the monomial basis of the polynomial space, say in the lexicographical ordering, with the derivatives taken in the same order is an upper triangular matrix with the non-zero diagonal entries $\alpha!$. So the determinant never vanishes. We have proved that Taylor interpolation is regular for any choice of the node.

It is, in fact, the only such multivariate Hermite interpolation, see [28].

Theorem 2. *In $\mathbb{R}^d$, $d \geqslant 2$, the only Hermite interpolation of type total or coordinate degree, which is regular for all choices of the nodes, is Taylor interpolation.*

This theorem is also true for more general polynomial spaces. The most general form I know is by Jia and Sharma, [24]. To formulate it, we need some terminology. Let $V \subset \Pi^d$ be a finite-dimensional space of polynomials. V is said to be scale invariant if $P(az) \in V$ for any $a \in \mathbb{R}$ and any $P \in V$. Also, for any polynomial $P = \sum_j a_j z^j$, the differential operator $P(D)$ is defined by

$$P(D)f = \sum_j a_j D^j f. \tag{28}$$

Now let Z be a node set. To each $z_q \in Z$, let there be associated a finite-dimensional space of polynomials $V_q \subset \Pi^d$. We choose any bases $P_{q,1},\dots,P_{q,r(q)}$ of V_q, $q = 1,\dots,m$ (then $r(q) = \dim V_q$), values $c_{q,i}$ and a set $A \in \mathbb{N}_0^d$. The *Abel-Goncharov* interpolation problem is to find a polynomial $Q \in \Pi_A$ (recall definition (24)) satisfying

$$P_{q,i}(D)Q(z_q) = c_{q,i}, \quad 1 \leqslant i \leqslant r(q),\ q = 1,\dots,m. \tag{29}$$

The choice of the bases does not affect the regularity of the interpolation.

Theorem 3. *Using the above notation, let V, V_q, $q = 1,\dots,m$, be scale invariant. Then Abel-Goncharov interpolation* (29) *is regular for any node set if*

$$V = \bigoplus_{q=1}^{m} V_q. \tag{30}$$

For a special case, Jia and Sharma prove more

Theorem 4. *Let $A \subset \mathbb{N}_0^d$. Let $A_q \subset A$, $q = 1,\dots,m$, $V = \Pi_A$, $V_q = \Pi_{A_q}$, $q = 1,\dots,m$. Then Abel–Goncharov interpolation* (29) *is regular for any node set Z if and only if A is the disjoint sum of the A_q, $q = 1,\dots,m$.*

This theorem implies Theorem 2. For example, for Hermite interpolation of type total degree, $\Pi_n^d = \Pi_A$ with $A = \{j \mid j \in \mathbb{N}_0^d, |j| \leqslant n\}$ while the derivatives to be interpolated derive from the polynomial spaces $V_q = \Pi_{A_q}$ with $A_q = \{j \mid j \in \mathbb{N}_0^d, |j| \leqslant k_q\}$. The same holds for Hermite interpolation of type coordinate degree. It also includes a similar theorem for multivariate Birkhoff interpolation.

In view of the these results, Jia and Sharma formulated the conjecture

Conjecture 5. *Let V and $\{V_q\}_{q=1}^{m}$ be scale-invariant subspaces of Π^d such that*

$$\bigcup_{q=1}^{m} V_q \subset V.$$

Then Abel-Goncharov interpolation (29) *is regular if and only if* (30) *holds.*

3.3. A.e. regular Hermite interpolation of type total degree in $\mathbb{R}^2$

From the previous section, we have seen that multivariate Hermite interpolation can, except for some very special cases, be at most regular for almost all choices of nodes. We denote this as being regular a.e. Not too much is known about even a.e. regular Hermite interpolation of type total degree. Gevorkian et al. [16] have proven

Theorem 6. *Bivariate Hermite interpolation of type total degree* (8) *is regular a.e. if there are at most 9 nodes with $k_q \geqslant 1$.*

Sauer and Xu [38] have proven

Theorem 7. *Multivariate Hermite interpolation of type total degree* (8) *in $\mathbb{R}^d$ with at most $d+1$ nodes having $k_q \geqslant 1$ is regular a.e. if and only if*

$$k_q + k_r < n$$

for $1 \leqslant q, r \leqslant m$, $q \neq r$.

The authors of Theorem 6 have also considered Hermite interpolation of type total degree from the point of view of algebraic geometry. These results will be discussed in Section 6. A conjecture due to them and Paskov [18,34], simultaneously does fit in here. Let us use the stenographic notation $\mathscr{N} = \{n; s_1, \ldots, s_m\}$ to stand for Hermite interpolation of type total degree on m nodes interpolating derivatives of order up to $k_q = s_q - 1$ at z_q by bivariate polynomials of degree n. We are using the standard notation of algebraic geometry here, working with orders of singularities s_q. Also, we will not be much concerned about the node set Z, since we are only interested in regularity a.e. In addition, we allow an s_q to be 0. This just means that there is no interpolation condition at that node. We also do not demand that the condition (7) requiring that the number of degrees of freedom in the interpolation space equals the number of functionals to be interpolated, holds.

With this freedom, we do not have interpolations n the strict sense of Equation 2 any more, so let us just call them *singularity schemes*. We introduce addition among singularity schemes just by vector addition. If $\mathscr{N} = \{n; s_1, \ldots, s_m\}$ and $\mathscr{R} = \{r; t_1, \ldots, t_m\}$, then

$$\mathscr{N} + \mathscr{R} = \{n + r; s_1 + t_1, \ldots, s_m + t_1\}. \tag{31}$$

We can add singularity schemes of different lengths by introducing zeros in the shorter of them.

The relevance of this addition is that if $Q_1 \in \Pi_n^2$ satisfies the homogeneous interpolation conditions of $\mathscr{N}$ and if $Q_2 \in \Pi_r^2$ satisfies the homogeneous interpolation conditions of $\mathscr{R}$, where $\mathscr{N}$ and $\mathscr{R}$ are of the same length and refer to the same nodes, then $Q_1 Q_2 \in \Pi_{n+r}^2$ satisfies the homogeneous interpolation conditions of $\mathscr{N} + \mathscr{R}$,

The conjecture is

Conjecture 8. *Let $\mathcal{N}$ correspond to a Hermite interpolation of type total degree (that is condition (7) holds). Then $\mathcal{N}$ is singular if and only if there are schemes $\mathcal{R}_i = \{r_i; t_{i,1}, \ldots, t_{i,m}\}, =1, \ldots, p$ satisfying*

$$\binom{r_i+2}{2} > \sum_{q=1}^{m} \binom{t_{i,q}+2}{2}, \quad i=1,\ldots,p,$$

such that

$$\mathcal{N} = \sum_{i=1}^{p} \mathcal{M}_i.$$

The sufficiency part of this theorem is easy to see. There are always nonzero polynomials Q_i satisfying the homogeneous interpolation conditions of $\mathcal{R}_i$, $i=1,\ldots,p$, since each of them can be found by solving a linear system of homogeneous equations with less equations than unknowns. By the above remark, $\Pi_{i=1}^{p} Q_i \in \Pi_n^2$ is a nonzero polynomial satisfying the homogeneous conditions for $\mathcal{N}$.

We have already seen an example of this. It is our favorite singular Hermite interpolation: the interpolation of first derivatives at two nodes in $\mathbb{R}^2$ by polynomials from Π_2^2. In the notation used here, this is the singularity scheme $\mathcal{N} = \{2; 2, 2\}$. It can be decomposed as $\mathcal{N} = \mathcal{R} + \mathcal{S}$, with $\mathcal{R} = \mathcal{S} = \{1; 1, 1\}$. More singular Hermite interpolations constructed using this idea can be found in [34; 28 Chapter 4].

There are essentially no other results for the a.e. regularity of general Hermite interpolation of total degree. More is known for uniform Hermite interpolation of total degree. The results which follow can be found in [38,28].

The simplest case of uniform Hermite interpolation is, of course, Lagrange interpolation in which partial derivatives of order zero are interpolated at each node. The number of nodes in $\dim \Pi_n^2$ for some n and it is regular a.e.

The next case is when all partial derivatives up to first order are interpolated at each node. Condition (7) is then

$$\binom{n+2}{2} = m \binom{1+2}{2} = 3m. \tag{32}$$

This equation has a solution for n and m if and only if $n = 1, 2 \bmod 3$.

Theorem 9. *For all n with $n = 1, 2 \bmod 3$, bivariate uniform Hermite interpolation of type total degree interpolating partial derivatives of order up to one is regular a.e., except for the two cases with $n=2$ (then $m=2$) and $n=4$ (then $m=5$). The two exceptional cases are singular.*

Note that our favorite singular Hermite interpolation is included. The smallest non-Taylor a.e. regular case is for $n=5$. The interpolation is then on 7 nodes.

The method of the proof of this and the following two theorems is to show that the Vandermonde determinant does not vanish identically by showing that one of its partial derivatives is a nonzero constant. The technique used to show this is the "coalescence" of nodes and, roughly speaking, tries to reduce the number of nodes of the interpolation until a Taylor interpolation is obtained.

If all partial derivatives up to second order are interpolated at each node, condition (7) becomes

$$\binom{n+2}{2} = m\binom{2+2}{2} = 6m. \tag{33}$$

This equation has a solution for n and m if and only if $n = 2, 7, 10, 11 \bmod 12$.

Theorem 10. *For all n with $n = 2, 7, 10, 11 \bmod 12$, bivariate uniform Hermite interpolation of type total degree interpolating partial derivatives of order up to two is regular a.e.*

The smallest non-Taylor a.e. regular case is for $n = 7$. The interpolation is then on 12 nodes. For third derivatives, we must have $n = 3, 14, 18, 19 \bmod 20$.

Theorem 11. *For all n with $n = 3, 14, 18, 19 \bmod 20$, bivariate uniform Hermite interpolation of type total degree interpolating partial derivatives of order up to three is regular a.e.*

The smallest non-Taylor a.e. regular case is for $n = 14$. The interpolation is then also on 12 nodes. For related results from algebraic geometry, see Section 6.

3.4. *A.e. regular bivariate Hermite interpolation of type coordinate degree in $\mathbb{R}^2$*

No theorems about the a.e. regularity of general Hermite interpolation of type coordinate degree in $\mathbb{R}^d$, as defined in (26), are known except for the relatively simple one given at the end of this subsection. But there are a few things known of the uniform case. If we want to interpolate all partial derivatives of order up to k_1 in x and to k_2 in y from $\Pi^2_{(n_1,n_2)}$, then

$$(n_1+1)(n_2+1) = m(k_1+1)(k_2+1) \tag{34}$$

must hold. The proofs of the the following theorems, all of which can be found in [28], are based on the same techniques as for the theorems on a.e. regularity of uniform Hermite interpolation of type total degree in the previous subsection.

Theorem 12. *If* (34) *is satisfied, and either k_1+1 divides n_1+1 or k_2+1 divides n_2+1, then bivariate uniform Hermite interpolation of type coordinate degree interpolating all partial derivatives of order up to k_1 in x and to k_2 in y from $\Pi^2_{(n_1,n_2)}$ is a.e. regular.*

This is more general than tensor product interpolation, since there one would have that both k_1+1 divides n_1+1 and k_2+1 divides n_2+1. If $n_1 = n_2 = n$ and $k_1 = k_2 = k$, which is a kind of (uniform)^2 Hermite interpolation, then (34) forces k to divide n and we have

Corollary 13. *Bivariate uniform Hermite interpolation of type coordinate degree interpolating all partial derivatives of order up to k in x and in y from $\Pi^2_{(n_1,n_2)}$ is a.e. regular.*

This corollary also holds in $\mathbb{R}^d$, but theorems like Theorem 12 in $\mathbb{R}^d$ require much more restrictive assumptions.

As for uniform Hermite interpolation of type total degree, interpolations involving only lower order derivatives can be taken care of completely.

Theorem 14. *For all combinations, except two, of k_1, k_2 with $0 \leqslant k_1, k_2 \leqslant 2$ and n_1, n_2 with $0 \leqslant k_1 \leqslant n_1$, $0 \leqslant k_2 \leqslant n_2$ satisfying* (34), *uniform bivariate Hermite interpolation of type coordinate degree interpolating all partial derivatives of order up to k_1 in x and to k_2 in y from $\Pi^2_{(n_1,n_2)}$ is a.e. regular. The two exceptional cases are $k_1 = 1$ and $k_2 = 2$ from $\Pi^2_{(2,3)}$ and the corresponding interpolation with x and y interchanged. These are singular.*

This theorem includes cases Theorem 12 does not cover. For example, interpolating partial derivatives of first order in x and second order in y at each of eight nodes from $\Pi^2_{(8,3)}$ is regular a.e. But Theorem 12 does not apply since neither 2 divides 9 nor does 3 divide 4.

We conclude this subsection with a theorem on non-uniform interpolation.

Theorem 15. *A bivariate Hermite interpolation of type coordinate degree interpolating all partial derivatives of order up to $k_{q,1}$ in x and to $k_{q,2}$ in y at z_q, $q = 1, \ldots, m$ from $\Pi^2_{(n_1,n_2)}$ is a.e. regular if the rectangle $(0, n_1 + 1) \times (n_2 + 1)$ is the disjoint union of the translates of the rectangles $(0, k_{q,1} + 1) \times (k_{q,2} + 1)$ $q = 1, \ldots, m$.*

This theorem does not hold in $\mathbb{R}^d$ for $d \geqslant 3$.

3.5. Singular Hermite interpolations in $\mathbb{R}^d$

The general trend of the results of this subsection will be that a Hermite interpolation in $\mathbb{R}^d$ will be singular if the number of nodes is small, typically $m \leqslant d + 1$. Of course, Taylor interpolations are excepted. Also Lagrange interpolation by linear polynomials ($m = d + 1$) is excluded. The theorems can all be found in [28,38].

Theorem 16. *Hermite interpolation of type total degree in $\mathbb{R}^d$, $d \geqslant 2$, is singular if the number of nodes satisfies $2 \leqslant m \leqslant d + 1$ except for the case of Lagrange interpolation which is a.e. regular.*

Implicitly, condition (7) is assumed to be satisfied.

The theorem includes our favorite singular Hermite interpolation. It is proved showing that the interpolation restricted to a certain hyperplane is not solvable.

One application of this theorem is a negative result related to the construction of finite elements. The statement is that there is no finite element interpolating all derivatives up to a given order (which may depend on the vertex) at each of the vertices of a tetrahedron in $\mathbb{R}^d$, $d \geqslant 2$, which interpolates from a complete space, say Π^d_n. The existence of such an element would have been desirable as it would have combined the highest degree approximation, $n + 1$, and global continuity available for a given amount of computational effort.

For interpolation of type coordinate degree, we have

Theorem 17. *Bivariate uniform Hermite interpolation of type coordinate degree interpolating all partial derivatives of order* k_1 *in* x *and to* k_2 *in* y *at either two or three nodes from* $\Pi^2_{(n_1,n_2)}$ *is singular unless either* p_1+1 *divides* n_1+1 *or* p_2+1 *divides* n_2+1.

The exceptional cases are regular by Theorem 12.

Singular uniform Hermite interpolation schemes on more than $d+1$ nodes in $\mathbb{R}^d$ are hard to find. Here are some for uniform Hermite interpolation of type total degree due to Sauer and Xu [38].

Theorem 18. *Uniform Hermite interpolation of type total degree interpolating all partial derivatives of order up to k at each of m nodes in* $\mathbb{R}^d$ *by polynomials from* Π^d_n *is singular if*

$$m < \frac{(n+1)\cdots(n+d)}{((n-1)/2+1)\cdots((n-1)/2+d)}.$$

In $\mathbb{R}^2$, the smallest example is covered by this theorem is the interpolation of all partial derivatives of up to first order at each of 5 nodes by quartic polynomials.

3.6. Conclusions

We have seen that Hermite interpolation of either total or coordinate degree type is regular if $m=1$, is singular if there are not too many nodes and, for uniform Hermite interpolation of type coordinate degree, there are no other exceptions if we are interpolating derivatives of order up to one, two or three. Really general theorems for a.e. regularity of interpolations with arbitrarily high derivatives are not known. The same holds for the results obtained by the methods of algebraic geometry, as we will see in Section 6.

In this section, we have assiduously ignored one of the essential components of interpolation. Most of the theorems of this section concerned a.e. regularity. To use these interpolations, one must find concrete node sets for which the interpolations are regular. Only for Lagrange interpolation are any systematic results are known. Otherwise, nothing is known. For this reason, special constructions of interpolations which include the node sets, such as those in Sections 5 and 6, are of importance, even if the interpolation spaces are not complete spaces Π^d_n.

4. Alternatives to classical multivariate Hermite interpolation

4.1. Lifting schemes

Our extension of univariate to "classical" multivariate Hermite interpolation, namely to multivariate Hermite interpolation of type total degree, has one glaring defect. It is not regular for any choice of nodes. It is in fact possible to get rid of this unfavorable property, but alas, at a price. The method to be introduced here, called "lifting", yields a multivariate interpolation which is formulated exactly as in the univariate case. In our context, it was introduced by Goodman [19] motivated by the first special cases, those of Kergin [25], Hakopian [21] and of Cavaretta et al. [5]. For a more complete survey, see the book [1] of Bojanov, Hakopian and Sahakian, or the thesis of Waldron [43] and the references therein.

Let us first formulate it for Lagrange interpolation: given m nodes $z_1,\dots,z_m$ in $\mathbb{R}^d$ and values $c_1,\dots,c_m$, there is a polynomial $P\in\Pi^d_{m-1}$ which interpolates the values

$$P(z_q)=c_q,\qquad q=1,\dots,m. \tag{35}$$

So what is the problem? The problem is that $\dim\Pi^d_{m-1}$ is much larger than m, so that Eq. (35) does not determine P uniquely. As we will see, there are many different ways to add additional interpolation conditions in order to make the interpolant unique.

But wait. Let us be more cautious. If the nodes can be chosen arbitrarily, can we be sure that there is at least one P in Π^d_{m-1} satisfying (35). Severi, [41], has already answered this question for us in the context of Hermite interpolation.

We will say that a Hermite interpolation (of type total degree) is *solvable* with polynomials from Π^d_n if given m, orders $\{k_q\}_{q=1}^m$ and values $c_{q,\alpha}$ for $0\leqslant|\alpha|\leqslant k_q$, $q=1,\dots,m$, there is, for any node set $Z=\{z_q\}_{q=1}^m$, a $P\in\Pi^d_n$ with

$$D^\alpha P(z_q)=c_{q,\alpha}\qquad 0\leqslant|\alpha|\leqslant k_q, q=1,\dots,m. \tag{36}$$

Theorem 19. *Let m and nonnegative integers $\{k_q\}_{q=1}^m$ be given. A necessary and sufficient condition that Hermite interpolation of type total degree be solvable with polynomials from Π^d_n is that*

$$n+1\geqslant\sum_{q=1}^m(k_q+1).$$

Thus, we see that Lagrange interpolation on m nodes is solvable from Π^d_{m-1}. Note also that the condition is the same for any dimension d. Actually, the case of Lagrange interpolation can be done directly. One can construct a Lagrange interpolating function from Π^d_{m-1} for a node by taking the product of $m-1$ hyperplanes passing through the other nodes but not through the given nodes.

We start with some definitions, Let $\Theta=\{\theta_1,\dots,\theta_m\}\in\mathbb{R}^d$ be a set of points with some of the nodes possibly repeated. Here and in the following, Θ denotes a point set with some of the points possibly repeated, while Z is our old notation of a point set with no points repeated. Given an integrable function f on $\mathbb{R}^d$ and a point set Θ, its *divided difference* $I_\Theta(f)$ of order $m-1$ is defined to be

$$I_\Theta(f)=\int_0^1\cdots\int_0^1 f(\theta_1+s_1(\theta_2-\theta_1)+\cdots+s_{m-1}(\theta_m-\theta_{m-1}))\,\mathrm{d}s_{m-1}\cdots\mathrm{d}s_1. \tag{37}$$

This is a direct generalization of the univariate divided difference (Eq. (16)) and leads to error estimates via remainder formulas (see [15, Section 2]).

For $z\in\mathbb{R}^d$ and f a continuously differentiable function on $\mathbb{R}^d$, the directional derivative, D_zf, of f in the direction z is denoted by

$$D_zf=z\cdot\nabla f.$$

To each $\lambda\in\mathbb{R}^d$, we associate the functional λ^* on $\mathbb{R}^d$ defined by $\lambda^*z=\lambda\cdot z$ (the Euclidian scalar product).

A *plane wave*, or ridge function, h in $\mathbb{R}^d$ is the composition of a functional and a univariate function $g:\mathbb{R}\to\mathbb{R}$

$$h(z)=(g\circ\lambda^*)(z)=g(\lambda^*z)\qquad\text{for } z\in\mathbb{R}^d. \tag{38}$$

Let $C^s(\mathbb{R}^d)$ be the space of s times continuously differentiable functions on $\mathbb{R}^d$.
Now the main definition

Definition 20. Let $s \in \mathbb{N}_0$ be given and L associate with each finite set $\Theta \subset \mathbb{R}$ of points (possibly repeated) a continuous linear map L_Θ

$$L_\Theta : C^s(\mathbb{R}) \to C^s(\mathbb{R}).$$

A continuous linear map

$$\mathscr{L}_\Theta : C^s(\mathbb{R}^d) \to C^s(\mathbb{R}^d)$$

is the *lift* of L to $\Theta \subset \mathbb{R}^d$ if it satisfies

$$\mathscr{L}_\Theta(g \circ \lambda^*) = (L_{\lambda^* \Theta} g)\lambda^* \tag{39}$$

for any $\lambda \in \mathbb{R}^d$ and any $g \in C^s(\mathbb{R}^d)$. Here $\lambda^* \Theta = \{\lambda^* \theta_1, \ldots, \lambda^* \theta_m\} \subset \mathbb{R}$ for $\Theta \subset \mathbb{R}^d$.

Since it is by no means clear that each map L has a lift, for example the univariate finite difference map has no lift, we say that L is *liftable* if it has a lift to each set Θ of point in $\mathbb{R}^d$. A lift, if it exists, is unique.

The maps we want to lift are Lagrange and Hermite maps. Given a node set $Z = \{z_1, \ldots, z_m\} \subset \mathbb{R}$ and $g \in C(\mathbb{R})$, the Lagrange map $L_Z g$ delivers the interpolant to g from Π^1_{m-1}. More generally, the Hermite map H_Θ based on the univariate Hermite interpolation given by (11) and (12) associates with each $g \in C^n(\mathbb{R})$, the Hermite interpolant from Π^d_n to the values $c_{q,\alpha} = D^\alpha g(z_q)$. Here, Z is the set of distinct points in Θ and k_q are their multiplicities. g was required to be in $C^n(\mathbb{R})$ with $n = \sum_{q=1}^m (k_q + 1) - 1$ since potentially all the points in Θ could coincide. Note also that the Hermite map becomes the Lagrange map when all the points of Θ are distinct.

The lift $\mathscr{L}$ of say the Lagrange map would necessarily associate to each node set $Z \subset \mathbb{R}^d$ and function $f \in C(\mathbb{R})$ a multivariate polynomial from Π^d_{m-1}. In fact, let $f \in C(\mathbb{R}^d)$. Then f can be approximated arbitrarily well by linear combinations of plane waves. Since, by definition, $\mathscr{L}_\Theta$ is continuous, $\mathscr{L}_\Theta(f)$ can be approximated arbitrarily well by linear combinations of functions of the form $(L_{\lambda^* \Theta} g) \circ \lambda^*$. But each $(L_{\lambda^* \Theta} g)$ is a univariate polynomial (in Π^1_{m-1}) and, consequently, so is $\mathscr{L}_\Theta(f)$ as their limit.

For the same reason, a lift $\mathscr{H}_\Theta$ of the Hermite map would also map $C^s(\mathbb{R}^d)$ to Π^d_n.

Before we formulate the general theorem on lifting Hermite maps, let us describe their precursors. Kergin (see [25]) first constructed a lift of the Hermite map but without using the concept of lifting.

Theorem 21. *Given a set of not necessarily distinct points $\Theta = \{\theta_1, \ldots, \theta_{n+1}\}$ from $\mathbb{R}^d$, there exists a unique linear map $\mathscr{H}_\Theta : C^n(\mathbb{R}^d) \to \Pi^d_n$ such that for each $f \in C^n(\mathbb{R}^d)$, each $Q \in \Pi^d_k$, $0 \leqslant k \leqslant n$ and each $J \subset \{1, \ldots, n+1\}$, there is a $z \in \operatorname{span}\{\theta_q\}_{q \in J}$ such that*

$$Q(D)(\mathscr{L}_\Theta(f) - f)(z) = 0.$$

Choosing $J = \{q\}$ in this theorem shows that $\mathscr{H}_\Theta(f)(\theta_q) = f(\theta_q)$, so that $\mathscr{H}_\Theta$ indeed interpolates function values. Similarly, if θ_q is repeated k_q times, then all partial derivatives of f up to order k_q are interpolated at θ_q.

Micchelli, respectively, Micchelli and Milman [29,30] showed how to construct the Kergin interpolant by showing that it interpolates the functionals

$$f(\theta_1), I_{\{\theta_1,\theta_2\}}(\partial f/\partial x_i), \qquad i=i,\dots,d,$$

$$\dots, \qquad I_{\{\theta_1,\dots,\theta_{n+1}\}}(D^\alpha f) \quad |\alpha|=n.$$

This is not such a nice representation of the functionals since, first of all, it is not clear from the formula that the interpolant is independent of the order in which the points θ_q are arranged and, secondly, it assumes an order of differentiability of f which is not really necessary. In fact, as was remarked in the original papers, it is independent of the choice of the nodes and, if the nodes are in general position, then Kergin's interpolant has a continuous extension from $C^n(\mathbb{R}^d)$ to $C^{d-1}(\mathbb{R}^d)$.

To elucidate this, let us look Lagrange interpolation at three noncollinear nodes $\{\theta_1,\theta_2,\theta_3\}$ in $\mathbb{R}^2$. The Kergin interpolant is from Π_2^2 whose dimension is 6. According to the representation given above, the functionals to be interpolated are $f(\theta_1)$, the average values of $\partial f/\partial x$ and $\partial f/\partial y$ along the line joining z_1 with z_2, and the average values of $\partial^2 f/\partial x^2$, $\partial^2 f/\partial x\partial y$ and $\partial^2 f/\partial y^2$ over the triangle formed by the nodes. In [28], it was shown that one obtains the same interpolation if one interpolates the functionals $f(z_q)$, $q=1,2,3$ and the average values of $\partial f/\partial n_i$ over the ith side of the triangle, $i=1,2,3$. Here, n_i is the normal to the ith side. This representation is symmetric in the nodes and one sees that only $f\in C^1(\mathbb{R}^2)$ is required instead of $C^2(\mathbb{R}^2)$. Waldron [44] gives explicit formulas for the general case.

Interpolating average values of functions or their derivatives is not as exotic as it may seem. For example, a well-known finite element, the Wilson brick (see [7]), interpolates average values of second derivatives.

Another generalization of univariate Hermite interpolation to $\mathbb{R}^d$ was given by Hakopian [21]. Let $\Theta\subset\mathbb{R}^d$ have $m\geqslant d+1$ nodes in general position, meaning that any $d+1$ of them form a nondegenerate tetrahedron. Then the problem of interpolating the functionals

$$I_{\tilde\Theta} \qquad \text{for all } \tilde\Theta\subset\Theta \text{ with } |\tilde\Theta|=d$$

for $f\in C(\mathbb{R}^d)$ with polynomials from Π_{m-d}^d is regular. Here $|\Theta|$ is the cardinality of Θ. Note that there are exactly

$$\binom{m}{d}$$

interpolation conditions, which is also the dimension of Π_{m-d}^d.

Take three noncollinear nodes z_1,z_2,z_3 in $\mathbb{R}^2$. Hakopians' interpolant interpolates the average value of a function over the three sides of the triangle formed by the nodes.

Goodman [19] showed that these two types of interpolations are the end points of a whole scale of univariate Hermite interpolations, which are all liftable.

Let $f\in C(\mathbb{R})$ and let $D^{-r}f$ be any function with $D^r(D^{-r})=f$. Then the generalized Hermite map $H_\Theta^{(r)}$ associated with r and the set $\Theta\subset\mathbb{R}$, with $|\Theta|=n+1$, is the map

$$H_\Theta^{(r)}: C^{n-r}(\mathbb{R})\to\Pi_{n-r}^1$$

given by

$$H_\Theta^{(r)}f=D^rH_\Theta D^{-r}f.$$

Theorem 22. *For any r, n with $0 \leqslant r \leqslant n$, $H_\Theta^{(r)}$, with $|\Theta| = n+1$, is liftable to $\mathscr{H}_\Theta^{(r)} : C^{n-r}(\mathbb{R}^d) \to \Pi_{n-r}^d$. The functionals interpolated are*

$$I_{\tilde{\Theta}}(P_\ell(D)f) \tag{40}$$

for all $\tilde{\Theta} \subset \Theta$ with $|\tilde{\Theta}| \geqslant r+1$ and where $\{P_\ell\}$ is a basis for the homogeneous d-variate polynomials of degree $|\tilde{\Theta}| - r - 1$.

If $r = 0$, we obtain the Kergin map. If $r = d-1$, $n \geqslant d-1$ and the points of Θ are in general position, we get the Hakopian map.

4.2. The least interpolant

Would it not be nice if given *any* functionals F_q, $q = 1,\ldots,m$, we could find a polynomial space V such that the equations

$$F_q P = c_q, \qquad q = 1,\ldots,m$$

have a unique solution $P \in V$ for any c_q? Note that the point of view has changed here. Given the functionals, we do not know which interpolation space to use, we derive it.

Well, if one is not too choosy, such a space almost always exists. All one needs is that the functionals be linearly independent in the space of all polynomials. Then a simple elimination argument shows that there are polynomials Q_r (dual polynomials) so that $F_q Q_r = \delta_{q,r}$. The linear span of these polynomials forms a possible space V.

The above construction is clearly not unique. One can require the interpolation space to have desirable properties. It turns out that the condition that the degree of the interpolant be as low as possible is a key concept, connecting the quest for good interpolation spaces with, among other things, Gröbner bases of polynomial ideals. This connection will be explained for Lagrange interpolation. Let $Z = \{z_1, \cdots, z_m\}$ be a set of nodes in $\mathbb{R}^d$ and V a space of polynomials from which Lagrange interpolation is regular. V is called degree reducing (the interpolant from V is degree reducing) if for any polynomial P, its interpolant Q satisfies $\deg Q \leqslant \deg P$.

A set of polynomials $\{P_\alpha | \alpha \in I \subset \mathbb{N}^d\}$ is called a Newton basis with respect to Z if Z can be indexed as $\{z_\alpha | \alpha \in I\}$ so that for any α, $\beta \in I$ with $|\alpha| \leqslant |\beta|$, one has $P_\alpha(z_\beta) = \delta_{\alpha,\beta}$ and for any n, there is a decomposition

$$\Pi_n^d = \operatorname{span}\{P_\alpha \mid |\alpha| \leqslant n\} \oplus \{Q \in \Pi_n^d \mid Q(z_q) = 0, q = 1,\ldots,m\}.$$

Sauer [35] shows that

Theorem 23. *A space of polynomials V has a Newton basis with respect to Z if and only if it is degree reducing for Z.*

Even degree reducing spaces are not unique except when $V = \Pi_n^d$ for some n. These ideas can be generalized to Hermite interpolation and to orderings of Π^d which are compatible with addition, see [15,35, Section 3].

A closely related subject is that of Gröbner bases and H-bases. Let $\prec$ be a total ordering of $\mathbb{N}^d$ which has 0 as its minimal element and is compatible with addition (in $\mathbb{N}^d$). We will use it to order the terms of a polynomial for which reason we call it a term order. Let I be an ideal of polynomials. A finite set $\mathscr{P}$ is called a Gröbner basis for I if any polynomial $Q \in I$ can be written as

$$Q = \sum_{P \in \mathscr{P}} Q_P P,$$

where the (term order) degree of any summand does not exceed the degree of Q. If the term order degree is replaced by total degree, then we have an H-basis.

Now suppose I is the ideal associated with the node set Z, i.e., $I = \{P \mid P(z_q) = 0,\ q = 1, \ldots, m\}$ and $\mathscr{P}$ a Gröbner or an H-basis for I. Given a set of values to interpolate, let Q be any polynomial which interpolates them. We really want an interpolant of lowest possible degree (in our chosen ordering), so if the degree of Q is too high, we can eliminate the term of highest degree with some polynomial multiple of one of the basis polynomials. Continuing, we reduce Q as much as possible, arriving at a "minimal degree interpolant" with respect to the order used. This also works the other way around. One orders monomials according to some term order. Then one uses Gaussian elimination to find an interpolant. During the elimination, one encounters zero columns. The linear combinations producing these columns are just the coefficients of a Gröbner or an H-basis polynomial. We again refer to the survey paper [15] for the precise formulations of the above sketchy ideas. Other sources are Sauer [35,36], Buchberger [3], and Groebner [20].

The least interpolant of de Boor and Ron [8], which we will define now, is such a degree reducing interpolant. It uses an ordering of degrees "in blocks" of total degree.

Definition 24. Let g be a real-analytic function on $\mathbb{R}^d$ (or at least analytic at $z = 0$)

$$g(z) = \sum_{|j|=0}^{\infty} a_j z^j.$$

Let j_0 be the smallest value of $|j|$ for which some $a_j \neq 0$. Then the *least term* of $g_\downarrow$ of g is

$$g_\downarrow = \sum_{j=j_0} a_j z^j.$$

Note that $g_\downarrow$ is a homogeneous polynomial of degree j_0.

Theorem 25. *Given a node set* $Z \subset \mathbb{R}^d$, *let*

$$\mathrm{Exp}_Z = \mathrm{span}\{\mathrm{e}^{z \cdot z_q} \mid q = 1, \ldots, m\}$$

and

$$\mathscr{P}_Z = \mathrm{span}\{g_\downarrow \mid g \in \mathrm{Exp}_Z\}.$$

Then Lagrange interpolation from $\mathscr{P}_Z$ *to values at* Z *is regular.*

The map $L : C(\mathbb{R}^d) \to \mathscr{P}_Z$ with Lf being the interpolant to the values of f on Z is called the least Lagrange interpolant. Note that $\dim \mathscr{P}_Z = \dim \mathrm{Exp}_Z$.

Let us now look at some examples in $\mathbb{R}^2$. First, let $Z = \{(0,0),\ (1,0),\ (0,1)\}$. We know that Lagrange interpolation from Π_1^2 at these nodes is regular. What does the least interpolant look like?

We have

$$\mathrm{Exp}_Z = \mathrm{span}\{\mathrm{e}^0,\ \mathrm{e}^x,\ \mathrm{e}^y\}$$
$$= \left\{ (a+b+c) + bx + cy + b\sum_{j=2}^{\infty}\frac{x^j}{j!} + c\sum_{j=2}^{\infty}\frac{y^j}{j!} \mid a,b,c\in\mathbb{R} \right\}$$

so that

$$\mathscr{P}_Z = \mathrm{span}\{1,x,y\} = \Pi_1^2.$$

If the nodes are collinear, say $Z=\{(0,0),\ (1,0),\ (2,0)\}$, we get

$$\mathrm{Exp}_Z = \mathrm{span}\{\mathrm{e}^0,\ \mathrm{e}^x,\ \mathrm{e}^y\}$$
$$= \left\{ (a+b+c) + (b+2c)x + (b+4c)x^2 + \sum_{j=3}^{\infty}(b+2^jc)\frac{x^j}{j!} \mid a,b,c\in\mathbb{R} \right\}$$

so that

$$\mathscr{P}_Z = \mathrm{span}\{1,x,x^2\},$$

which is the correct *univariate* space for Lagrange interpolation on three nodes on a line.

For the general case, we require that the functionals F_q be of the form $P(D)(z_q)$, where $P\in\Pi^d$ is a polynomial and $z_q\in\mathbb{R}^d$. Then the formal power series associated with a functional F is

$$g_F(z) = F(\mathrm{e}^{y\cdot z}) := \sum_{|j|=0}^{\infty}\frac{F(y^j)}{j!}z^j.$$

The power series is called formal, since there is no convergence assumption. For the point evaluation functional $F(f)=f(z_q)$, we get $g_{F_q}(z)=\mathrm{e}^{z\cdot z_q}$.

Now we can write down the interpolation space matching the functionals $\{F_q\}$

Theorem 26. *Let $\mathscr{F}=\{F_1,\ldots,F_m\}$ be functionals defined on Π^d. Let*

$$\mathscr{G}_{\mathscr{F}} = \mathrm{span}\{g_{F_1},\ldots,g_{F_m}\}$$

and

$$\mathscr{P}_{\mathscr{F}} = \mathrm{span}\{g_{\downarrow} \mid g\in\mathscr{G}_{\mathscr{F}}\}.$$

Then for any values c_q, $q=1,\ldots,m$, there is exactly one $p\in\mathscr{P}_{\mathscr{F}}$ with

$$F_q(P) = c_q, \qquad q=1,\ldots,m.$$

As an example, let us take our favorite singular Hermite interpolation. The functionals are evaluation of function values and the two first partial derivatives at z_1 and z_2. To simplify computations, we take $Z=\{(0,0),\ (1,0)\}$.

To evaluate the power series, we use

$$\frac{\partial}{\partial x}\,\mathrm{e}^{xx_q+yy_q} = x\mathrm{e}^{z\cdot z_q},$$
$$\frac{\partial}{\partial y}\,\mathrm{e}^{xx_q+yy_q} = y\mathrm{e}^{z\cdot z_q},$$

so that the power series are

$$e^0, xe^0, ye^0, e^x, xe^x, ye^x.$$

Doing the calculation as above, we get

$$\mathscr{P}_{\mathscr{F}} = \text{span}\{1, x, y, x^2, xy, x^3\}.$$

You can check for yourself that our Hermite interpolation from this space is regular. If we had taken nodes not lying on the coordinate axes, $\mathscr{P}_{\mathscr{F}}$ would no longer have a monomial basis.

Now that we have seen that it is relatively easy to construct the least interpolant, let us see what we are buying. By construction, we can now carry out Lagrange interpolation on any number of nodes without worrying whether the number matches a $\dim \Pi_n^d$. Other good properties of the least Lagrange interpolation on Z are that one can show that one has affine invariance in the sense that $\mathscr{P}_{aZ+z_0} = \mathscr{P}_Z$, for any real a and $z_0 \in \mathbb{R}$. Also one has transformation rules as for a change of variables: if A is a nonsingular $d \times d$ matrix, then $\mathscr{P}_{AZ} = \mathscr{P}_Z \cdot A^{\mathrm{T}}$.

One of its most important properties is that the least interpolant is degree reducing. From this, it follows that if Lagrange interpolation on Z is regular for some complete space Π_n^d, then $\mathscr{P}_Z = \Pi_n^d$. Formulating this differently, if $Z \subset \mathbb{R}^d$, $m = \dim \Pi_n^d$ for some j and the functionals are the point evaluation functionals, then $\mathscr{P}_Z = \Pi_n^d$ a.e., since Lagrange interpolation is regular a.e.

This raises the question of the behavior of the spaces $\mathscr{P}_Z$ when Z changes or what happens to $\mathscr{P}_Z$ when two nodes coalesce. If two of the nodes coalesce along a straight line, it can be shown that the least Lagrange interpolant converges to the Hermite interpolant which replaces one of the functional evaluations at the coalesced node with the directional derivative in the direction of the line.

But the first question does raise some computational issues. If $m = \dim \Pi_n^d$ for some j, then the set of Z for which $\mathscr{P}_Z = \Pi_n^d$ is open in $\mathbb{R}^{dm}$. If Z moves to a location in which Lagrange interpolation is singular, $\mathscr{P}_Z$ must change discontinuously. Thus, around this location, computation of a basis for $\mathscr{P}_Z$ is unstable. This is not to say that classical polynomial interpolation is better in this regard. It is even worse: the interpolant just does not exist for those Z. That this problem can be solved computationally is show by de Boor and Ron in [9]. They have have also developed a MATLAB program implementing these ideas.

5. Explicit interpolation schemes

5.1. Introduction

In this section, we will discuss two types of interpolation schemes. The first is to find concrete locations of points for which classical Lagrange interpolation is regular. The second is to construct Hermite interpolations, including the node sets, guaranteed to be regular. These are usually not of the classical type, but they have certain advantages. For example, some of them can be computed in a Newton like way.

5.2. Explicit Lagrange interpolation schemes

Chung and Yao, [6], give a quasi-constructive description of locations of nodes in $\mathbb{R}^d$ for which Lagrange interpolation is regular. They satisfy the

Definition 27. A node set $Z = \{z_1, \ldots, z_m\} \subset \mathbb{R}^d$ with $m = \dim \Pi_n^d$ for some n is said to satisfy the *General Condition* if to each node z_q, there exist n hyperplanes $H_{q,1}, \ldots, H_{q,n}$ such that

(a) z_q does not lie on any of these hyperplanes,
(b) all other nodes lie on at least one of them.

A node set satisfying the general condition is said to be a *natural lattice*. It can easily be shown that Lagrange interpolation from Π_n^d on a natural lattice is regular since the n functions

$$\lambda_q(z) := \frac{\prod_{i=1}^n H_{q,i}(z)}{\prod_{i=1}^n H_{q,i}(z_q)}$$

are Lagrange fundamental functions. Here $H_{q,i}(z) = 0$ are the hyperplanes required by the General Condition. This is very much in the spirit of the univariate construction in Eq. (14). This idea has been extended to Hermite interpolation by Busch, see [4].

Finding regular Lagrange interpolations via natural lattices is not really constructive. It is actually just a sufficient condition for regularity. However, it has been the motivation for several explicit constructions. Let us first look at triangular arrays

$$Z_n^d := \{j \mid j \in \mathbb{N}_0^d, |j| \leqslant n\}.$$

With a little bit of work, one can convince oneself that each Z_n^d satisfies the general condition (do not forget the hyperplanes perpendicular to the direction $(1, \ldots, 1)$!)

A nicer example in $\mathbb{R}^2$, worked out in great detail by Sauer and Xu, [39], starts with $2r+1$ points equi-distributed on the circumference of the unit disk. We number the points $z_1, \ldots, z_{2r+1}$ and connect z_q with z_{q+r} by a straight line. Here, $q+r$ is taken modulo $2r+1$. Let $Z^{(r)}$ be the set of intersections of these lines within and on the circumference of the circle. Then it can be shown that

Theorem 28. *The node set $Z^{(r)}$ described above contains exactly $r(2r+1) = \dim \Pi_{2r-1}^2$ points. It is a natural lattice and, consequently, Lagrange interpolation on $Z^{(r)}$ from Π_{2r-1}^2 is regular.*

Due to the concrete location of the nodes, they lie on r concentric circles, Sauer and Xu are able to give compact formulas for the fundamental functions and a point-wise bound for the error of the interpolant. Bos [2] has some similar constructions which we discuss in Section 4.

Now, let us look at a really constructive approach to Lagrange interpolation given by Gasca and Maeztu [11]. Let there be given $n+1$ distinct lines $\ell_i(z) = 0$, $i = 0, \ldots, n$. To each line ℓ_i, we associate lines $\ell_{i,j}$, $j = 0, \ldots, r(i)$ each of which intersect ℓ_i at exactly one point $z_{i,j}$. In addition, it is assumed that the node $z_{i,j}$ does not lie on any of the lines $\ell_0, \ldots, \ell_{i-1}$, $1 \leqslant i \leqslant n$. Sets of lines satisfying these conditions are called admissible.

We set

$$Z = \{z_{i,j} | 0 \leqslant j \leqslant r(i), 0 \leqslant i \leqslant n\}. \tag{41}$$

Now we must define the space from which we will interpolate. Let

$$P_{0,0}(z)=1, \tag{42}$$

$$P_{0,j}(z)=\ell_{0,0}\cdots\ell_{0,j-1}(z), \quad 1\leqslant j\leqslant r(0), \tag{43}$$

$$P_{i,0}(z)=\ell_0\cdots\ell_{i-1}(z), \quad 1\leqslant i\leqslant n, \tag{44}$$

$$P_{i,j}(z)=\ell_0\cdots\ell_{i-1}(z)\ell_{i,0}\cdots\ell_{i,j-1}(z), \quad 1\leqslant j\leqslant r(i), 1\leqslant i\leqslant n. \tag{45}$$

Then we set

$$V=\operatorname{span}\{P_{i,j}\,|\,0\leqslant j\leqslant r(i), 0\leqslant i\leqslant n\}. \tag{46}$$

Theorem 29. *Lagrange interpolation on Z given by* (41) from *V given by* (46) *is regular.*

How to prove this theorem becomes clear once we recognize that (42) and the following equations can be given recursively by

$$P_{i,0}(z)=P_{i-1,0}(z)\ell_{i-1}(z), \quad 1\leqslant i\leqslant n,$$

$$P_{i,j}(z)=P_{i,j-1}(z)\ell_{i,j-1}(z), \quad 1\leqslant j\leqslant r(i), 1\leqslant i\leqslant n.$$

If we order the nodes of Z lexicographically, that is $(i,j)<(i',j')$ if $i<i'$ or if $i=i'$, then $j<j'$, it is easy to see from the above recursive construction that

$$P_{i',j'}(z_{i,j})=0 \quad \text{if } (i,j)<(i',j').$$

If Z is admissible, $P_{i,j}(z_{i,j})\neq 0$. Thus the interpolant can be constructed exactly as in the univariate Newton construction of Section 2.1.

Many people have considered the special case that the lines ℓ_i are taken parallel to the x-axis. Then the lines $\ell_{i,j}$ can be taken parallel to the y-axis.

What does the interpolation space look like? First, we must note that many different admissible choices of lines can lead to the same node set. Thus we can have many different spaces for the same functionals. In one special, but very important case, V is independent of the lines. This is the case when we would like to have $V=\Pi_n^2$ for some n. To achieve this, we choose $r(i)=n-i$, $0\leqslant i\leqslant n$ and the lines so that the admissibility condition is satisfied. Then $\dim V=\dim \Pi_n^2$. But, by Theorem 29, all the polynomials in (42) are linearly independent and, from (42), it can be seen that each of them is of degree not exceeding n. Thus $V=\Pi_n^2$ as desired, independently of the choice of lines.

Comparing their results with those of Chung and Yao, Gasca and Maeztu have made the conjecture.

Conjecture 30. *Let Z be a natural lattice in $\mathbb{R}^2$ for some n. Then one of the lines involved in the definition of the general condition contains exactly $n+1$ nodes of Z.*

Note that no line can contain more that $n+1$ nodes, since then Lagrange interpolation will not be regular.

If the conjecture were true, then we could remove those $n+1$ nodes obtaining a smaller node set $Z^{(1)}$, which satisfies the general condition with respect to Π_{n-1}^2. Continuing in this way, we could conclude that any natural lattice is a special case of the above construction of Gasca and Maeztu for obtaining interpolation spaces with $V = \Pi_n^2$.

Gasca and Maeztu's construction has a straightforward generalization to $\mathbb{R}^d$, leading to a Newton formula for Lagrange interpolation in $\mathbb{R}^d$. Their method also includes Hermite interpolation. This will be presented in the following subsection.

The final type of construction we mention here can be subsumed under the concept of "Boolean sum". A detailed exposition of this class of constructions, which can also be used for trigonometric polynomials, can be found in a book by Delvos and Shempp [10].

The simplest example of such interpolations is called Biermann interpolation. It is based on univariate interpolation and the boolean sum of two commuting projectors. Let L and M be two commuting projectors. Then their *boolean sum* $L \oplus M$ is given by

$$L \oplus M = L + M - LM.$$

The projectors we use are those of univariate Lagrange interpolation

$$(L_n f)(x) = \sum_{q=1}^{n+1} f(x_q)\lambda_{q,n}(x) \quad \text{for } f \in C(\mathbb{R}), \tag{47}$$

where $\lambda_{q,n}$ are the Lagrange fundamental functions for interpolation from Π_n^1 on $X = \{x_1, \ldots, x_{n+1}\}$. These projectors are extended to $C(\mathbb{R}^2)$ by just applying them to one of the variables

$$(L_n f)(x, y) = \sum_{q=1}^{n+1} f(x_q, y)\lambda_{q,n}(x) \quad \text{for } f \in C(\mathbb{R}^2). \tag{48}$$

By M_s, we denote the same projector, but now applied to the y variable and based on the node set $Y = \{y_1, \ldots, y_{s+1}\}$. Then L_n and M_s commute and

$$(L_n M_s f)(x, y) = \sum_{p=1}^{n+1} \sum_{q=1}^{s+1} f(x_p, y_q)\lambda_{p,n}(x)\lambda_{q,s}(y) \quad \text{for } f \in C(\mathbb{R}^2).$$

Choose now increasing integer sequences $1 \leqslant j_1 < \cdots < j_r$; $1 \leqslant l_1 < \cdots < l_r$ and nodes $X = \{x_1, \ldots, x_{j_r+1}\}$, $Y = \{y_1, \ldots, y_{l_r+1}\}$. Then the Biermann projector is defined by

$$B_r = L_{j_1} M_{l_r} \oplus L_{j_2} M_{l_{r-1}} \oplus \cdots L_{j_r} M_{l_1}. \tag{49}$$

The interpolation space is defined similarly

$$V = \Pi_{(j_1, l_r)} + \Pi_{(j_2, l_{r-1})} + \cdots \Pi_{(j_r, l_1)}. \tag{50}$$

Here the sum of two subspaces U and V is defined by $U + V = \{u + v \mid u \in U,\ v \in V\}$.

Theorem 31. *Lagrange interpolation from V given in* (50) *on the node set*

$$Z = \{(x_i, y_j) \mid 1 \leqslant i \leqslant j_m + 1,\ 1 \leqslant j \leqslant l_{r+1-m} + 1,\ 1 \leqslant m \leqslant r\}$$

is regular. The interpolation is given explicitly by

$$(B_r f)(x, y) = f(x_i, y_j)$$

where B_r is the Biermann projector (49).

The interpolation space has a monomial basis, so if we write $V = \Pi_A$, then A resembles (optically) the node set. Both have the same cardinality

$$\sum_{m=1}^{r} (k_m + 1)(l_{r+1-m} - l_{r-m}).$$

The special case that $k_m = l_m = m - 1$ for $m = 1, \ldots, r+1$ turns out to be the Lagrange interpolation from Π_r^2 on the triangular array given in Section 2.2. The Biermann projector is defined similarly in higher dimensional euclidean spaces.

One of the nice features of these kinds of interpolations is that the error of interpolation can be expressed as a product of the univariate errors.

5.3. Explicit Hermite interpolation schemes

Gasca and Maeztus technique for Lagrange interpolation given in the previous subsection can be used to obtain Hermite interpolation simply by dropping the assumption the lines need be distinct, and that the intersection $z_{i,j}$ of ℓ_i and $\ell_{i,j}$ not lie on any of the lines "preceding" it in the sense mentioned in the previous subsection. The definition of the interpolating space remains the same as before (42), (46), but with the lines repeated according to their multiplicities. But the functionals to be interpolated change. This is to be Hermite interpolation after all. To describe them, we use the notation previously introduced. Let t_i be lines orthogonal to ℓ_i and $t_{i,j}$ be lines orthogonal to $\ell_{i,j}$. We define the numbers

$$a_i = \begin{cases} 0 & \text{if } j = 0, \\ \text{the number of lines among} & \\ \{\ell_0, \ldots, \ell_{i-1}\} \text{ coincident with } \ell_i & \text{if } 1 \leqslant i \leqslant n, \end{cases} \tag{51}$$

$$b_i = \begin{cases} 0 & \text{if } i = 1, \\ \text{the number of lines among} & \\ \{\ell_0, \ldots, \ell_{i-1}\} \text{ that contain } z_{i,j} & \text{if } 1 \leqslant i \leqslant n, \end{cases} \tag{52}$$

$$c_{i,j} = \begin{cases} 0 & \text{if } j = 0, \\ \text{the number of lines among} & \\ \{\ell_{i,0}, \ldots, \ell_{i,j-1}\} \text{ that contain } z_{i,j} & \text{if } 1 \leqslant j \leqslant r(i). \end{cases} \tag{53}$$

The functionals to be interpolated are

$$D_{i,j} f = D_{t_i}^{a_i} D_{t_{i,j}}^{b_i + c_{i,j}} f(z_{i,j}), \quad 1 \leqslant i \leqslant n, \ 1 \leqslant j \leqslant r(i), \tag{54}$$

where, as before, $D_t f$ is the directional derivative of f in the direction t.

Theorem 32. *The interpolation of the functionals* (54) *from the space spanned by the polynomials* (42) *is regular.*

For another approach which yields regular interpolation schemes similar in flavor to those just discussed, see Gevorkian et al. [17]. There $V = \Pi_n^d$ and they obtain their interpolation conditions by projections onto the intersection of families of hyperplanes.

If we start with a Hermite interpolation, then it seems clear that one obtains another by subjecting all components, the node set and the interpolation space to an affine transformation. This was carried out systematically by Gasca and Mühlbach in [12–14]. There starting with a node set lying on a Cartesian grid, they apply projectivities to obtain new schemes. For these new schemes, one can find Newton-type interpolation formulas and formulas for the error of interpolation resembling the univariate ones. These projectivities allow mapping "finite" points to "infinity" and thus one can obtain nodes lying on pencils of rays. This extends an approach used by Lee and Philips [26].

5.4. Node sets on algebraic varieties

The interpolation schemes presented in the previous subsections were very much concerned with lines and hyperplanes. In this subsection, we look at Lagrange interpolation on node sets restricted to algebraic varieties. The presentation is taken from Bos [2]. By *algebraic variety* or algebraic manifold, we mean sets of the form

$$W = \{z \in \mathbb{R}^d \mid P(z) = 0,\ P \in \mathscr{P}\}, \tag{55}$$

where $\mathscr{P}$ is a collection of polynomials from Π^d. Given a point set $E \subset \mathbb{R}^d$, the *ideal* I_E associated with E is

$$I_E = \{P \in \Pi^d \mid P(z) = 0,\ z \in E\}. \tag{56}$$

If E is a variety, then we say that I_E is the *ideal of the variety*. An ideal is called *principal* if there is a $Q \in \Pi^d$ such that

$$I = \{PQ \mid P \in \Pi^d\}. \tag{57}$$

Finally, given a point set $E \subset \mathbb{R}^d$,

$$N_n^d(E) = \dim(\Pi_n^d|_E).$$

Much of the following is based on the

Lemma 33. *Let W be an algebraic variety whose ideal I_W is principal being represented by the polynomial Q. Then*

$$N_n^d(W) = \binom{n+d}{d} - \binom{n - \deg Q + d}{d}.$$

In the following, we fix n, d and want to interpolate from Π_n^d.

Let $W_1, \dots, W_N$ be algebraic varieties whose ideals are principal and are represented by the polynomials $Q_1, \dots, Q_N$ having degrees $n_1, \dots, n_N$. Assume that these polynomials are pairwise relatively prime and, in addition,

$$n_1 + \cdots + n_{N-1} < n, \tag{58}$$

$$n_1 + \cdots + n_N \geqslant n. \tag{59}$$

Set $s_i = n - n_1 - \cdots - n_{i-1}$ and let Z_i be an arbitrary set of $N_{s_i}^d$ points on W_i, $i = 1,\ldots,N$ such that all of these points are distinct. If $n_1 + \cdots + n_N > n$, set

$$Z = \bigcup_{i=1}^{N} Z_i \tag{60}$$

and if $n_1 + \cdots + n_N = n$, put

$$Z = \bigcup_{i=1}^{N} Z_i \cup \{a\}, \tag{61}$$

where a does not lie on any of the W_i. With this choice, there are always a total of $\dim \Pi_n^d$ nodes in Z.

The reason for choosing the nodes this way is that the regularity of Lagrange interpolation on Z can be decomposed to the Lagrange regularity on each of the varieties.

Theorem 34. *Let Z, Z_i, $i = 1,\ldots,N$, be chosen as above. If Lagrange interpolation from $\Pi_{s_i}^d|_{W_i}$ on Z_i is regular for $i = 1,\ldots,N$, then Lagrange interpolation from Π_n^d on Z is regular.*

This theorem is proved by repeated application of Lemma 33.

For a simple example of this technique, we consider varieties which are concentric circles in $\mathbb{R}^2$. Take distinct radii R_i, $i = 1,\ldots,N$, and

$$W_i = \{(x, y) \mid x^2 + y^2 = R_i^2\}.$$

Then each I_{W_i} is a principal ideal with $P_i = x^2 + y^2 - R_i^2$, so $n_i = 2$. We fix n and want to interpolate from Π_n^2. By (58) and (59), we must choose $N = [(n + 1)/2]$, where $[a]$ is the integer part of a. Then $n_1 + \cdots + n_N > n$ if n is odd and $n_1 + \cdots + n_N = n$ if n is even. It follows that $s_i = n - 2(i - 1)$ and that, by Lemma 33, that

$$N_{s_i}^d(W_i) = \binom{s_i + 2}{2} - \binom{s_i - 2 + 2}{2} = 2s_i + 1.$$

But Lagrange interpolation by algebraic polynomials of degree s_i on $2s_i + 1$ nodes on a circle centered at the origin is the same as interpolation by trigonometric polynomials of the same degree, which is regular. So taking an additional point $(0, 0)$ in n is even, Theorem 34 allows us to conclude that Lagrange interpolation from Π_n^d on a node set consisting of $2n - 4i + 5$, $i = 1,\ldots,N$ nodes lying respectively on $N = [(n + 1)/2]$ concentric circles, is regular.

The Lagrange interpolation by Sauer and Xu mentioned in Section 5.2 is also has its nodes lying on concentric circles. In that construction, there are $r(r + 1)$ nodes when interpolating from Π_{2r-1}^2. These nodes are distributed over r circles, with $r + 1$ nodes on each of them. In the example of Bos, the number of nodes on the circles differs from circle to circle, but their location on the circles can be chosen at will. Another difference is that the circles given by Sauer and Xu have the predetermined radii

$$R_j = \frac{\cos j\pi/(2j + 1)}{\cos r\pi/(2j + 1)}, \quad j = 1,\ldots,r.$$

Despite the resemblance, it does not seem that there is any connection between these schemes.

As we have seen in this and the previous section, the study of polynomial ideals is quite useful for multivariate interpolation. More about them can be found in Möller [32,33], Xu [45] and the references given in Section 4.

6. Related topics

6.1. Introduction

In this section, we will look at the theory of singularities, i.e., at the investigation of the set of polynomials having zeros of given multiplicities. The last subject will be the results of Vassiliev on the minimal dimension an interpolation space must have in order to be able to solve a Lagrange interpolation for any choice of nodes.

6.2. Singularities

In concordance with the notation generally used, we will speak of singularities of a given order of a function. A function f defined on $\mathbb{R}^d$ has a *singularity of order s* at z if

$$D^{\alpha} f(z) = 0 \qquad \text{for } |\alpha| < s.$$

On the other hand, we consider polynomials in euclidian and not in projective spaces, so that some of our notation does differ for example from the survey paper of Miranda [31], from which some of this material was taken. Let $Z = \{z_1, \ldots, z_m\}$ be a set of nodes in $\mathbb{R}^d$ and $\{s_1, \ldots, s_m\} \subset \mathbb{N}^d$. Then $\mathscr{L}_n^{(d)}(-\sum_{q=1}^m s_q z_q)$ stands for the subspace of Π_n^d of polynomials having a singularity of order s_q at z_q for $q = 1, \ldots, m$. $\mathscr{L}_n^{(d)}(-\sum_{q=1}^m s_q z_q)$ could consist of just the zero polynomial or be very large, since there is no connection between the number of singularities and the degree of the polynomials. The *virtual dimension* $\mathrm{v}_n^{(d)}$ of $\mathscr{L}_n^{(d)}(-\sum_{q=1}^m s_q z_q)$ is

$$\mathrm{v}_n^{(d)}\left(-\sum_{q=1}^m s_q z_q\right) = \binom{d+n}{d} - \sum_{q=1}^m \binom{s_q - 1}{d}.$$

Intuitively, if we take Π_n^d and subject it to the

$$\sum_{q=1}^m \binom{s_q - 1}{d}$$

conditions that the polynomials have the given singularities on Z, then we expect to reduce its dimension by exactly this number, unless this number is larger than the dimension of Π_n^d, in which case, we expect to get dimension 0. Thus, we define the *expected dimension* $\mathrm{e}_n^{(d)}$ by

$$\mathrm{e}_n^{(d)}\left(-\sum_{q=1}^m s_q z_q\right) = \max\left\{0, \mathrm{v}_n^{(d)}\left(-\sum_{q=1}^m s_q z_q\right)\right\}.$$

What does this mean for an interpolation scheme? There

$$\binom{d+n}{d} = \sum_{q=1}^m \binom{s_q - 1}{d}, \tag{62}$$

so that the expected dimension is always 0. If the true dimension is not 0, then there is a polynomial in Π_n^d which satisfies the homogeneous interpolation conditions. Thus the interpolation is singular.

As we have seen from Theorem 2, for each Hermite interpolation (except the Taylor interpolation) there are always node sets Z for which the dimension of $\mathscr{L}_n^{(d)}(-\sum_{q=1}^m s_q z_q)$ is nonzero, i.e., for which the interpolation is singular. This is nothing special. On the other hand, it is a rather special situation if the dimension of $\mathscr{L}_n^{(d)}(-\sum_{q=1}^m s_q z_q)$ is *always* larger than the expected dimension. Thus we say that the system of homogeneous equations described by $(n; s_1, \ldots, s_m)$, whose solution yields $\mathscr{L}_n^{(d)}(-\sum_{q=1}^m s_q z_q)$ if solved for the node set Z, is *special* if

$$\inf_Z \dim \mathscr{L}_n^{(d)}\left(-\sum_{q=1}^m s_q z_q\right) > \mathrm{e}_n^{(d)}\left(-\sum_{q=1}^m s_q z_q\right). \tag{63}$$

An example is the system $(2; 2, 2)$, which is our favorite singular Hermite interpolation. It is special. We have

$$\inf_Z \dim \mathscr{L}_2^{(2)}\left(-\sum_{q=1}^2 2z_q\right) = 1$$

while

$$\mathrm{e}_2^{(2)}\left(-\sum_{q=1}^2 2z_q\right) = 0.$$

Nodes for which the minimum in (63) is attained are said to be in *general position*. This is not quite the usual definition, but is equivalent. Be aware of the fact that we have already used "general position" with another meaning.

So, the concept of special systems is wider than that of singular Hermite interpolations of type total degree. A special system which happens to be an interpolation, i.e., for which (62) holds, is a singular interpolation scheme.

It has been a problem of long standing in algebraic geometry to determine, or to characterize all special systems. They have come up with the following conjecture in $\mathbb{R}^2$.

Conjecture 35. *Let $(n; s_1, \ldots, s_m)$ be a special system in $\mathbb{R}^2$ and P a solution of it for some node set in general position. Then there is a (nonzero) polynomial Q such that Q^2 divides P. The polynomial Q is a solution of a system $(r; t_1, \ldots, t_m)$ with the same nodes and with*

$$r^2 - \sum_{q=1}^m s_q = -1. \tag{64}$$

Hirschowitz [22,23] has verified this conjecture for some special cases

Theorem 36. *Let $n, d \geqslant 2$. The linear system $(n; 2, \ldots, 2)$ with m nodes in $\mathbb{R}^d$ is special exactly in the following cases*:

(a) *for $n = 2$: if and only if $2 \leqslant m \leqslant d$*,

(b) *for $n \geqslant 3$: if and only if $(d, n.m)$ is one of $(2, 4, 5)$, $(3, 4, 9)$, $(4, 3, 7)$ or $(4, 4, 14)$.*

We have already seen some of these. Theorem 16, formulated in the terminology of this section, states that interpolations $(n; s_1, \ldots, s_m)$ in $\mathbb{R}^d$ are special if $2 \leqslant m \leqslant d+1$. One must exercise care when comparing the two results, since the above theorem also includes non-interpolation schemes. For example, the systems described in a) are interpolating if and only if $m = 1 + d/2$, i.e., only for spaces of even dimension. In those cases, they are, indeed, covered by Theorem 16.

Of the special schemes in (b), $(2,4,5)$, $(4,3,7)$ and $(4,4,14)$ are interpolations. The scheme $(3,4,9)$ has 36 interpolation conditions, while the dimension of the interpolation space is 35. The singularity of $(2,4,5)$ and $(4,4,14)$ follow from Theorem 18.

There are many other results in the literature. For example, Hirschowitz [22], treats the case $(n; 3, \ldots, 3)$ exhaustively.

Let us now return to the condition $n^2 - \sum_q s_q^2 = -1$ of Conjecture 35. It derives from the calculus on schemes we used in Section 3.3: addition is

$$(n_1; s_{1,1}, \ldots, s_{1,m}) + (n_2; s_{2,1}, \ldots, s_{2,m}) = (n_1 + n_2; s_{1,1} + s_{2,1}, \ldots, s_{1,m} + s_{2,m})$$

and an "inner product" is

$$\langle (n_1; s_{1,1}, \ldots, s_{1,m}), (n_2; s_{2,1}, \ldots, s_{2,m}) \rangle = n_1 n_2 - \sum_{q=1}^{m} s_{1,q} s_{2,q}.$$

Thus if we set $\mathcal{N} = (n; s_1, \ldots, s_m)$, then

$$n^2 - \sum_{q=1}^{m} s_q^2 = \langle \mathcal{N}, \mathcal{N} \rangle.$$

Many facts about interpolation schemes can be expressed in terms of this calculus. For example, if $\mathcal{N}_1$ and $\mathcal{N}_2$ are special then so is $\mathcal{N}_1 + \mathcal{N}_2$. In fact, if P_i are polynomials satisfying $\mathcal{N}_i$, $i = 1, 2$, then, as we have seen before, $P_1 \cdot P_2$ satisfies $\mathcal{N}_1 + \mathcal{N}_2$. Or, Bezout's theorem (in $\mathbb{R}^2$) can be expressed in the following way: If P_i are polynomials satisfying $\mathcal{N}_i$, $i = 1, 2$ which are relatively prime, then

$$\langle \mathcal{N}_1, \mathcal{N}_2 \rangle \geqslant 0.$$

Here the schemes are not assumed to be special.

We can also reformulate Hirschowitzs' conjecture

Conjecture 37. *Let $\mathcal{N}$ be a special interpolation scheme in $\mathbb{R}^2$. Then there is another scheme $\mathcal{R}$ satisfying condition* (64) *such that*

$$\langle \mathcal{N}, \mathcal{R} \rangle = -2.$$

More details about this calculus can be found in Miranda [31].

It is interesting to note, that Gevorkian Hakopian and Sahakian have a series of papers of singular Hermite interpolations, some of which we have already referred to, in which they use exactly this calculus, but with a slightly different notation. They have also formulated a conjecture about singular interpolations which is based on schemes which have fewer interpolation conditions than the dimension of the space being interpolated from. They say that $(n; s_1, \ldots, s_1)$ belongs to the *less condition* class (LC), if (in $\mathbb{R}^2$)

$$\binom{n+2}{2} > \sum_{q=1}^{m} \binom{s_q+2}{2}.$$

As we have seen, the importance of these "less" schemes is that there always exists a polynomial satisfying the homogeneous conditions, i.e., $\mathscr{L}_n^{(2)}(-\sum_{q=1}^m s_q z_q)$ has dimension at least one for each node set.

Their conjecture, reformulated in terms of special schemes, is

Conjecture 38. *Let $\mathcal{N}$ be an interpolation scheme in $\mathbb{R}^2$. Then $\mathcal{N}$ is special if and only if there are schemes $\mathcal{M}_i$, $i=1,\ldots,r$ belonging to LC, such that*

$$\mathcal{N} = \sum_{i+1}^{r} \mathcal{M}_i.$$

This conjecture is not true in $\mathbb{R}^3$: $\mathcal{N}=(4;3,3,1,\ldots,1)$ with 15 1's is a singular interpolation scheme in $\mathbb{R}^3$ which is not decomposable into "less" schemes.

No one seems to have integrated the results from these two sources.

6.3. Lagrange interpolation spaces of minimal dimension

In this subsection, we stay with Lagrange interpolation but allow interpolating spaces consisting of arbitrary functions. What we want to do is to fix the number of nodes and then find one interpolation space, such that Lagrange interpolation on that number of nodes is always solvable, no matter where the nodes are located. If the number of nodes is equal to the dimension of a complete space of polynomials (in $\mathbb{R}^d$ for $d \geqslant 2$), then it is clear that we cannot choose that space itself. But perhaps there is another possibly non-polynomial space of the same dimension that does the trick. If not, then we allow the dimension of the space from which we will interpolate to increase until we have found one that works. The theorem of Severi (Theorem 19) tells us that if we want to do Lagrange interpolation on m nodes in $\mathbb{R}^d$, then we can do it with Π_{m-1}^d. Its dimension is

$$\binom{m-1+d}{d.}$$

If we allow noncomplete spaces of polynomials, then one can get away with smaller dimension. In [28], it is shown that it can be done with a space of dimension roughly $m \ln m$ in $\mathbb{R}^2$. Vassiliev [42] has considered these questions in a more general context.

Let M be a topological space, V a finite dimensional subspace of $C(M)$ (not necessarily polynomial). V is called m-interpolating (over M) if the Lagrange interpolation problem: find $P \in V$ such that

$$f(z_q) = c_q, \qquad q=1,\ldots,m \tag{65}$$

is solvable for any choice of nodes z_q from M and values c_q. Before, in the context of Severis' theorem, we called the problem "solvable" in this case. Here $\dim V = m$ is not assumed and, in general, not possible. Now we want to find the space of lowest possible dimension, and define

$$I(V,m) = \min_Z \dim V,$$

where the minimum is taken over all spaces which are m-interpolating and over all node sets Z of size m.

For example, $I(\mathbb{R},m)=m-1$ and a space of minimal dimension is Π^1_{m-1}. A first interesting result by Vassiliev is

Theorem 39.

$$2m - b(m) \leqslant I(\mathbb{R}^2,m) \leqslant 2m-1,$$

where $b(m)$ is the number of ones in the binary representation of m.

It immediately follows that, for example, that $I(\mathbb{R}^2,2^n)=2\cdot 2^n-1$, since $b(2^n)=1$. For $m=3$, $b(m)=2$ and the lower bound is the true value. A space of minimal dimension is $V=\mathrm{span}\{1,x,y,x^2+y^2\}$. V is 3-interpolating. In fact, $V=\mathrm{span}\{1,\Re z^t, \Im z^t \mid 1\leqslant t\leqslant m-1\}$, where $z=x+iy$, provides the upper bound in the theorem. The lower bound is the difficult one to prove.

If $M=S^1$ is the unit circle in $\mathbb{R}^2$, then $I(M,m)=m$ if m is odd and $I(M,m)=m+1$ if m is even, for we can take the space of trigonometric polynomials of degree $[m/2]$ in both cases.

Theorem 40. *For any d-dimensional manifold M, we have*

$$I(M,m) \leqslant m(d+1).$$

In the case of the unit circle, the theorem predicts $2m$ instead of the correct answer $2[m/2]+1$.

I suppose this theory has more conjectures than proven theorems, so, appropriately, we close with another conjecture of Vassiliev

Conjecture 41. *If d is a power of 2, then*

$$I(\mathbb{R}^d,m) \geqslant m + (d-1)(m-b(m)).$$

References

[1] B.D. Bojanov, H.A. Hakopian, A.A. Sahakian, Spline Functions and Multivariate Interpolations, Kluwer Academic Publishers, Dordrecht, 1993.
[2] L. Bos, On certain configurations of points in $\Re^n$ which are unisolvent for polynomial interpolation, J. Approx. Theory 64 (1991) 271–280.
[3] B. Buchberger, Ein algorithmisches Kriterium für die Lösbarkeit eines algebraischen Gleichungssystems, Aeq. Math. 4 (1970) 373–383.
[4] J.R. Busch, Osculatory interpolation in $\Re^n$, SIAM J. Numer. Anal. 22 (1985) 107–113.
[5] A.L. Cavaretta, C.A. Micchelli, A. Sharma, Multivariate interpolation and the Radon transform, Part I, Math. Z. 174 (1980), 263–269; Part II, in: R.A. DeVore, K. Scherer (Eds.), Quantitative Approximation, Academic Press, New York, 1980, pp. 49–62.
[6] C.K. Chung, T.H. Yao, On lattices admitting unique Lagrange interpolations, SIAM J. Numer. Anal. 14 (1977) 735–741.

[7] P.G. Ciarlet, The Finite Element Method for Elliptic Problems, North-Holland, New York, 1978.
[8] C. de Boor, A. Ron, The least solution for the polynomial interpolation problem, Math. Zeit. 210 (1992) 347–378.
[9] C. de Boor, A. Ron, Computational aspects of polynomial interpolation in several variables, Math. Comp. 58 (1992) 705–727.
[10] F.-J. Delvos, W. Schempp, Boolian Methods in Interpolation and Approximation, Longman Scientific & Technical, Harlow, 1989.
[11] M. Gasca, J.I. Maeztu, On Lagrange and Hermite interpolation in $\Re^n$, Numer. Math. 39 (1982) 1–14.
[12] M. Gasca, G. Mühlbach, Multivariate polynomial interpolation under projectivities, Part I: Lagrange and Newton interpolation formulas, Numer. Algebra 1 (1991) 375–400.
[13] M. Gasca, G. Mühlbach, Multivariate polynomial interpolation under projectivities, Part II: Neville–Aitken formulas, Numer. Algor. 2 (1992) 255–278.
[14] M. Gasca, G. Mühlbach, Multivariate polynomial interpolation under projectivities III, Remainder formulas, Numer. Algor. 8 (1994) 103–109.
[15] M. Gasca, T. Sauer, Multivariate polynomial interpolation, Adv. Comp. Math., to appear.
[16] H. Gevorkian, H. Hakopian, A. Sahakian, On the bivariate Hermite interpolation, Mat. Zametki 48 (1990) 137–139 (in Russian).
[17] H. Gevorkian, H. Hakopian, A. Sahakian, J. Approx. Theory 80 (1995) 50–75.
[18] H. Gevorkian, H. Hakopian, A. Sahakian, Bivariate Hermite interpolation and numerical curves, J. Approx. Theory 85 (1996) 297–317.
[19] T.N.T. Goodman, Interpolation in minimal semi-norm, and multivariate B-splines, J. Approx. Theory 37 (1983) 212–223.
[20] W. Gröbner, Algebraische Geometrie I, II, Bibliographisches Institut Mannheim, Mannheim, 1968, 1970.
[21] H.A. Hakopian, Les différences divisées de plusieurs variables et les interpolations multidimensionnelles de types Lagrangien et Hermitien, C. R. Acad. Sci. Paris 292 (1981) 453–456.
[22] A. Hirschowitz, La méthode d'Horace pour l'interpolation á plusieurs variables, Man. Math. 50 (1985) 337–388.
[23] A. Hirschowitz, Une conjecture pour la cohomologie des diviseurs sur les surfaces rationelles génériques, J. Reine Angew. Math. 397 (1989) 208–213.
[24] R.-Q. Jia, A. Sharma, Solvability of some multivariate interpolation problems, J. Reine Angew. Math. 421 (1991) 73–81.
[25] P. Kergin, A natural interpolation of C^K functions, J. Approx. Theory 29 (1980) 29–278.
[26] S.L. Lee, G.M. Phillips, Construction of lattices for Lagrange interpolation in projective spaces, Constructive Approx. 7 (1991) 283–297.
[27] G.G. Lorentz, K. Jetter, S.D. Riemenschneider, Birkhoff Interpolation, Addison-Wesley, Reading, MA, 1983.
[28] R.A. Lorentz, Multivariate Birkhoff Interpolation, Springer, Heidelberg, 1992.
[29] C. Micchelli, A constructive approach to Kergin interpolation in $\mathbb{R}^k$: multivariate B-splines and Lagrange interpolation, Rocky Mountain J. Math. 10 (1979) 485–497.
[30] C.A. Micchelli, P. Milman, A formula for Kergin interpolation in R^k, J. Approx. Theory 29 (1980) 294–296.
[31] N. Miranda, Linear systems of plane curves, Notices AMS 46 (1999) 192–202.
[32] H.M. Möller, Linear abhängige Punktfunktionale bei zweidimensionalen Interpolations- und Approximations-problemen, Math. Zeit. 173 (1980) 35–49.
[33] H.M. Möller, Gröbner bases and numerical analysis, in: B. Buchberger, F. Winkler (Eds.), Groebner Bases and Applications, Cambridge University Press, Cambridge, 1998, pp. 159–179.
[34] S.H. Paskov, Singularity of bivariate interpolation, J. Approx. Theory 75 (1992) 50–67.
[35] T. Sauer, Polynomial interpolation of minimal degree, Numer. Math. 78 (1997) 59–85.
[36] T. Sauer, Gröbner bases, H-bases and interpolation, Trans. AMS, to appear.
[37] T. Sauer, Y. Xu, On multivariate Lagrange interpolation, Comp. Math. 64 (1995) 1147–1170.
[38] T. Sauer, Y. Xu, On multivariate Hermite interpolation, Adv. Comp. Math. 4 (1995) 207–259.
[39] T. Sauer, Y. Xu, Regular points for Lagrange interpolation on the unit disc, Numer. Algorithm 12 (1996) 287–296.
[40] L. Schumaker, Fitting surfaces to scattered data, in: G.G. Lorentz, C.K. Chui, L.L. Schumaker (Eds.), Approximation Theory 2, Academic Press, Boston, 1976.
[41] F. Severi, E. Löffler, Vorlesungen über Algebraische Geometrie, Teubner, Berlin, 1921.

[42] V.A. Vassiliev, Complements of discriminants of smooth maps: topology and applications, Transl. Math. Mon. 98 (1992) 209–210; translated from Funkt. Analiz i Ego Pril. 26 (1992) 72–74.

[43] S. Waldron, L_p error bounds for multivariate polynomial interpolation schemes, Ph. D. Thesis, Department of Mathematics, University of Wisconsin, 1995.

[44] S. Waldron, Mean value interpolation for points in general position, Technical Report, Dept. of Math., Univ of Auckland, 1999.

[45] Y. Xu, Polynomial interpolation in several variables, cubature formulae, and ideals, Adv. Comp. Math., to appear.

ELSEVIER

Journal of Computational and Applied Mathematics 122 (2000) 203–222

JOURNAL OF COMPUTATIONAL AND APPLIED MATHEMATICS

www.elsevier.nl/locate/cam

Interpolation by Cauchy–Vandermonde systems and applications

G. Mühlbach

Institut für Angewandte Mathematik, Universität Hannover, Postfach 6009, 30060 Hannover, Germany

Received 30 April 1999; received in revised form 22 September 1999

Abstract

Cauchy–Vandermonde systems consist of rational functions with prescribed poles. They are complex ECT-systems allowing Hermite interpolation for any dimension of the basic space. A survey of interpolation procedures using CV-systems is given, some equipped with short new proofs, which generalize the well-known formulas of Lagrange, Neville–Aitken and Newton for interpolation by algebraic polynomials. The arithmetical complexitiy is $\mathcal{O}(N^2)$ for N Hermite data. Also, inversion formulas for the Cauchy–Vandermonde matrix are surveyed. Moreover, a new algorithm solving the system of N linear Cauchy–Vandermonde equations for multiple nodes and multiple poles recursively is given which does not require additional partial fraction decompositions. As an application construction of rational B-splines with prescribed poles is discussed.

MSC: 41A05; 65D05

Keywords: Prescribed poles; Cauchy–Vandermonde systems; Interpolation algorithms

1. Cauchy–Vandermonde systems; definitions and notations

Let

$$\mathscr{B} = (b_1, b_2, \ldots) \tag{1}$$

be a given sequence of not necessarily distinct points of the extended complex plane $\bar{\mathbb{C}} = \mathbb{C} \cup \{\infty\}$. With $\mathscr{B}$ we associate a system

$$\mathscr{U} = (u_1, u_2, \ldots) \tag{2}$$

of basic rational functions defined by

$$u_j(z) = \begin{cases} z^{\nu_j(b_j)} & \text{if } b_j = \infty, \\ (z - b_j)^{-(\nu_j(b_j)+1)} & \text{if } b_j \in \mathbb{C}. \end{cases} \tag{3}$$

0377-0427/00/$ - see front matter
PII: S0377-0427(00)00364-2

Here

$$\nu_j(b) \text{ is the multiplicity of } b \text{ in the sequence } (b_1,\dots,b_{j-1}). \tag{4}$$

The system $\mathscr{U}$ has been called [18,17,15,13,12,5] the *Cauchy–Vandermonde system* (CV-system for brevity) associated with the pole sequence $\mathscr{B}$. By $\mathbb{N}$ we denote the set of positive integers. For any fixed $N \in \mathbb{N}$ to the initial section of $\mathscr{B}$

$$\mathscr{B}_N = (b_1,\dots,b_N) \tag{5}$$

there corresponds the basis

$$U_N = (u_1,\dots,u_N) \tag{6}$$

of the N-dimensional *Cauchy–Vandermonde space* $\mathscr{U}_N{:=}\operatorname{span} U_N$. Indeed, for every $N \in \mathbb{N}$ U_N is an *extended complete Čebyšev system* on $\mathbb{C} \setminus \{b_1,\dots,b_N\}$. This follows from

Proposition 1. *For any system*

$$\mathscr{A} = (a_1,a_2,\dots) \tag{7}$$

of not necessarily distinct complex numbers a_i such that $\mathscr{A}$ and $\mathscr{B}$ have no common point, for every $N \in \mathbb{N}$ and for every complex function f which is defined and sufficiently often differentiable at the multiple nodes of the initial section

$$\mathscr{A}_N = (a_1,\dots,a_N) \tag{8}$$

of $\mathscr{A}$ there is a unique $p \in \mathscr{U}_N$ that satisfies the interpolation conditions

$$\left(\frac{\mathrm{d}}{\mathrm{d}z}\right)^{\mu_i(a_i)} (p-f)(a_i) = 0, \quad i = 1,\dots,N. \tag{9}$$

Here

$$\mu_i(a) \text{ is the multiplicity of } a \text{ in the sequence } (a_1,\dots,a_{i-1}). \tag{10}$$

We also express the interpolation conditions by saying that u agrees with f at $a_1,\dots,a_N$ counting multiplicities. There is a simple proof due to Walsh [20] reducing it to interpolation by algebraic polynomials. Before repeating this proof we will introduce some notations to simplify formulas to be derived later. We sometimes will assume that the systems $\mathscr{A}_N$ of nodes and $\mathscr{B}_N$ of poles are *consistently ordered*, i.e.,

$$\mathscr{A}_N = (\underbrace{\alpha_1,\dots,\alpha_1}_{m_1},\underbrace{\alpha_2,\dots,\alpha_2}_{m_2},\dots,\underbrace{\alpha_p,\dots,\alpha_p}_{m_p}), \tag{11}$$

$$\mathscr{B}_N = (\underbrace{\beta_0,\dots,\beta_0}_{n_0},\underbrace{\beta_1,\dots,\beta_1}_{n_1},\dots,\beta_{q-1},\underbrace{\beta_q,\dots,\beta_q}_{n_q}) \tag{12}$$

corresponding with

$$U_N = \left(1,z,\dots,z^{n_0-1},\frac{1}{z-\beta_1},\dots,\frac{1}{(z-\beta_1)^{n_1}},\dots,\frac{1}{z-\beta_q},\dots,\frac{1}{(z-\beta_q)^{n_q}}\right), \tag{13}$$

where $\alpha_1,\dots,\alpha_p$ and $\beta_0,\beta_1,\dots,\beta_q$ are pairwise distinct and $m_1+\dots+m_p=N$, $m_i \geqslant 0$ and $n_0+n_1+\dots+n_q=N$, $n_0 \geqslant 0$, $n_i \geqslant 1$ for $i \in \{1,\dots,q\}$, respectively. In most cases, we will assume $\beta_0=\infty$ to be in front of the other poles corresponding with (13), in some considerations we assume $\beta_0 = \infty$ at the end of the pole sequence (12),

$$\mathscr{B}_N = (\underbrace{\beta_1,\ldots,\beta_1}_{n_1},\ldots,\beta_{q-1},\underbrace{\beta_q,\ldots,\beta_q}_{n_q},\underbrace{\beta_0,\ldots,\beta_0}_{n_0},) \tag{14}$$

corresponding with

$$U_N = \left(\frac{1}{z-\beta_1},\ldots,\frac{1}{(z-\beta_1)^{n_1}},\ldots,\frac{1}{z-\beta_q},\ldots,\frac{1}{(z-\beta_q)^{n_q}},1,z,\ldots,z^{n_0-1}\right). \tag{15}$$

Notice that $p, m_1, \ldots, m_p$ as well as $q, n_0, \ldots, n_q$ do depend on N. Of course, there is no loss of generality in assuming that the nodes and poles are ordered consistently. This only means reordering the system U_N keeping it to be an extended complete Čebyšev system on $\mathbb{C} \setminus \{b_1,\ldots,b_N\}$ and reordering Eq. (9) according to a permutation of $\mathscr{A}_N$ to get the node system consistently ordered.

Another simplification results from adopting the following notation. If $(\gamma_1,\ldots,\gamma_k)$ is a sequence in $\bar{\mathbb{C}}$, then

$$\gamma_j^* := \begin{cases} \gamma_j & \text{if } \gamma_j \in \mathbb{C},\ \gamma_j \neq 0, \\ 1 & \text{if } \gamma_j = \infty \text{ or } \gamma_j = 0, \end{cases} \tag{16}$$

$$\prod_{j=1}^{k}{}^{*} \gamma_j := \prod_{j=1}^{k} \gamma_j^*, \tag{17}$$

i.e., the symbol $\prod^*$ means that each factor equal to ∞ or to zero is replaced by a factor 1.

Proof of Proposition 1. Let $A_N(z) = (z-a_1)\cdot\ldots\cdot(z-a_N)$ be the *node polynomial* and $B_N = (z-b_1)^* \cdot \ldots \cdot (z-b_N)^*$ be the *pole polynomial* associated with the systems $\mathscr{A}_N$ and $\mathscr{B}_N$, respectively. Let φ be the polynomial of degree $N-1$ at most that interpolates $B_N f$ at the nodes of $\mathscr{A}_N$ counting multiplicites.Then $p{:=}\varphi/B_N$ satisfies (9). Indeed, $p \in \mathscr{U}_N$ follows from the partial fraction decomposition theorem and, since B_N and A_N are prime, Leibniz' rule combined with an induction argument yields

$$\left(\frac{\mathrm{d}}{\mathrm{d}z}\right)^{\mu_i(a_i)} (f-p)(a_i) = 0, \quad i=1,\ldots,N \Leftrightarrow \left(\frac{\mathrm{d}}{\mathrm{d}z}\right)^{\mu_i(a_i)} (B_n f - \varphi)(a_i) = 0, \quad i=1,\ldots,N. \qquad \square$$

2. Interpolation by Cauchy–Vandermonde systems

With the node sequence (7) there is naturally associated a sequence

$$\mathscr{L} = (L_1, L_2, \ldots), \quad u \mapsto L_i = \left(\frac{\mathrm{d}}{\mathrm{d}z}\right)^{\mu_i(a_i)} u(a_i) \tag{18}$$

of Hermite-type linear functionals where $\mu_i(a)$ is defined by (10). For interpolation by algebraic polynomials there are well-known classical formulas connected with the names of Lagrange, Newton, Neville and Aitken expressing the interpolant in terms of the nodes and interpolation data

$$w_i = \left(\frac{\mathrm{d}}{\mathrm{d}z}\right)^{\mu_i(a_i)} f(a_i), \quad i=1,\ldots,N. \tag{19}$$

Since CV-systems in many aspects are close to algebraic polynomials it should be expected that there are similar interpolation formulas for CV-systems. Such formulas are given in the next three subsections.

2.1. Lagrange's interpolation formula

The basic Lagrange functions ℓ_j $(j=1,\ldots,N)$ for the N-dimensional CV-space $\mathscr{U}_N$ are uniquely determined by the conditions of biorthogonality

$$\langle L_i, \ell_j \rangle = \delta_{i,j}, \quad i,j=1,\ldots,N. \tag{20}$$

For certain purposes it is important that we are able to change easily between the one-index enumeration of Hermite functionals (18) corresponding to the one-index enumeration (7) of the nodes and the two-index enumeration

$$\langle L_i, f \rangle = \left(\frac{\mathrm{d}}{\mathrm{d}z}\right)^{\rho} f(\alpha_r), \quad r=1,\ldots,p, \ \rho=0,\ldots,m_r-1, \tag{21}$$

where we assume that $\mathscr{A}_N$ is consistently ordered as in (11). This is done by the one-to-one mapping $\varphi=\varphi_N$:

$$(r,\rho) \overset{\varphi}{\mapsto} i = \varphi(r,\rho) = m_1+\cdots+m_{r-1}+\rho+1. \tag{22}$$

Similarly, between the enumeration of the CV functions (2) corresponding to the one-index enumeration (1) of the poles and the two-index enumeration

$$u_j = u_{s,\sigma}, \quad s=0,\ldots,q, \ \sigma=1,\ldots,n_s, \tag{23}$$

where

$$u_{s,\sigma}(z) = \begin{cases} \dfrac{1}{(z-b_s)^{\sigma}} & s=1,\ldots,q, \ \sigma=1,\ldots,n_s, \\ z^{\sigma-1} & s=0, \ \sigma=1,\ldots,n_0 \end{cases} \tag{24}$$

corresponding to the consistently ordered pole system (15), there is the one-to-one mapping $\psi=\psi_N$:

$$(s,\sigma) \overset{\psi}{\mapsto} j = \psi(s,\sigma) = n_1+\cdots+n_{s-1}+\sigma. \tag{25}$$

In order to give a Lagrange-type formula the following notation is needed:

$$B_N(z) = \prod_{j=1}^{N}{}^{*}(z-b_j) = \prod_{t=1}^{q}(z-\beta_t)^{n_t},$$

$$\omega_\ell(z) = \prod_{\substack{s=1 \\ s\neq\ell}}^{p}(z-\alpha_s)^{m_s}, \quad \ell=1,\ldots,p,$$

$$v_{\ell,\lambda}(z) = \frac{1}{\lambda!}(z-\alpha_\ell)^{\lambda}, \quad \ell=1,\ldots,p, \ \lambda=0,\ldots,m_\ell-1,$$

$$P_{\ell,\lambda}(z) = \sum_{i=0}^{m_\ell-\lambda-1} \frac{1}{i!}\left(\frac{\mathrm{d}}{\mathrm{d}z}\right)^i \left(\frac{B_N}{\omega_\ell}\right)(\alpha_\ell)(z-\alpha_\ell)^i,$$

$$d_\ell^i(\cdot) = \left(\frac{\mathrm{d}}{\mathrm{d}z}\right)^i (\cdot)_{z=\alpha_\ell}, \quad \ell = 1,\ldots,p, \; i = 0,\ldots,m_\ell - 1.$$

Proposition 2. *Assume that node system* (8) *and pole system* (5) *when consistently ordered are identical with* (11) *and* (14), *respectively. Given a function f that is defined on $\mathscr{A}_N$ and sufficiently often differentiable at the multiple nodes, the interpolant $p \in \mathscr{U}_N$ of f at $\mathscr{A}_N$ admits the Lagrange-type representation,*

$$p = p\begin{bmatrix} u_1,\ldots,u_N \\ a_1,\ldots,a_N \end{bmatrix} f = p_1^N f = \sum_{i=1}^{N} \left(\frac{\mathrm{d}}{\mathrm{d}z}\right)^{\mu_i(a_i)} f(a_i)\ell_i = \sum_{\ell=1}^{p}\sum_{\lambda=0}^{m_\ell-1}\left(\frac{\mathrm{d}}{\mathrm{d}z}\right)^{\lambda} f(\alpha_\ell)\omega_\ell^\lambda, \tag{26}$$

where the Lagrange-type basis functions are

$$\ell_i(z) = \ell_{\varphi(\ell,\lambda)}(z) = \omega_\ell^\lambda(z) = \frac{\omega_\ell(z)}{B_N(z)} P_{\ell,\lambda}(z) v_{\ell,\lambda}(z). \tag{27}$$

Observe that in case all poles are at infinity formula (26) reduces to the well-known Lagrange–Hermite interpolation formula [2] for interpolation by algebraic polynomials.

The *proof* [12] is simple. One only has to check that the functions $\omega_\ell^\lambda \in \mathscr{U}_N$ are biorthogonal to the functionals d_s^σ which can be verified by repeatedly using the Leibniz' formula. □

It is another simple task to find the coefficients $A_{s,\sigma}^{\ell,\lambda}$ of the partial fraction decomposition in

$$\omega_\ell^\lambda = \sum_{s=0}^{q}\sum_{\sigma=1}^{n_s} A_{s,\sigma}^{\ell,\lambda} u_{s,\sigma},$$

$$A_{s,\sigma}^{\ell,\lambda} = \begin{cases} \dfrac{D_s^{n_s-\sigma}[\omega_\ell P_{\ell,\lambda} v_{\ell,\lambda}]}{(n_s-\sigma)!\prod_{t=1, t\neq s}^{q}(\beta_s-\beta_t)^{n_t}}, & s = 1,\ldots,q, \; \sigma = 1,\ldots,n_s, \\[2ex] \dfrac{D_0^{n_0-\sigma}}{(n_0-\sigma)!}\left[\dfrac{\omega_\ell P_{\ell,\lambda} v_{\ell,\lambda}}{B_N z^{n_0-1}}\right], & s = 0, \; \sigma = 1,\ldots,n_0. \end{cases} \tag{28}$$

Here the differentiation D_s^σ is defined by $D_s^\sigma(\cdot) = (\mathrm{d}/\mathrm{d}z)^\sigma(\cdot)_{z=\beta_s}$.

By somewhat tedious but elementary calculations it is possible to express the coefficients (28) solely in terms of the nodes and poles [12]. If one knows the coefficients $c_{j,t} = A_{\psi^{-1}(t)}^{\varphi^{-1}(j)}$ of the expansion

$$\ell_j = \sum_{t=1}^{N} c_{j,t} u_t, \quad j = 1,\ldots,N, \tag{29}$$

it is easy to give an explicit formula of the inverse of the *Cauchy–Vandermonde matrix*

$$V := V\begin{pmatrix} u_1,\ldots,u_N \\ L_1,\ldots,L_N \end{pmatrix} = (\langle L_i, u_j\rangle)_{i=1,\ldots,N}^{j=1,\ldots,N}. \tag{30}$$

In fact, since for $j = 1,\dots,N$,

$$\ell_j = \frac{1}{\det V}\begin{vmatrix} \langle L_1,u_1\rangle & \cdots\cdots\cdots & \langle L_1,u_N\rangle \\ \vdots & & \vdots \\ \langle L_{j-1},u_1\rangle & \cdots\cdots\cdots & \langle L_{j-1},u_N\rangle \\ u_1 & \cdots\cdots\cdots & u_N \\ \langle L_{j+1},u_1\rangle & \cdots\cdots\cdots & \langle L_{j+1},u_N\rangle \\ \vdots & & \vdots \\ \langle L_N,u_1\rangle & \cdots\cdots\cdots & \langle L_N,u_N\rangle \end{vmatrix} \tag{31}$$

the adjoint V_{adj} of V equals

$$V_{\mathrm{adj}} = (\det V)C \tag{32}$$

where C has entries $c_{j,t}$ defined by (29). Hence

$$V^{-1} = C^{\top}. \tag{33}$$

It is remarkable [6] that in case of q simple finite poles and a pole at infinity of multiplicity $n_0, q+n_0 = N$, and N simple nodes the inverse of V can be factorized as

$$V^{-1} = \begin{pmatrix} D_1 & 0 \\ 0 & H(s) \end{pmatrix} V^{\top} D_2, \tag{34}$$

where D_1, D_2 are diagonal matrices of dimensions q and N, respectively, and where $H(s)$ is a triangular Hankel matrix of the form

$$H(s) = \begin{pmatrix} s_1 & s_2 & s_3 & \dots & s_{n_q} \\ s_2 & s_3 & \dots & \cdot & \\ s_3 & \dots & \cdot & & \\ \cdot & \cdot & & & \\ \cdot & & & & \\ s_{n_q} & & & & \end{pmatrix}.$$

2.2. *The Neville–Aitken interpolation formula*

In [7] a Neville-Algorithm is given which computes the whole triangular field

$$(p_i^k f)_{i=1,\dots,N-k+1}^{k=1,\dots,N}, \qquad p_i^k f = p\begin{bmatrix} u_1,\dots,u_k \\ a_i,\dots,a_{i+k-1} \end{bmatrix} f \in \mathscr{U}_k \tag{35}$$

of interpolants recursively where $p_i^k f$ agrees with the function f on $\{a_i,\dots,a_{i+k-1}\}$. In [7] this algorithm was derived from the general Neville–Aitken algorithm [11,3] via explicit formulas for the Cauchy–Vandermonde determinants [17]. In [5] we were going the other way around and have given a different proof of the Neville–Aitken algorithm which is purely algebraic. In this survey we will derive the algorithm by a simple direct argument.

Proposition 3. *Let* $k \in \mathbb{N}$,

$$\mathscr{A}_{k+1} = (a_1, \dots, a_k, a_{k+1}) = (\underbrace{\alpha_1, \dots, \alpha_1}_{m_1}, \alpha_2, \dots, \alpha_{p-1}, \underbrace{\alpha_p, \dots, \alpha_p}_{m_p}) \in \mathbb{C}^{k+1},$$

with $\alpha_1, \dots, \alpha_p$ *pairwise distinct and* $m_1 + \cdots + m_p = k+1$, $\mathscr{A}_k = (a_1, \dots, a_k)$, $\mathscr{A}'_k = (a_2, \dots, a_{k+1})$ *and* $a_1 \neq a_{k+1}$. *Let* $\mathscr{U}_{k+1} = (u_1, \dots, u_{k+1})$ *be a CV-system associated with the pole system* $\mathscr{B}_{k+1} = (b_1, \dots, b_{k+1})$. *Suppose* $\mathscr{A}_{k+1} \cap \mathscr{B}_{k+1} = \emptyset$. *Let* $p_1 \in \mathscr{U}_k$ *interpolate* f *at* $\mathscr{A}_k$ *and* $p_2 \in \mathscr{U}_k$ *interpolate* f *at* $\mathscr{A}'_k$ *and let* $p_3 \in \mathscr{U}_{k+1}$ *interpolate* f *at* $\mathscr{A}_{k+1}$.

(i) *If* $b_{k+1} \in \mathbb{C}$ *then*

$$p_3(z) = \frac{p_1(z)(z - a_{k+1})(b_{k+1} - a_1) - p_2(z)(z - a_1)(b_{k+1} - a_{k+1})}{(a_{k+1} - a_1)(z - b_{k+1})}. \tag{36}$$

(ii) *If* $b_{k+1} = \infty$ *then*

$$p_3(z) = \frac{p_1(z)(z - a_{k+1}) - p_2(z)(z - a_1)}{a_1 - a_{k+1}}. \tag{37}$$

Proof. (i) Call the right-hand side of (36) $\tilde{p}_3$. It belongs to $\mathscr{U}_{k+1}$ in view of

$$\frac{z - a_{k+1}}{z - b_{k+1}} = 1 + \frac{b_{k+1} - a_{k+1}}{z - b_{k+1}} \quad \text{and} \quad \frac{z - a_1}{z - b_{k+1}} = 1 + \frac{b_{k+1} - a_1}{z - b_{k+1}}$$

by the partial fraction decomposition theorem. Obviously, $\tilde{p}_3$ interpolates f at $\mathscr{A}_{k+1}$ if all nodes are simple since the weights add to one and each of the unknown values $p_1(a_{k+1}), p_2(a_1)$ has factor 0. It is a consequence of Leibniz' rule that this holds true also in case of multiple nodes. In fact,

$$\begin{aligned}
&\left(\frac{\mathrm{d}}{\mathrm{d}z}\right)^{\mu} \tilde{p}_3|_{z=\alpha_i} \\
&= \frac{b_{k+1} - a_1}{a_{k+1} - a_1} \left(\frac{\mathrm{d}}{\mathrm{d}z}\right)^{\mu} \left[p_1(z)\frac{z - a_{k+1}}{z - b_{k+1}}\right]_{z=\alpha_i} - \frac{b_{k+1} - a_{k+1}}{a_{k+1} - a_1} \left(\frac{\mathrm{d}}{\mathrm{d}z}\right)^{\mu} \left[p_2(z)\frac{z - a_1}{z - b_{k+1}}\right]_{z=\alpha_i} \\
&= \frac{b_{k+1} - a_1}{a_{k+1} - a_1} \sum_{\lambda=0}^{\mu} \binom{\mu}{\lambda} \left(\frac{\mathrm{d}}{\mathrm{d}z}\right)^{\lambda} p_1(z)|_{z=\alpha_i} \left(\frac{\mathrm{d}}{\mathrm{d}z}\right)^{\mu-\lambda} \frac{z - a_{k+1}}{z - b_{k+1}}|_{z=\alpha_i} \\
&\quad - \frac{b_{k+1} - a_{k+1}}{a_{k+1} - a_1} \sum_{\lambda=0}^{\mu} \binom{\mu}{\lambda} \left(\frac{\mathrm{d}}{\mathrm{d}z}\right)^{\lambda} p_2(z)|_{z=\alpha_i} \left(\frac{d}{dz}\right)^{\mu-\lambda} \frac{z - a_1}{z - b_{k+1}}|_{z=\alpha_i} \\
&= \frac{b_{k+1} - a_1}{a_{k+1} - a_1} \sum_{\lambda=0}^{\mu} \binom{\mu}{\lambda} \left(\frac{\mathrm{d}}{\mathrm{d}z}\right)^{\lambda} f(z)|_{z=\alpha_i} \left(\frac{\mathrm{d}}{\mathrm{d}z}\right)^{\mu-\lambda} \frac{z - a_{k+1}}{z - b_{k+1}}|_{z=\alpha_i} \\
&\quad - \frac{b_{k+1} - a_{k+1}}{a_{k+1} - a_1} \sum_{\lambda=0}^{\mu} \binom{\mu}{\lambda} \left(\frac{\mathrm{d}}{\mathrm{d}z}\right)^{\lambda} f(z)|_{z=\alpha_i} \left(\frac{\mathrm{d}}{\mathrm{d}z}\right)^{\mu-\lambda} \frac{z - a_1}{z - b_{k+1}}|_{z=\alpha_i} \\
&= \left(\frac{\mathrm{d}}{\mathrm{d}z}\right)^{\mu} \left[\frac{b_{k+1} - a_1}{a_{k+1} - a_1} f(z) \frac{z - a_{k+1}}{z - b_{k+1}} - \frac{b_{k+1} - a_{k+1}}{a_{k+1} - a_1} f(z) \frac{z - a_1}{z - b_{k+1}}\right]_{z=\alpha_i} \\
&= \left(\frac{\mathrm{d}}{\mathrm{d}z}\right)^{\mu} f(z)|_{z=\alpha_i}
\end{aligned}$$

since the weights add to one. This is evident for all $i = 1, \dots, p$ and $\mu = 0, \dots, m_p - 1$ except for $i = 1$ and $\mu = m_1 - 1$ or $i = p$ and $\mu = m_p - 1$. If $i = 1$ and $\mu = m_1 - 1$ then the unknown derivative

$(\mathrm{d}/\mathrm{d}z)^{m_1-1}p_2(z)|_{z=\alpha_1}$ has the factor $(z-a_1)/(z-b_{k+1})|_{z=\alpha_1}$ which vanishes. Similarly, for $i=p$ and $\mu=m_p-1$ the unknown derivative $(\mathrm{d}/\mathrm{d}z)^{m_p-1}p_1(z)|_{z=\alpha_p}$ has the factor $(z-a_{k+1})/(z-b_{k+1})|_{z=\alpha_p}$ which vanishes.

(ii) Obviously, the right-hand side of (37) belongs to $\mathscr{U}_{k+1}$ and satisfies the interpolation conditions as is shown by the same argument used in the proof of (i). □

Remarks.

1. Letting $b_{k+1}\to\infty$ in (36) gives an alternative proof of (37). Observe that (37) is the classical Neville–Aitken recursion for interpolation by polynomials. It has to be used anytime a pole ∞ is inserted.
2. $p_3(z)$ is a convex combination of $p_1(z)$ and $p_2(z)$ if a_1, a_{k+1}, z, b_{k+1} are real and $a_1\leqslant z\leqslant a_{k+1}<b_{k+1}$ or $b_{k+1}<a_1\leqslant z\leqslant a_{k+1}$.
3. Proposition 2 constitutes an algorithm of arithmetical complexity $\mathcal{O}(N^2)$ to compute the triangular field (35) recursively from initializations $p_i^\ell f\in\mathscr{U}_\ell$ $(\ell\geqslant 1)$ which are generalized Taylor interpolants: $p_i^\ell f$ agrees with f at $(a_i,\dots,a_{i+\ell-1})$ where all nodes are identical $a_i=\cdots=a_{i+\ell-1}=:a$. From (26) and (27) immediately

$$p_i^\ell(f)=\sum_{\lambda=0}^{l-1}\left(\frac{\mathrm{d}}{\mathrm{d}z}\right)^{\lambda}f(a)\omega_1^\lambda \tag{38}$$

with

$$\omega_1^\lambda=\frac{1}{B_\ell(z)}\sum_{\mu=0}^{\ell-\lambda-1}\frac{(\mathrm{d}/\mathrm{d}z)^\mu B_\ell(a)}{\mu!}(z-a)^{\mu+\lambda}\in\mathscr{U}_\ell \tag{39}$$

are derived.

2.3. Newton's formula and the interpolation error

Given a complex function f which is defined and sufficiently often differentiable at the multiple nodes of system (8) then (9) constitutes a system of linear equations for the coefficients c_j of the generalized polynomial

$$p=pf=:\sum_{j=1}^{N}c_ju_j\in\mathscr{U}_N, \tag{40}$$

where the coefficient

$$c_j=c_{1,j}^N(f)=\begin{bmatrix}u_1,\dots,u_N & f\\ a_1,\dots,a_N & j\end{bmatrix}$$

will be referred to as the jth *divided difference* of f with respect to the systems U_N and $\mathscr{A}_N$. In [6,9] for consistently ordered poles which are assumed to be simple if finite and for simple nodes algorithms for solving system (9) recursively are derived whose arithmetical complexity is $\mathcal{O}(N^2)$.

In this section we are going to derive Newton's formula for the interpolant obtained in [14] and a procedure to compute the divided differences

$$\begin{bmatrix}u_1,\dots,u_{k+1} & f\\ a_i,\dots,a_{i+k} & k+1\end{bmatrix}$$

recursively [14], see also [10,16]. For the latter a new short proof is given. This way we will establish an algorithm to compute the interpolant (40) in the general case of multiple nodes and multiple poles in Newton's form recursively whose arithmetical complexity again is $\mathcal{O}(N^2)$. Later, in Proposition 6 of Section 4, we will derive an algorithm solving the linear system (9) recursively in the general case of multiple poles and multiple nodes avoiding the additional partial fraction decomposition.

Proposition 4. *If* $p_1^k f =: \sum_{j=1}^k c_{1,j}^k(f)u_j \in \mathcal{U}_k$ *interpolates* f *at* $\mathcal{A}_k = (a_1,\dots,a_k)$ *and* $p_1^{k+1} f =: \sum_{j=1}^{k+1} c_{1,j}^{k+1}(f)u_j \in \mathcal{U}_{k+1}$ *interpolates* f *at* $\mathcal{A}_{k+1} = (a_1,\dots,a_k,a_{k+1})$, *then*

$$p_1^{k+1} f = p_1^k f + c_{1,k+1}^{k+1}(f) r_1^k u_{k+1}, \tag{41}$$

where

$$r_1^k u_{k+1}(z) = u_{k+1}(z) - p_1^k u_{k+1}(z) = \frac{A_k(z)}{B_{k+1}(z)}\frac{B_k(b_{k+1})}{A_k(b_{k+1})} \tag{42}$$

with A_k *the node polynomial associated with* $\mathcal{A}_k$, $A_k(b_{k+1}) := \prod_{j=1}^k {}^*(b_{k+1} - a_j)$ *and with* B_k, B_{k+1} *the pole polynomials associated with the pole systems* $\mathcal{B}_k$ *and* $\mathcal{B}_{k+1}$, *respectively. Furthermore,*

$$c_{1,k+1}^{k+1}(f) = \left[\begin{matrix} u_1,\dots,u_{k+1} \\ a_1,\dots,a_{k+1}\end{matrix}\,\middle|\,\begin{matrix} f \\ k+1\end{matrix}\right]$$

with

$$\left[\begin{matrix} u_1,\dots,u_{k+1} \\ a_1,\dots,a_{k+1}\end{matrix}\,\middle|\,\begin{matrix} f \\ k+1\end{matrix}\right] = \begin{cases} \dfrac{\left[\begin{matrix} u_1,\dots,u_k \\ a_2,\dots,a_{k+1}\end{matrix}\,\middle|\,\begin{matrix} f \\ k\end{matrix}\right] - \left[\begin{matrix} u_1,\dots,u_k \\ a_1,\dots,a_k\end{matrix}\,\middle|\,\begin{matrix} f \\ k\end{matrix}\right]}{\frac{a_{k+1}-a_1}{(a_{k+1}-b_{k+1})^*}\cdot\frac{B_k(b_{k+1})}{A_k(b_{k+1})}\cdot\frac{A'_{k-1}(b_k)}{B_{k-1}(b_k)}} & \text{if } a_1 \neq a_{k+1}, \\[2ex] \det V\begin{pmatrix} u_1,\dots,u_{k-1},f \\ a,\dots,a,a\end{pmatrix} \Big/ \det V\begin{pmatrix} u_1,\dots,u_{k-1},u_k \\ a,\dots,a,a\end{pmatrix} \end{cases} \tag{43}$$

if in the second case all nodes are identical $a_1 = \cdots = a_{k+1} = a$. *Here* $A'_{k-1}(b_{k+1}) := \prod_{j=2}^k {}^*(b_{k+1} - a_j)$. *For any* $z \in \mathbb{C}\setminus\mathcal{B}_k$

$$r_1^k f(z) = f(z) - p_1^k f(z) = [a_1,\dots,a_k,z](B_k f)\frac{A_k(z)}{B_k(z)} \tag{44}$$

with

$$[a_1,\dots,a_k,z](B_k f) = \sum_{i=1}^k [a_1,\dots,a_i]B_k[a_i,\dots,a_k,z]f + [a_1,\dots,a_k,z]B_k f(z) \tag{45}$$

denoting the ordinary divided difference where $[a_1,\dots,a_k,z]B_k = 0$ *iff at least one pole is at infinity. Moreover, if* $b_{k+1} \in \bar{\mathbb{C}}$ *is chosen arbitrarily,*

$$r_1^k f(z) = \left[\begin{matrix} u_1,\dots,u_{k+1} \\ a_1,\dots,a_k,z\end{matrix}\,\middle|\,\begin{matrix} f \\ k+1\end{matrix}\right]\frac{A_k(z)B_k(b_{k+1})}{B_{k+1}(z)A_k(b_{k+1})}. \tag{46}$$

Proof. Consider linear system (9) for the coefficients of $p_1^{k+1} f$:

$$V\begin{pmatrix} u_1,\dots,u_{k+1} \\ L_1,\dots,L_{k+1}\end{pmatrix}\begin{pmatrix} c_{1,1}^{k+1}(f) \\ \vdots \\ c_{1,k+1}^{k+1}(f)\end{pmatrix} = \begin{pmatrix} \langle L_1, f\rangle \\ \vdots \\ \langle L_{k+1}, f\rangle\end{pmatrix}$$

bordered by the equation

$$\sum_{j=1}^{k+1} c_{1,j}^{k+1}(f)u_j(z) + \xi = 0$$

thus introducing a new unknown $\xi = -p_1^{k+1} f(z)$ where $z \notin \{a_1, \dots, a_{k+1}\}$ is arbitrary. The new system reads

$$\begin{pmatrix} V\begin{pmatrix} u_1,\dots,u_{k+1} \\ L_1,\dots,L_{k+1} \end{pmatrix} & \mathbf{0} \\ \langle L,u_1\rangle \dots \langle L,u_{k+1}\rangle & 1 \end{pmatrix} \begin{pmatrix} \boldsymbol{c} \\ \xi \end{pmatrix} = \begin{pmatrix} \boldsymbol{b} \\ 0 \end{pmatrix}, \tag{47}$$

where $\langle L,u_j\rangle := u_j(z)$ $(j = 1,\dots,k+1)$, $\boldsymbol{c} = (c_{1,1}^{k+1}(f),\dots,c_{1,k+1}^{k+1}(f))^\top$ and $\boldsymbol{b} = (\langle L_i, f\rangle)_{i=1,\dots,k+1}$ with $\langle L,f\rangle := 0$. By block elimination of the unknowns $c_1,\dots,c_k$ in the last equation of the bordered system using

$$V\begin{pmatrix} u_1,\dots,u_k \\ L_1,\dots,L_k \end{pmatrix}$$

as pivot we get the equation

$$\alpha_1 c_{1,k+1}^{k+1}(f) + \xi = \beta_1.$$

Here α_1, β_1 are certain Schur complements:

$$\alpha_1 = \det V\begin{pmatrix} u_1,\dots,u_k,u_{k+1} \\ L_1,\dots,L_k,L \end{pmatrix} \Big/ \det V\begin{pmatrix} u_1,\dots,u_k \\ L_1,\dots,L_k \end{pmatrix} = \langle L, r_1^k\, u_{k+1}\rangle$$

$$= \langle L,u_{k+1}\rangle - (\langle L,u_1\rangle,\dots,\langle L,u_k\rangle) V\begin{pmatrix} u_1,\dots,u_k \\ L_1,\dots,L_k \end{pmatrix}^{-1} \begin{pmatrix} \langle L_1,u_{k+1}\rangle \\ \vdots \\ \langle L_k,u_{k+1}\rangle \end{pmatrix}, \tag{48}$$

similarly,

$$\beta_1 = V\begin{pmatrix} u_1,\dots,u_k,f \\ L_1,\dots,L_k,L \end{pmatrix} \Big/ V\begin{pmatrix} u_1,\dots,u_k \\ L_1,\dots,L_k \end{pmatrix} = -p_1^k f(z).$$

Since z is arbitrary this yields Eq. (41). It holds trivially for $z \in \{a_1,\dots,a_k\}$.

The proof of (42) starts from the obvious representation

$$\det V\begin{pmatrix} u_1,\dots,u_k,u_{k+1} \\ L_1,\dots,L_k,L \end{pmatrix} = e\frac{A_k(z)}{B_{k+1}(z)}, \tag{49}$$

where A_k is the node polynomial associated with $\mathscr{A}_k$ and B_{k+1} is the pole polynomial associated with $\mathscr{B}_{k+1}$ and where e must be a constant. To determine e consider the partial fraction decomposition of $A_k(z)/B_{k+1}(z) = \sum_{j=1}^{k+1} d_j u_j(z)$. It is easy to see that in any case

$$d_{k+1} = \frac{A_k(b_{k+1})}{B_k(b_{k+1})}. \tag{50}$$

By comparing coefficients of u_{k+1} on both sides of (49) we find

$$\det V\begin{pmatrix} u_1,\dots,u_k \\ L_1,\dots,L_k \end{pmatrix} = e d_{k+1}. \tag{51}$$

Now (42) follows from (48)–(51).

To prove the remainder formulas (44) and (46) consider the node system $\tilde{\mathscr{A}}_{k+1}:=(a_1,\dots,a_k,z)$ with $z\in\mathbb{C}\backslash\mathscr{B}_k$. From Walsh's proof of Proposition 1 we see that for $z\notin\{a_1,\dots,a_k\}$ the interpolation error is

$$r_1^k f(z)=p_1^{k+1}f(z)-p_1^k f(z)=[a_1,\dots,a_k,z](B_k f)\frac{A_k(z)}{B_k(z)}. \tag{52}$$

If $z\in\{a_1,\dots,a_k\}$ (44) holds trivially. Eq. (45) results by application of Leibniz' rule for ordinary divided differences. Let $b_{k+1}\in\bar{\mathbb{C}}$ be arbitrary. By applying (41) to $\tilde{\mathscr{A}}_{k+1}$ with $z\in\mathbb{C}\setminus(\mathscr{B}_k\cup\mathscr{A}_k)$ (46) results. Again, (46) holds trivially, if $z\in\mathscr{A}_k$.

By comparison of (46) and (44) the following relation between ordinary and generalized divided differences obtains

$$\left[\begin{matrix}u_1,\dots,u_{k+1}\\a_1,\dots,a_k,z\end{matrix}\middle|\begin{matrix}f\\k+1\end{matrix}\right]=(z-b_{k+1})^*\frac{A_k(b_{k+1})}{B_k(b_{k+1})}[a_1,\dots,a_k,z](B_k f). \tag{53}$$

Clearly, Eq. (53) holding for all $z\notin\{a_1,\dots,a_k\}$ remains true for arbitrary $z=a_{k+1}\in\mathbb{C}\setminus\mathscr{B}_{k+1}$ by continuity of ordinary divided differences as functions of the nodes showing that the generalized divided differences on the left-hand side share this property. Moreover, using the well-known recurrence relation for ordinary divided differences from (53) with $z=:a_{k+1}$ yields

$$\left[\begin{matrix}u_1,\dots,u_{k+1}\\a_1,\dots,a_{k+1}\end{matrix}\middle|\begin{matrix}f\\k+1\end{matrix}\right]=(a_{k+1}-b_{k+1})^*\frac{A_k(b_{k+1})}{B_k(b_{k+1})}\frac{[a_2,\dots,a_{k+1}](B_k\cdot f)-[a_1,\dots,a_k](B_k f)}{a_{k+1}-a_1}. \tag{54}$$

Leibniz' rule for ordinary divided differences gives

$$\begin{aligned}[a_2,\dots,a_{k+1}](B_k f)&=[a_2,\dots,a_{k+1}]((z-b_k)B_{k-1}f)\\&=(a_{k+1}-b_k)^*[a_2,\dots,a_{k+1}](B_{k-1}f)+1[a_2,\dots,a_k](B_{k-1}f),\end{aligned}$$

$$\begin{aligned}[a_1,\dots,a_k](B_k f)&=[a_1,\dots,a_k]((z-b_k)B_{k-1}f)\\&=(a_1-b_k)^*[a_1,\dots,a_k](B_{k-1}f)+1[a_2,\dots,a_k](B_{k-1}f)\end{aligned}$$

and by subtraction

$$\begin{aligned}&\frac{[a_2,\dots,a_{k+1}](B_k f)-[a_1,\dots,a_k](B_k f)}{a_{k+1}-a_1}\\&\quad=\frac{(a_{k+1}-b_k)^*[a_2,\dots,a_{k+1}](B_{k-1}f)-(a_1-b_k)^*[a_1,\dots,a_k](B_{k-1}f)}{a_{k+1}-a_1}.\end{aligned}$$

Now (43) follows if the last expression is inserted on the right-hand side of (54). □

Example. Given $\mathscr{A}_5=(0,0,1,1,-2)$ and $\mathscr{B}_5=(\infty,\infty,-1,-1,2)$ corresponding with $U_5=(u_1,u_2,u_3,u_4,u_5)$ with

$$u_1(z)=1,\quad u_2(z)=z,\quad u_3(z)=\frac{1}{z+1},\quad u_4(z)=\frac{1}{(z+1)^2},\quad u_5(z)=\frac{1}{z-2}.$$

Given of a function f the interpolation data

$$f(0)=\tfrac{3}{2},\quad f'(0)=-2,\quad f(1)=\tfrac{3}{4},\quad f'(1)=-\tfrac{3}{8},\quad f(-2)=-\tfrac{9}{8},$$

find $p\in\mathscr{U}_5$ that agrees with f at $\mathscr{A}_5$.

According to (43) we easily compute the table of divided differences of f:

z_i	$\left[\begin{array}{c\|c} u_1 & f \\ \cdot & 1 \end{array}\right]$	$\left[\begin{array}{cc\|c} u_1 & u_2 & f \\ \cdot & \cdot & 2 \end{array}\right]$	$\left[\begin{array}{ccc\|c} u_1 & u_2 & u_3 & f \\ \cdot & \cdot & \cdot & 3 \end{array}\right]$	$\left[\begin{array}{cccc\|c} u_1 & u_2 & u_3 & u_4 & f \\ \cdot & \cdot & \cdot & \cdot & 4 \end{array}\right]$
0	$\left[\begin{array}{c\|c} u_1 & f \\ 0 & 1 \end{array}\right] = \frac{3}{2}$			
0		$\left[\begin{array}{cc\|c} u_1 & u_2 & f \\ 0 & 0 & 2 \end{array}\right] = -2$		
1	$\left[\begin{array}{c\|c} u_1 & f \\ 1 & 1 \end{array}\right] = \frac{3}{4}$	$\left[\begin{array}{cc\|c} u_1 & u_2 & f \\ 0 & 1 & 2 \end{array}\right] = -\frac{3}{4}$	$\left[\begin{array}{ccc\|c} u_1 & u_2 & u_3 & f \\ 0 & 0 & 1 & 3 \end{array}\right] = \frac{5}{2}$	
1		$\left[\begin{array}{cc\|c} u_1 & u_2 & f \\ 1 & 1 & 2 \end{array}\right] = -\frac{3}{8}$	$\left[\begin{array}{ccc\|c} u_1 & u_2 & u_3 & f \\ 0 & 0 & 1 & 3 \end{array}\right] = \frac{3}{2}$	$\left[\begin{array}{cccc\|c} u_1 & u_2 & u_3 & u_4 & f \\ 0 & 0 & 1 & 1 & 4 \end{array}\right] = 2$
-2	$\left[\begin{array}{c\|c} u_1 & f \\ -2 & 1 \end{array}\right] = -\frac{9}{4}$	$\left[\begin{array}{cc\|c} u_1 & u_2 & f \\ 1 & -2 & 2 \end{array}\right] = 1$	$\left[\begin{array}{ccc\|c} u_1 & u_2 & u_3 & f \\ 1 & 1 & -2 & 3 \end{array}\right] = \frac{11}{6}$	$\left[\begin{array}{cccc\|c} u_1 & u_2 & u_3 & u_4 & f \\ 0 & 1 & 1 & -2 & 4 \end{array}\right] = -\frac{1}{6}$

$$\left[\begin{array}{ccccc|c} u_1 & u_2 & u_3 & u_4 & u_5 & f \\ 0 & 0 & 1 & 1 & -2 & 5 \end{array}\right] = \frac{13}{27}.$$

From Newton's formula (41) we get the interpolant in Newton's form

$$p(z) = p_1^5 f(z) = \frac{3}{2} - 2z + \frac{5}{2}\frac{z^2}{z+1} + 2\frac{-z^2(z-1)}{2(z+1)^2} + \frac{13}{27}\frac{9}{4}\frac{z^2(z-1)^2}{(z+1)^2(z-2)},$$

which, by additional partial fraction decompositions, equals

$$p_1^5 f(z) = -\frac{1}{6} + \frac{7}{12}z + \frac{73}{54}\frac{1}{z+1} + \frac{5}{9}\frac{1}{(z+1)^2} + \frac{13}{27}\frac{1}{z-2}.$$

In Section 4 we will present an alternative method computing the interpolant avoiding the additional partial fraction decompositions.

3. Cauchy–Vandermonde determinants

In this section we give a new short proof of the explicite formula of the Cauchy–Vandermonde determinant [17,7,15] in terms of the nodes and poles. It will be derived as a simple consequence of Proposition 4.

Proposition 5. *For consistently ordered node and pole systems as in* (11) *and* (12) *that have no common points*

$$\det V\begin{pmatrix} u_1,\ldots,u_N \\ L_1,\ldots,L_N \end{pmatrix} = \operatorname{mult}(\mathscr{A}_N)\frac{\prod_{\substack{k,j=1\\k>j}}^{N}{}^*(a_k-a_j)\prod_{\substack{k,j=1\\k\geqslant j}}^{N}{}^*(a_k-b_j)}{\prod_{\substack{k,j=1\\k>j}}^{N}{}^*(b_k-b_j)\prod_{\substack{k,j=1\\k>j}}^{N}{}^*(b_k-a_j)} \tag{55}$$

with

$$\operatorname{mult}(\mathscr{A}_N)=\prod_{k=1}^{N}\mu_k(a_k)! \tag{56}$$

where $\mu_k(a)$ *is defined by* (10) *and use is made of notations* (16) *and* (17).

Proof. From (48) and (42) for $k+1=N$ we get

$$r_1^{N-1}u_N(z)=\frac{\det V\begin{pmatrix}u_1,\dots,u_{N-1},u_N\\ L_1,\dots,L_{N-1},L\end{pmatrix}}{\det V\begin{pmatrix}u_1,\dots,u_{N-1}\\ L_1,\dots,L_{N-1}\end{pmatrix}}$$
$$=\frac{(z-a_1)\cdot\ldots\cdot(z-a_{N-1})}{(z-b_1)^*\cdot\ldots\cdot(z-b_{N-1})^*(z-b_N)^*}\,\frac{(b_N-b_1)^*\cdot\ldots\cdot(b_N-b_{N-1})^*}{(b_N-a_1)^*\cdot\ldots\cdot(b_N-a_{N-1})^*}.$$

As a consequence,

$$\det V\begin{pmatrix}u_1,\dots,u_{N-1},u_N\\ L_1,\dots,L_{N-1},L_N\end{pmatrix}=\det V\begin{pmatrix}u_1,\dots,u_{N-1}\\ L_1,\dots,L_{N-1}\end{pmatrix}\left(\frac{\mathrm{d}}{\mathrm{d}z}\right)^{\mu_N(a_N)} r_1^{N-1}u_N(z)\big|_{z=a_N}.$$

By Leibniz' rule

$$\left(\frac{\mathrm{d}}{\mathrm{d}z}\right)^{\mu_N(a_N)}\frac{(z-a_1)\cdot\ldots\cdot(z-a_{N-1})}{(z-b_1)^*\cdot\ldots\cdot(z-b_{N-1})^*(z-b_N)^*}\bigg|_{z=a_N}=\mu_N(a_N)!\,\frac{\prod_{j=1}^{N-1}{}^*(a_N-a_j)}{\prod_{j=1}^{N}{}^*(a_N-b_j)}.$$

Hence, we have got a recursion for the determinants considered. Since $\det V\binom{u_1}{a_1}=\mu_1(a_1)!\cdot 1/(a_1-b_1)^*$ an induction argument proves (55). □

4. Solution of linear CV-systems

In this section we will present a new method solving the system of linear equations (9) recursively where no additional partial fraction decomposition is needed. Its proof is based upon Proposition 3.

Proposition 6. *Let* $k\in\mathbb{N}$ *and let the CV-systems* $(u_1,\dots,u_k)$ *and* $(u_1,\dots,u_k,u_{k+1})$ *correspond to the pole systems*

$$\mathscr{B}_k=(b_1,\dots,b_k)=(\underbrace{\beta_0,\dots,\beta_0}_{n_0},\underbrace{\beta_1,\dots,\beta_1}_{n_1},\dots,\underbrace{\beta_q,\dots,\beta_q}_{n_q})$$

and $\mathscr{B}_{k+1}=(b_1,\dots,b_k,b_{k+1})$, *respectively, where it is assumed that* $\mathscr{B}_k$ *is consistenly ordered with* $\beta_0=\infty,\beta_1,\dots,\beta_q\in\mathbb{C}$ *pairwise distinct and* $n_0+n_1+\dots+n_q=k$. *We set for* $r=0,\dots,q$,

$$j_r{:=}n_0+\dots+n_r.$$

Let $\mathscr{A}_k=(a_1,\dots,a_k)\in\mathbb{C}^k$ *and* $\mathscr{A}_{k+1}=(a_1,\dots,a_k,a_{k+1})\in\mathbb{C}^{k+1}$ *be arbitrary node systems with*

$$\alpha_2{:=}a_{k+1}\neq a_1=:\alpha_1.$$

By $\mathscr{A}'_k$ we denote the node system $\mathscr{A}'_k=(a_2,\dots,a_{k+1})$. Given a function f defined and sufficiently often differentiable at the multiple nodes of $\mathscr{A}_{k+1}$. Let

$$p_1^k f =: \sum_{j=1}^{k} c_{1,j}^k(f)u_j \in \mathscr{U}_k \qquad \textit{interpolate } f \textit{ at } \mathscr{A}_k,$$

$$p_2^k f =: \sum_{j=1}^{k} c_{2,j}^k(f)u_j \in \mathscr{U}_k \qquad \textit{interpolate } f \textit{ at } \mathscr{A}'_k, \textit{ and}$$

$$p_1^{k+1} f =: \sum_{j=1}^{k+1} c_{1,j}^{k+1}(f)u_j \in \mathscr{U}_{k+1} \ \textit{interpolate } f \textit{ at } \mathscr{A}_{k+1}.$$

For simplicity, the argument f of the divided differences in the following formulas will be dropped.

(i) *If $\beta:=b_{k+1}\in\mathbb{C}\setminus\{b_1,\dots,b_k\}$ corresponding to $u_{k+1}(z)=1/(z-\beta)$ then*

$$c_{1,j}^{k+1}=\frac{c_{1,j}^k(\beta-\alpha_1)-c_{2,j}^k(\beta-\alpha_2)}{\alpha_2-\alpha_1}+\frac{(\beta-\alpha_1)(\beta-\alpha_2)}{\alpha_2-\alpha_1}\sum_{\lambda=j+1}^{j_0}(c_{1,\lambda}^k-c_{2,\lambda}^k)\beta^{\lambda-j-1}, \quad j=1,\dots,j_0, \tag{57}$$

$$c_{1,j}^{k+1}=\frac{c_{1,j}^k(\beta-\alpha_1)(\alpha_2-b_j)-c_{2,j}^k(\beta-\alpha_2)(\alpha_1-b_j)}{(\alpha_2-\alpha_1)(\beta-b_j)} -\frac{(\beta-\alpha_1)(\beta-\alpha_2)}{\alpha_2-\alpha_1}\sum_{\lambda=j+1}^{j_{r+1}}\frac{c_{1,\lambda}^k-c_{2,\lambda}^k}{(\beta-b_j)^{\lambda-j+1}} \quad j_r<j\leqslant j_{r+1};\ r=0,\dots,q-1, \tag{58}$$

$$c_{1,k+1}^{k+1}=\frac{(\beta-\alpha_1)(\beta-\alpha_2)}{\alpha_2-\alpha_1}\left(\sum_{\lambda=1}^{j_0}(c_{1,\lambda}^k-c_{2,\lambda}^k)\beta^{\lambda-1}+\sum_{r=0}^{q-1}\sum_{\lambda=j_r+1}^{j_{r+1}}\frac{c_{1,\lambda}^k-c_{2,\lambda}^k}{(\beta-b_\lambda)^{\lambda-j_r}}\right). \tag{59}$$

(ii) *If $\beta:=b_{k+1}=\beta_i\in\mathbb{C}$ corresponding with $u_{k+1}(z)=1/(z-\beta_i)^{n_i+1}$ then*

$$c_{1,j}^{k+1}=\frac{c_{1,j}^k(\beta-\alpha_1)-c_{2,j}^k(\beta-\alpha_2)}{\alpha_2-\alpha_1}+\frac{(\beta-\alpha_1)(\beta-\alpha_2)}{\alpha_2-\alpha_1}\sum_{\lambda=j+1}^{j_0}(c_{1,\lambda}^k-c_{2,\lambda}^k)\beta^{\lambda-j-1}, \quad j=1,\dots,j_0, \tag{60}$$

$$c_{1,j}^{k+1}=\frac{c_{1,j}^k(\beta-\alpha_1)(\alpha_2-b_j)-c_{2,j}^k(\beta-\alpha_2)(\alpha_1-b_j)}{(\alpha_2-\alpha_1)(\beta-b_j)} +\frac{(\beta-\alpha_1)(\beta-\alpha_2)}{\alpha_2-\alpha_1}\sum_{\lambda=j+1}^{j_{r+1}}\frac{c_{1,\lambda}^k-c_{2,\lambda}^k}{(\beta-b_j)^{\lambda-j+1}}, \quad j_r<j\leqslant j_{r+1},\ r=0,\dots,i-2,i,i+1,\dots,q-1, \tag{61}$$

$$c_{1,j_{i-1}+1}^{k+1}=\frac{c_{1,j_{i-1}+1}^k(\beta-\alpha_1)-c_{2,j_{i-1}+1}^k(\beta-\alpha_2)}{\alpha_2-\alpha_1} +\frac{(\beta-\alpha_1)(\beta-\alpha_2)}{\alpha_2-\alpha_1}\left[\sum_{\lambda=1}^{j_0}(c_{1,\lambda}^k-c_{2,\lambda}^k)\beta^{\lambda-1}+\sum_{\substack{r=0\\ r\neq i-1}}^{q-1}\sum_{\lambda=j_r+1}^{j_{r+1}}\frac{c_{1,\lambda}^k-c_{2,\lambda}^k}{(\beta-b_\lambda)^{\lambda-j_r}}\right], \tag{62}$$

$$c_{1,j}^{k+1} = \frac{c_{1,j}^k(\beta-\alpha_1) - c_{2,j}^k(\beta-\alpha_2)}{\alpha_2-\alpha_1} + \frac{(\beta-\alpha_1)(\beta-\alpha_2)}{\alpha_2-\alpha_1}(c_{1,j-1}^k - c_{2,j-1}^k), \quad j = j_{i-1}+2,\dots,j_i, \tag{63}$$

$$c_{1,k+1}^{k+1} = \frac{(\beta-\alpha_1)(\beta-\alpha_2)}{\alpha_2-\alpha_1}(c_{1,j_i}^k - c_{2,j_i}^k). \tag{64}$$

(iii) *If* $\beta := b_{k+1} = \infty$ *corresponding with* $u_{k+1}(z) = z^{n_0}$ *then*

$$c_{1,1}^{k+1} = \frac{c_{1,1}^k\alpha_2 - c_{2,1}^k\alpha_1 - \sum_{\lambda=0}^{q-1}(c_{1,j_\lambda+1}^k - c_{2,j_\lambda+1}^k)}{\alpha_2-\alpha_1}, \tag{65}$$

$$c_{1,j}^{k+1} = \frac{c_{1,j}^k\alpha_2 - c_{2,j}^k\alpha_1 - (c_{1,j-1}^k - c_{2,j-1}^k)}{\alpha_2-\alpha_1}, \quad j = 2,\dots,j_0, \tag{66}$$

$$c_{1,j}^{k+1} = \frac{c_{1,j}^k\alpha_2 - c_{2,j}^k\alpha_1 - (c_{1,j}^k - c_{2,j}^k)b_j - (c_{1,j+1}^k - c_{2,j+1}^k)}{\alpha_2-\alpha_1}, \quad j_0 < j < k,\ j \neq j_1, j_2, \dots, j_q, \tag{67}$$

$$c_{1,j}^{k+1} = \frac{c_{1,j}^k\alpha_2 - c_{2,j}^k\alpha_1 - (c_{1,j}^k - c_{2,j}^k)b_j}{\alpha_2-\alpha_1}, \quad j = j_i,\ i = 1,\dots,q, \tag{68}$$

$$c_{1,k+1}^{k+1} = -\frac{c_{1,j_0}^k - c_{2,j_0}^k}{\alpha_2-\alpha_1}. \tag{69}$$

Proof. According to (36) if $\beta := b_{k+1} \in \mathbb{C}$ we have

$$\begin{aligned} p_1^{k+1} &= \sum_{j=1}^{k+1} c_{1,j}^{k+1} u_j \\ &= \left(\sum_{j=1}^{k} c_{1,j}^k u_j\right)\left(1 + \frac{\beta-\alpha_2}{z-\beta}\right)\frac{\beta-\alpha_1}{\alpha_2-\alpha_1} - \left(\sum_{j=1}^{k} c_{2,j}^k u_j\right)\left(1 + \frac{\beta-\alpha_1}{z-\beta}\right)\frac{\beta-\alpha_2}{\alpha_2-\alpha_1} \\ &= \frac{1}{\alpha_2-\alpha_1}\sum_{j=1}^{k}(c_{1,j}^k(\beta-\alpha_1) - c_{2,j}^k(\beta-\alpha_2))u_j \\ &\quad + \frac{(\beta-\alpha_1)(\beta-\alpha_2)}{\alpha_2-\alpha_1}\sum_{j=1}^{k}(c_{1,j}^k - c_{2,j}^k)u_j\frac{1}{z-\beta}. \end{aligned} \tag{70}$$

If $\beta = \infty$ according to (37) we have

$$\begin{aligned} p_1^{k+1} &= \sum_{j=1}^{k+1} c_{1,j}^{k+1} u_j = \frac{\sum_{j=1}^k c_{1,j}^k u_j(z-\alpha_2) - \sum_{j=1}^k c_{2,j}^k u_j(z-\alpha_1)}{\alpha_1-\alpha_2} \\ &= \frac{\sum_{j=1}^k (c_{1,j}^k\alpha_2 - c_{2,j}^k\alpha_1)u_j - \sum_{j=1}^k (c_{1,j}^k - c_{2,j}^k)u_j z}{\alpha_2-\alpha_1}. \end{aligned} \tag{71}$$

Now by partial fraction decomposition

$$z^{\nu}\frac{1}{z-\beta}=\sum_{\lambda=0}^{\nu-1}\beta^{\nu-\lambda-1}z^{\lambda}+\beta^{\nu}\frac{1}{z-\beta}, \tag{72}$$

$$\frac{1}{(z-b)^{\nu+1}}\frac{1}{z-\beta}=\sum_{\lambda=0}^{\nu}\frac{-1}{(\beta-b)^{\lambda+1}}\frac{1}{(z-b)^{\nu+1-\lambda}}+\frac{1}{(\beta-b)^{\nu+1}}\frac{1}{z-\beta}, \tag{73}$$

$$\frac{1}{(z-b)^{\nu+1}}z=\frac{z-b+b}{(z-b)^{\nu+1}}=\frac{1}{(z-b)^{\nu}}+b\frac{1}{(z-b)^{\nu+1}}. \tag{74}$$

Eq. (73) is readily verified by multiplying both sides by $(z-b)^{\nu+1}(z-\beta)$ and making use of the finite geometric series. Eq. (72) follows from

$$z^{\nu}\frac{1}{z-\beta}=\frac{(z-\beta+\beta)^{\nu}}{z-\beta}=\sum_{\mu=0}^{\nu}\binom{\nu}{\mu}\beta^{\nu-\mu}(z-\beta)^{\mu-1}$$
$$=\frac{\beta^{\nu}}{z-\beta}+\sum_{\mu=1}^{\nu}\binom{\nu}{\mu}\beta^{\nu-\mu}\sum_{\lambda=0}^{\mu-1}\binom{\mu-1}{\lambda}z^{\lambda}(-\beta)^{\mu-\lambda-1}.$$

Here the second sum can be extended over $\lambda=0,\dots,\nu-1$ since the binomial coefficients $\binom{\mu-1}{\lambda}$ vanish for the extra summands. By interchanging the two summations we obtain

$$z^{\nu}\frac{1}{z-\beta}=\frac{\beta^{\nu}}{z-\beta}+\sum_{\lambda=0}^{\nu-1}z^{\lambda}\beta^{\nu-\lambda-1}(-1)^{\lambda-1}\sum_{\mu=1}^{\nu}\binom{\nu}{\mu}\binom{\mu-1}{\lambda}(-1)^{\mu}.$$

The second sum equals

$$(-1)^{\nu}\sum_{\mu=0}^{\nu}\binom{\nu}{\mu}\binom{\mu-1}{\lambda}(-1)^{\nu-\mu}+(-1)^{\lambda+1}=(-1)^{\lambda+1}$$

since the sum in the last equation is the forward difference $\Delta_1^{\nu}f(0)=f^{(\nu)}(\xi)=0$, where

$$f(x)=\binom{x-1}{\lambda}=\frac{(x-1)(x-2)\cdots(x-\lambda)}{\lambda!}$$

is a polynomial of degree λ and $\lambda\leqslant\nu-1<\nu$.

Let now $\nu:=\nu_j(b_j)$ denote the multiplicity of b_j in $(b_1,\dots,b_{j-1})$. Consider case (i): $\beta=b_{k+1}\in\mathbb{C}$, $\beta\notin\{b_1,\dots,b_k\}$. Then from (72) and (73) (for simplicity we drop the argument z)

$$u_j\frac{1}{z-\beta}=\sum_{\lambda=0}^{j-2}\beta^{\lambda}u_{j-1-\lambda}+\beta^{j-1}u_{k+1},\quad 1\leqslant j\leqslant j_0, \tag{75}$$

$$u_j\frac{1}{z-\beta}=\sum_{\lambda=0}^{\nu}\frac{-1}{(\beta-b_j)^{\lambda+1}}u_{j-\lambda}+\frac{1}{(\beta-b_j)^{\nu+1}}u_{k+1},\quad j_0<j\leqslant k. \tag{76}$$

In case (ii): $\beta = b_{k+1} = \beta_i \in \mathbb{C}$, from (72) and (73)

$$u_j \frac{1}{z-\beta} = \sum_{\lambda=0}^{j-2} \beta^\lambda u_{j-1-\lambda} + \beta^{j-1} u_{j_{i-1}+1}, \quad 1 \leqslant j \leqslant j_0 \tag{77}$$

$$u_j \frac{1}{z-\beta} = \sum_{\lambda=0}^{\nu} \frac{-1}{(\beta - b_j)^{\lambda+1}} u_{j-\lambda} + \frac{1}{(\beta - b_j)^{\nu+1}} u_{j_{i-1}+1}, \quad j_0 < j \leqslant j_{i-1} \text{ or } j_i < j \leqslant k, \tag{78}$$

$$u_j \frac{1}{z-\beta} = u_{j+1}, \quad j_{i-1} < j < j_i, \tag{79}$$

$$u_j \frac{1}{z-\beta} = u_{k+1}, \quad j = j_i. \tag{80}$$

In case (iii): $\beta = b_{k+1} = \infty$ we have

$$u_j z = u_{j+1}, \quad j = 1, \ldots, j_0 - 1, \tag{81}$$

$$u_j z = z^{n_0} = u_{k+1}, \quad j = j_0, \tag{82}$$

$$u_j z = u_1 + b_j u_j, \quad j = j_i + 1, i = 0, \ldots, q-1, \tag{83}$$

$$u_j z = u_{j-1} + b_j u_j \quad j_0 + 1 < j \leqslant k, j \neq j_i + 1, i = 0, \ldots, q-1. \tag{84}$$

Eqs. (81) and (82) are obvious and (83) and (84) follow from (74). The rest of the proof consists in comparing coefficients. □

Remark. (i) The arithmetical complexity for computing $(c_{1,j}^{k+1})_{j=1,\ldots,k+1}$ from $(c_{1,j}^k)_{j=1,\ldots,k}$ and $(c_{2,j}^k)_{j=1,\ldots,k}$ according to (57)–(64) in cases (i) or (ii) is $\mathcal{O}(k+1+\sum_{r=0}^q (n_r-1)^2)$ and $\mathcal{O}(k+1)$ in case (iii).

(ii) It should be noticed that the first term in (58) resp. (61) is the recursion (36) with p_1, p_2 replaced by $c_{1,j}^k, c_{2,j}^k$ and with $z = b_j$. Similarly, the first term in (57), (60), (62) and (63) is recursion (37) with p_1, p_2 replaced by $c_{1,j}^k, c_{2,j}^k$ and with $z = b_j$.

Consider once more the *example* given in Section 2.3. According to Proposition 5 we compute the triangular field of solutions

$$p_i^k = p\begin{bmatrix} u_1, \ldots, u_k \\ a_i, \ldots, a_{i+k} \end{bmatrix} =: \sum_{j=1}^k c_{i,j}^k \cdot u_j, \quad k = 1, \ldots, 5, \ i = 1, \ldots, 6-k.$$

The initializations which are certain generalized Taylor polynomials of f are computed according to Proposition 4, or alternatively, according to (38) and (39).

a_i	$p\begin{bmatrix} u_1 \\ \cdot \end{bmatrix}$	$p\begin{bmatrix} u_1 & u_2 \\ \cdot & \cdot \end{bmatrix}$	$p\begin{bmatrix} u_1 & u_2 & u_3 \\ \cdot & \cdot & \cdot \end{bmatrix}$
0	$p\begin{bmatrix} u_1 \\ 0 \end{bmatrix} = \frac{3}{2} \cdot u_1$		
0		$p\begin{bmatrix} u_1 & u_2 \\ 0 & 0 \end{bmatrix} = \frac{3}{2} - 2 \cdot z$	
1	$p\begin{bmatrix} u_1 \\ 1 \end{bmatrix} = \frac{3}{4} \cdot u_1$	$p\begin{bmatrix} u_1 & u_2 \\ 0 & 1 \end{bmatrix} = \frac{3}{2} - \frac{3}{4} \cdot z$	$p\begin{bmatrix} u_1 & u_2 & u_3 \\ 0 & 0 & 1 \end{bmatrix} = -2 + \frac{1}{2} \cdot z + \frac{5}{2} \frac{1}{z+1}$
1		$p\begin{bmatrix} u_1 & u_2 \\ 0 & 1 \end{bmatrix} = \frac{9}{8} - \frac{3}{8} \cdot z$	$p\begin{bmatrix} u_1 & u_2 & u_3 \\ 0 & 1 & 1 \end{bmatrix} = \frac{3}{2} \frac{1}{z+1}$
−2	$p\begin{bmatrix} u_1 \\ -2 \end{bmatrix} = -\frac{9}{8} \cdot u_1$	$p\begin{bmatrix} u_1 & u_2 \\ 1 & -2 \end{bmatrix} = -\frac{1}{4} + z$	$p\begin{bmatrix} u_1 & u_2 & u_3 \\ 1 & 1 & -2 \end{bmatrix} = -\frac{1}{4} + \frac{1}{12} \cdot z + \frac{11}{6} \frac{1}{z+1}$

$$p\begin{bmatrix} u_1 & u_2 & u_3 & u_4 \\ 0 & 0 & 1 & 1 \end{bmatrix} = 2 - \frac{1}{2} z - \frac{5}{2} \frac{1}{z+1} + 2 \frac{1}{(z+1)^2},$$

$$p\begin{bmatrix} u_1 & u_2 & u_3 & u_4 \\ 0 & 1 & 1 & -2 \end{bmatrix} = -\frac{1}{6} + \frac{1}{24} z + \frac{11}{6} \frac{1}{z+1} - \frac{1}{6} \frac{1}{(z+1)^2},$$

$$p\begin{bmatrix} u_1 & u_2 & u_3 & u_4 & u_5 \\ 0 & 0 & 1 & 1 & -2 \end{bmatrix} = -\frac{1}{6} + \frac{7}{12} z + \frac{73}{54} \frac{1}{z+1} + \frac{5}{9} \frac{1}{(z+1]^2} + \frac{13}{27} \frac{1}{z-2}.$$

For a theory of convergence of rational interpolants with prescribed poles to analytic functions as $N \to \infty$ confer [1].

5. Applications

CV-systems have been used to construct rational B-splines with prescribed poles. A. Gresbrand [8] has found a recursion fomula for such splines that reduces to de Boor's recursion when all poles are at infinity. Given a weakly increasing sequence $t = (t_j)_{j=0}^{m+1}$ of knots in $[a, b]$ where $t_0 = a$ and $t_{m+1} = b$ are simple knots and $t_i < t_{i+n-1}$ for all i. The extended knot sequence is $t_{\text{ext}} = (t_j)_{j=-n+1}^{m+n}$ where a and b are repeated precisely n times each. Given a pole sequence $(b_1, \ldots, b_n) \in \bar{\mathbb{R}} \setminus [a, b]$ that is consistently ordered with $b_1 = \infty$, then for $j = 0, \ldots, m$ define

$$B_0^1 := \chi_{[t_0, t_1]} \quad \text{and} \quad B_j^1 := \chi_{(t_j, t_{j+1}]}$$

with χ_S denoting the characteristic function of a set S and for $k = 2, \ldots, n$ and for $j = -k+2, \ldots, m$ define

$$\lambda_j^k(x) := \frac{x - t_j}{t_{j+k-1} - t_j} \frac{1}{(k-1)(x - b_k)^*} \frac{\operatorname{perm}\,(t_{j+i} - b_{\ell+1})_{i=1,\ldots,k-1}^{\ell=1,\ldots,k-1}}{\operatorname{perm}\,(t_{j+i} - b_{\ell+1})_{i=1,\ldots,k-2}^{\ell=1,\ldots,k-2}},$$

$$\mu_j^k(x) := \frac{t_{j+k-}1 - x}{t_{j+k-1} - t_j} \frac{1}{(k-1)(x - b_k)^*} \frac{\operatorname{perm}\,(t_{j+i-1} - b_{\ell+1})_{i=1,\ldots,k-1}^{\ell=1,\ldots,k-1}}{\operatorname{perm}\,(t_{j+i} - b_{\ell+1})_{i=1,\ldots,k-2}^{\ell=1,\ldots,k-2}},$$

where the permanent of a matrix $A=(a_{i,j})\in\mathbb{K}^{n\times n}$ is defined as

$$\operatorname{perm} A=\sum_{\sigma\in S_n}\prod_{i=1}^{n} a_{i,\sigma(i)}.$$

Here S_n denotes the symmetric group of all permutations of order n. Then for $k=2,\ldots,n$ and $j=-k+1,\ldots,m$

$$B_j^k(x):=\lambda_j^k(x)B_j^{k-1}(x)+\mu_{j+1}^k(x)B_{j+1}^{k-1}(x) \tag{85}$$

are rational B-splines of order k with prescribed poles $(b_1,\ldots,b_k)$, i.e., when restricted to any knot interval then B_j^k belongs to the CV-space $\mathscr{U}_k$. Gresbrand [8] has proved that

$$B_j^k(x)=\frac{\operatorname{perm}(t_{j+i}-b_{\ell+1})_{i=1,\ldots,k-1}^{\ell=1,\ldots,k-1}}{(k-1)!B_k(x)}N_j^k(x), \tag{86}$$

where B_k is the pole polynomial associated with the pole system $(b_1,\ldots,b_k)$ and $N_j^k(x)=(t_{j+k}-t_j)[t_j,\ldots,t_{j+k}](\cdot-x)_+^{k-1}$ is the ordinary polynomial B-spline function of order k with knots $t_j,\ldots,t_{j+k}$.

The rational B-splines with prescribed poles (84) share many properties with the de Boor B-splines [8]:

1. They can be computed recursively by a de Boor like algorithm, see (83).
2. $\operatorname{supp} B_j^k=\operatorname{supp} N_j^k=[t_j,t_{j+k}]$.
3. B_j^k has precisely the same smoothness as N_j^k iff all poles are chosen in the exterior of $[a,b]$.
4. B_j^k is nonnegative.
5. The B_j^k form a partition of unity.
6. There are knot insertion algorithms.
7. There is a simple connection with NURBS.

The prescribed poles can serve as additional shape-controlling parameters. Given a knot sequence t and a controll polygon corresponding to t_{ext} by suitably choosing the poles "corners" of the B-spline curve can be generated which are more or less sharp while maintaining the smoothness properties controlled by the knot sequence. The splines (84) are an example of Čebyševian splines which when restricted to any knot interval belong to the same CV-space $\mathscr{U}_k$. In other words for each knot interval the spline curve has the same poles outside of $[a,b]$. Clearly, one can consider also the more general case where in each knot interval $(t_i,t_{i+1}]$ individually for the spline curve poles are prescribed outside $[t_i,t_{i+1}]$. Not surprisingly, then the computation is more laborous, but we expect also in the general case existence of a recursive procedure [4].

We conclude with mentioning another application. Recently, interpolants from CV-spaces have been proved useful for approximation of transfer functions of infinite-dimensional dynamical systems [19].

References

[1] A. Ambroladze, H. Wallin, Rational interpolants with prescribed poles, theory and practice, Complex Variables 34 (1997) 399–413.

[2] I. Berezin, N. Zhidkov, Computing Methods, Vol. 1, Addison-Wesley, Reading, MA, 1965.

[3] C. Brezinski, The Mühlbach–Neville–Aitken algorithm and some extensions, BIT 20 (1980) 444–451.

[4] B. Buchwald, Computation of rational B-splines with prescribed poles, in preparation.

[5] C. Carstensen, G. Mühlbach, The Neville–Aitken formula for rational interpolants with prescribed poles, J. Comput. Appl. Math. 26 (1992) 297–309.

[6] T. Finck, G. Heinig, K. Rost, An inversion formula and fast algorithm for Cauchy–Vandermonde matrices, Linear Algebra Appl. 183 (1995) 179–191.

[7] M. Gasca, J.J. Martinez, G. Mühlbach, Computation of rational interpolants with prescribed poles, J. Comput. Appl. Math. 26 (1989) 297–309.

[8] A. Gresbrand, Rational B-splines with prescribed poles, Numer. Algorithms 12 (1996) 151–158.

[9] G. Heinig, K. Rost, Recursive solution of Cauchy–Vandermonde systems of equations, Linear Algebra Appl. 218 (1995) 59–72.

[10] G. Mühlbach, A recurrence formula for generalized divided differences and some applications, J. Approx. Theory 9 (1973) 165–172.

[11] G. Mühlbach, Neville-Aitken algorithms for interpolation of Čebyšev systems in the sense of Newton and in a generalized sense of Hermite, in: A.G. Law, B.N. Sahney (Eds.), Theory of Approximation and Applications, Academic Press, New York, 1976, pp. 200–212.

[12] G. Mühlbach, On Hermite interpolation by Cauchy–Vandermonde systems: the Lagrange formula, the adjoint and the inverse of a Cauchy–Vandermonde matrix, J. Comput. Appl. Math. 67 (1996) 147–159.

[13] G. Mühlbach, Linear and quasilinear extrapolation algorithms, in: R. Vichnevetsky, I. Vignes (Eds.), Numerical Mathematics and Applications, Elsevier Science Publishers, IMACS, Amsterdam, 1986, pp. 65–71.

[14] G. Mühlbach, On interpolation by rational functions with prescribed poles with applications to multivariate interpolation, J. Comput. Appl. Math. 32 (1990) 203–216.

[15] G. Mühlbach, Computation of Cauchy–Vandermonde determinants, J. Number Theory 43 (1993) 74–81.

[16] G. Mühlbach, The recurrence relation for generalized divided differences with respect to ECT-systems, Numer. Algorithms 22 (1999) 317–326.

[17] G. Mühlbach, L. Reimers, Linear extrapolation by rational functions, Exponentials and logarithmic functions, J. Comput. Appl. Math. 17 (1987) 329–344.

[18] L. Reimers, Lineare Extrapolation mit Tschebyscheff-Systemen rationaler und logarithmischer Funktionen, Dissertation, Universität Hannover, 1984.

[19] A. Ribalta Stanford, G. López Lagomasino, Approximation of transfer functions of infinite dimensional dynamical systems by rational interpolants with prescribed poles, Report 433, Institut für Dynamische Systeme, Universität Bremen, 1998.

[20] J. Walsh, Interpolation and approximation by rational functions in the complex domain, Amer. Math. Soc. Colloq. Publ., Vol. 20, American Mathematical Society, Providence, RI, 1960.

ELSEVIER

Journal of Computational and Applied Mathematics 122 (2000) 223–230

JOURNAL OF COMPUTATIONAL AND APPLIED MATHEMATICS

www.elsevier.nl/locate/cam

The E-algorithm and the Ford–Sidi algorithm

Naoki Osada

Tokyo Woman's Christian University, Zempukuji, Suginamiku, Tokyo 167-8585, Japan

Received 26 April 1999; received in revised form 15 February 2000

Abstract

The E-algorithm and the Ford–Sidi algorithm are two general extrapolation algorithms. It is proved that the E-algorithm and the Ford–Sidi algorithm are mathematically (although not operationally) equivalent. A slightly more economical algorithm is given. Operation counts are discussed. © 2000 Elsevier Science B.V. All rights reserved.

Keywords: Extrapolation method; Acceleration of convergence; E-algorithm; Ford–Sidi algorithm

1. Introduction

Let (s_n) be a sequence and $(g_j(n))$, $j=1,2,\ldots$, be known auxiliary sequences. Suppose that there exist unknown constants (c_j), $j=1,\ldots,k$, such that

$$s_{n+i} = T_k^{(n)} + \sum_{j=1}^{k} c_j g_j(n+i), \quad i=0,\ldots,k. \tag{1}$$

If the system of linear equations (1) is nonsingular, then by Cramer's rule $T_k^{(n)}$ can be expressed as the ratio of two determinants

$$T_k^{(n)} = \begin{vmatrix} s_n & \cdots & s_{n+k} \\ g_1(n) & \cdots & g_1(n+k) \\ & \cdots & \\ g_k(n) & \cdots & g_k(n+k) \end{vmatrix} \Bigg/ \begin{vmatrix} 1 & \cdots & 1 \\ g_1(n) & \cdots & g_1(n+k) \\ & \cdots & \\ g_k(n) & \cdots & g_k(n+k) \end{vmatrix}. \tag{2}$$

Many known sequence transformations which are used to accelerate the convergence are of the form $(s_n) \mapsto (T_k^{(n)})$. (For a review, see [2].) Two famous recursive algorithms are known to compute $T_k^{(n)}$. One is the E-algorithm proposed by Schneider [6], Håvie [5] and Brezinski [1], independently.

E-mail address: osada@twcu.ac.jp (N. Osada)

PII: S0377-0427(00)00366-6

Schneider and Håvie derived it using Gaussian elimination while Brezinski using Sylvester's determinantal identity. The other is the Ford–Sidi algorithm [4], which requires a smaller number of arithmetic operations than the E-algorithm. Ford and Sidi derived their algorithm using Sylvester's identity.

Ford and Sidi [4] mentioned the main difference of the recursion of the E-algorithm and that of the Ford–Sidi algorithm. Brezinski and Redivo Zaglia [3] derived the E-algorithm and the Ford–Sidi algorithm using annihilation difference operators and gave the relations between the two algorithms.

In this paper we show that the E-algorithm and the Ford–Sidi algorithm are mathematically equivalent in the following sense: two algorithms are computing the same quantities but in a different way, and the recurrence relations of the E-algorithm can be derived from those of the Ford–Sidi algorithm, and vice versa.

In Section 2 we review the recurrence relations and the number of operation counts of the two algorithms. In Section 3 we show that the two algorithms are mathematically equivalent. In Section 4 we give an efficient implementation for the Ford–Sidi algorithm. This implementation is slightly more economical than the original implementation for the Ford–Sidi algorithm.

2. The E-algorithm and the Ford–Sidi algorithm — review

Throughout this paper let $s=(s_n)$ be any sequence to be transformed and $g_j=(g_j(n))$, $j=1,2,\ldots,$ be any auxiliary sequences. We denote 1 by the constant sequence with $1_n = 1$, for $n = 0,1,\ldots$. Assume that all denominators are nonzero.

2.1. The E-algorithm

The E-algorithm is defined as follows. For $n=0,1,\ldots,$ the quantities $E_k^{(n)}$ and $g_{k,j}^{(n)}$ are defined by

$$E_0^{(n)} = s_n,$$

$$g_{0,j}^{(n)} = g_j(n), \quad j = 1,2,\ldots,$$

$$E_k^{(n)} = \frac{E_{k-1}^{(n)} g_{k-1,k}^{(n+1)} - E_{k-1}^{(n+1)} g_{k-1,k}^{(n)}}{g_{k-1,k}^{(n+1)} - g_{k-1,k}^{(n)}}, \quad k = 1,2,\ldots, \tag{3}$$

$$g_{k,j}^{(n)} = \frac{g_{k-1,j}^{(n)} g_{k-1,k}^{(n+1)} - g_{k-1,j}^{(n+1)} g_{k-1,k}^{(n)}}{g_{k-1,k}^{(n+1)} - g_{k-1,k}^{(n)}}, \quad k = 1,2,\ldots; \ j = k+1,\ldots . \tag{4}$$

The recurrence relations (3) and (4) are called the main rule and the auxiliary rule of the E-algorithm, respectively. Brezinski [1] proved the following theorem using Sylvester's determinantal identity.

Theorem 1. *For $n=0,1,\ldots,$ and $k=1,2,\ldots,$ $E_k^{(n)}$ and $g_{k,j}^{(n)}$ are represented as*

$$E_k^{(n)} = \begin{vmatrix} s_n & \cdots & s_{n+k} \\ g_1(n) & \cdots & g_1(n+k) \\ & \cdots & \\ g_k(n) & \cdots & g_k(n+k) \end{vmatrix} \Bigg/ \begin{vmatrix} 1 & \cdots & 1 \\ g_1(n) & \cdots & g_1(n+k) \\ & \cdots & \\ g_k(n) & \cdots & g_k(n+k) \end{vmatrix},$$

$$g_{k,j}^{(n)} = \begin{vmatrix} g_j(n) & \cdots & g_j(n+k) \\ g_1(n) & \cdots & g_1(n+k) \\ & \cdots & \\ g_k(n) & \cdots & g_k(n+k) \end{vmatrix} \Bigg/ \begin{vmatrix} 1 & \cdots & 1 \\ g_1(n) & \cdots & g_1(n+k) \\ & \cdots & \\ g_k(n) & \cdots & g_k(n+k) \end{vmatrix}, \quad j > k,$$

respectively.

If we set $c_k^{(n)} = g_{k-1,k}^{(n)}/(g_{k-1,k}^{(n+1)} - g_{k-1,k}^{(n)})$, then (3) and (4) become

$$E_k^{(n)} = E_{k-1}^{(n)} - c_k^{(n)}(E_{k-1}^{(n+1)} - E_{k-1}^{(n)}), \quad k > 0, \tag{5}$$

$$g_{k,j}^{(n)} = g_{k-1,j}^{(n)} - c_k^{(n)}(g_{k-1,j}^{(n+1)} - g_{k-1,j}^{(n)}), \quad k > 0, j > k, \tag{6}$$

respectively.

For given $s_0,\ldots,s_N$, the computation of $E_k^{(n-k)}$, $0 \leqslant n \leqslant N$, $0 \leqslant k \leqslant n$, requires $\frac{1}{3}N^3 + O(N^2)$ $g_{k,j}^{(n)}$'s. The number of operation counts for the E-algorithm, as mentioned in [1], is $\frac{5}{3}N^3 + O(N^2)$, while that with the implementation using (5) and (6) becomes $N^3 + O(N^2)$. More precisely, the latter is $N^3 + \frac{5}{2}N^2 + \frac{3}{2}N$.

We remark that Ford and Sidi [4] implemented the E-algorithm by rewriting in the forms

$$E_k^{(n)} = \frac{E_{k-1}^{(n+1)} - c_k^{(n)}E_{k-1}^{(n)}}{d_k^{(n)}}, \quad k > 0, \tag{7}$$

$$g_{k,j}^{(n)} = \frac{g_{k-1,j}^{(n+1)} - c_k^{(n)}g_{k-1,j}^{(n)}}{d_k^{(n)}}, \quad k > 0, \ j > k, \tag{8}$$

where $c_k^{(n)} = g_{k-1,k}^{(n+1)}/g_{k-1,k}^{(n)}$, and $d_k^{(n)} = 1 - c_k^{(n)}$. The implementation using (7) and (8) requires exactly the same number of arithmetic operations as that using (5) and (6). However, one can avoid the loss of significant figures by using (5) and (6).

2.2. The Ford–Sidi algorithm

The Ford–Sidi algorithm is defined as follows. Let $u = (u_n)$ be one of sequences $s = (s_n)$, 1, or $g_j = (g_j(n))$. The quantities $\psi_k^{(n)}(u)$ are defined by

$$\psi_0^{(n)}(u) = \frac{u_n}{g_1(n)}, \tag{9}$$

$$\psi_k^{(n)}(u) = \frac{\psi_{k-1}^{(n+1)}(u) - \psi_{k-1}^{(n)}(u)}{\psi_{k-1}^{(n+1)}(g_{k+1}) - \psi_{k-1}^{(n)}(g_{k+1})}, \quad k > 0. \tag{10}$$

Ford and Sidi [4] proved the following theorem using Sylvester's determinantal identity.

Theorem 2. *For $n = 0, 1, \ldots,$ and $k = 1, 2, \ldots,$ $\psi_k^{(n)}(u)$ are represented as*

$$\psi_k^{(n)}(u) = \begin{vmatrix} u_n & \cdots & u_{n+k} \\ g_1(n) & \cdots & g_1(n+k) \\ & \cdots & \\ g_k(n) & \cdots & g_k(n+k) \end{vmatrix} \Bigg/ \begin{vmatrix} g_{k+1}(n) & \cdots & g_{k+1}(n+k) \\ g_1(n) & \cdots & g_1(n+k) \\ & \cdots & \\ g_k(n) & \cdots & g_k(n+k) \end{vmatrix}.$$

By Theorems 1 and 2, $T_k^{(n)}$ can be evaluated by

$$T_k^{(n)} = \frac{\psi_k^{(n)}(s)}{\psi_k^{(n)}(1)}. \tag{11}$$

Following the implementation by Ford and Sidi [4], the computation of $T_k^{(n-k)}$, $0 \leqslant n \leqslant N$, $0 \leqslant k \leqslant n$, requires $\frac{1}{3}N^3 + \frac{3}{2}N^2 + \frac{7}{6}N$ subtractions, and $\frac{1}{3}N^3 + \frac{5}{2}N^2 + \frac{25}{6}N + 2$ divisions, a total of $\frac{2}{3}N^3 + 4N^2 + \frac{16}{3}N + 2$ arithmetic operations.

Remark. (1) The Ford–Sidi algorithm requires g_{k+1} for computing $\psi_k^{(n)}$. However, for computing $T_k^{(n)}$ the Ford–Sidi algorithm does not require g_{k+1}. The reason is as follows: Let $a_1, \ldots, a_{k+1}$ be any sequence such that

$$\begin{vmatrix} a_1 & \cdots & a_{k+1} \\ g_1(n) & \cdots & g_1(n+k) \\ & \cdots & \\ g_k(n) & \cdots & g_k(n+k) \end{vmatrix} \neq 0.$$

Let

$$\bar{\psi}_k^{(n)}(u) = \begin{vmatrix} u_1 & \cdots & u_{k+1} \\ g_1(n) & \cdots & g_1(n+k) \\ & \cdots & \\ g_k(n) & \cdots & g_k(n+k) \end{vmatrix} \Bigg/ \begin{vmatrix} a_1 & \cdots & a_{k+1} \\ g_1(n) & \cdots & g_1(n+k) \\ & \cdots & \\ g_k(n) & \cdots & g_k(n+k) \end{vmatrix}.$$

Then by (11) we have

$$T_k^{(n)} = \frac{\bar{\psi}_k^{(n)}(s)}{\bar{\psi}_k^{(n)}(1)}.$$

Using the above trick, when $s_0, \ldots, s_N$, $g_1, \ldots, g_N$ are given, we can determine all the $T_k^{(n)}$, $0 \leqslant n + k \leqslant N$, by the Ford–Sidi algorithm.

(2) Moreover, neither the value of g_{k+1} nor sequence $a_1,\ldots,a_{k+1}$ is required in computing $T_k^{(n)}$, when we use

$$T_k^{(n)} = \frac{\psi_{k-1}^{(n+1)}(s) - \psi_{k-1}^{(n)}(s)}{\psi_{k-1}^{(n+1)}(1) - \psi_{k-1}^{(n)}(1)},$$

which is derived from (10) and (11). The implementation of this fact is just the efficient Ford–Sidi algorithm described in Section 4 of this paper.

3. Mathematical equivalence of the E-algorithm and the Ford–Sidi algorithm

3.1. *The Ford–Sidi algorithm is derived from the E-algorithm*

Let $u=(u_n)$ be any sequence. Suppose that the sequence transformations $E_k : u=(u_n) \mapsto E_k(u) = (E_k^{(n)}(u))$ are defined by

$$E_0^{(n)}(u) = u_n, \tag{12}$$

$$E_k^{(n)}(u) = \frac{E_{k-1}^{(n)}(u)E_{k-1}^{(n+1)}(g_k) - E_{k-1}^{(n+1)}(u)E_{k-1}^{(n)}(g_k)}{E_{k-1}^{(n+1)}(g_k) - E_{k-1}^{(n)}(g_k)}, \quad k>0. \tag{13}$$

Suppose the sequence transformations $\psi_k : u=(u_n) \mapsto \psi_k(u) = (\psi_k^{(n)}(u))$ are defined by

$$\psi_k^{(n)}(u) = \frac{E_k^{(n)}(u)}{E_k^{(n)}(g_{k+1})}, \quad k=0,1,\ldots\,. \tag{14}$$

Theorem 3. *The quantities* $\psi_k^{(n)}(u)$ *defined by* (14) *satisfy* (9) *and* (10).

Proof. It follows from (12) and (14) that $\psi_0^{(n)}(u) = u_n/g_1(n)$.

Using the mathematical induction on k, it can be easily proved that $E_k^{(n)}(1)=1$. Thus, we have

$$\psi_k^{(n)}(1) = \frac{1}{E_k^{(n)}(g_{k+1})}, \quad k=0,1,\ldots\,. \tag{15}$$

By (14), we have

$$\psi_k^{(n)}(g_j) = \frac{E_k^{(n)}(g_j)}{E_k^{(n)}(g_{k+1})}, \quad k=0,1,\ldots; \quad j=k+1,k+2,\ldots\,. \tag{16}$$

From (13) and (16), we obtain

$$\begin{aligned}
\psi_{k-1}^{(n+1)}(g_{k+1}) - \psi_{k-1}^{(n)}(g_{k+1}) &= \frac{E_{k-1}^{(n+1)}(g_{k+1})}{E_{k-1}^{(n+1)}(g_k)} - \frac{E_{k-1}^{(n)}(g_{k+1})}{E_{k-1}^{(n)}(g_k)} \\
&= \frac{E_{k-1}^{(n)}(g_{k+1})E_{k-1}^{(n+1)}(g_k) - E_{k-1}^{(n+1)}(g_{k+1})E_{k-1}^{(n)}(g_k)}{E_{k-1}^{(n+1)}(g_k) - E_{k-1}^{(n)}(g_k)}\,\frac{E_{k-1}^{(n)}(g_k) - E_{k-1}^{(n+1)}(g_k)}{E_{k-1}^{(n+1)}(g_k)E_{k-1}^{(n)}(g_k)} \\
&= E_k^{(n)}(g_{k+1})\left(\frac{1}{E_{k-1}^{(n+1)}(g_k)} - \frac{1}{E_{k-1}^{(n)}(g_k)}\right).
\end{aligned} \tag{17}$$

By dividing the both sides of (13) by $E_k^{(n)}(g_{k+1})$, we have

$$\frac{E_k^{(n)}(u)}{E_k^{(n)}(g_{k+1})} = \frac{E_{k-1}^{(n)}(u)/E_{k-1}^{(n)}(g_k) - E_{k-1}^{(n+1)}(u)/E_{k-1}^{(n+1)}(g_k)}{E_k^{(n)}(g_{k+1})(1/E_{k-1}^{(n)}(g_k) - 1/E_{k-1}^{(n+1)}(g_k))},$$

therefore from (14) and (17), we obtain

$$\psi_k^{(n)}(u) = \frac{\psi_{k-1}^{(n+1)}(u) - \psi_{k-1}^{(n)}(u)}{\psi_{k-1}^{(n+1)}(g_{k+1}) - \psi_{k-1}^{(n)}(g_{k+1})}. \qquad \square$$

We note that Brezinski and Redivo Zaglia [3] derived the relations (14)–(16) from their definitions of $E_k^{(n)}(u)$ and $\psi_k^{(n)}(u)$.

3.2. The E-algorithm is derived from the Ford–Sidi algorithm

Suppose that the sequence transformations ψ_k satisfy (9) and (10) for any sequence $u = (u_n)$. Let the sequence transformations E_k be defined by

$$E_k^{(n)}(u) = \frac{\psi_k^{(n)}(u)}{\psi_k^{(n)}(1)}. \tag{18}$$

Theorem 4. *The $E_k^{(n)}(u)$ defined by (18) satisfies (12) and (13).*

Proof. Since (10), we have

$$\psi_k^{(n)}(g_{k+1}) = 1.$$

Hence, by the definition (18), we obtain

$$E_k^{(n)}(g_{k+1}) = \frac{1}{\psi_k^{(n)}(1)}. \tag{19}$$

By (18) and (10),

$$E_k^{(n)}(u) = \frac{\psi_k^{(n)}(u)}{\psi_k^{(n)}(1)} = \frac{\psi_{k-1}^{(n+1)}(u) - \psi_{k-1}^{(n)}(u)}{\psi_{k-1}^{(n+1)}(1) - \psi_{k-1}^{(n)}(1)}$$

$$= \frac{E_{k-1}^{(n+1)}(u)/E_{k-1}^{(n+1)}(g_k) - E_{k-1}^{(n)}(u)/E_{k-1}^{(n)}(g_k)}{1/E_{k-1}^{(n+1)}(g_k) - 1/E_{k-1}^{(n)}(g_k)}$$

$$= \frac{E_{k-1}^{(n+1)}(g_k)E_{k-1}^{(n)}(u) - E_{k-1}^{(n)}(g_k)E_{k-1}^{(n+1)}(u)}{E_{k-1}^{(n+1)}(g_k) - E_{k-1}^{(n)}(g_k)}. \qquad \square$$

By Theorems 3 and 4, we consider that the E-algorithm and the Ford–Sidi algorithm are mathematically equivalent.

4. An efficient implementation for the Ford–Sidi algorithm

Let $\psi_k^{(n)}(u)$ be defined by (9) and (10). By (11) and (10), $T_k^{(n)}$ in Eq. (2) is represented as

$$T_k^{(n)} = \frac{\psi_{k-1}^{(n+1)}(s) - \psi_{k-1}^{(n)}(s)}{\psi_{k-1}^{(n+1)}(1) - \psi_{k-1}^{(n)}(1)}, \quad k > 0. \tag{20}$$

Using (20), the Ford–Sidi algorithm is implemented as follows.

$\{$read $s_0,\ g_1(0)\}$
$\psi_0^{(0)}(s){:=}s_0/g_1(0);\ \psi_0^{(0)}(1){:=}1/g_1(0);\ T_0^{(0)}{:=}s_0;$
$\{$save $T_0^{(0)},\ \psi_0^{(0)}(s),\ \psi_0^{(0)}(1)\}$
$\{$read $s_1,\ g_1(1)\}$
$\psi_0^{(1)}(s){:=}s_1/g_1(1);\ \psi_0^{(1)}(1){:=}1/g_1(1);\ T_0^{(1)}{:=}s_1;$
$TN{:=}\psi_0^{(1)}(s) - \psi_0^{(0)}(s);\ TD{:=}\psi_0^{(1)}(1) - \psi_0^{(0)}(1);\ T_1^{(0)}{:=}TN/TD;$
$\{$save $T_0^{(1)},\ T_1^{(0)},\ \psi_0^{(1)}(s),\ \psi_0^{(1)}(1),\ TN,\ TD,$ discard $\psi_0^{(0)}(s),\ \psi_0^{(0)}(1)\}$
for $n{:=}2$ **to** N **do**
begin
  $\{$read $s_n,\ g_j(n),\ 1 \leqslant j \leqslant n-1,\ g_n(m),\ 0 \leqslant m \leqslant n\}$
  for $j{:=}2$ **to** $n-1$ **do** $\psi_0^{(n)}(g_j){:=}g_j(n)/g_1(n);$
  for $m{:=}0$ **to** n **do** $\psi_0^{(m)}(g_n){:=}g_n(m)/g_1(m);$
  for $k{:=}1$ **to** $n-2$ **do for** $m{:=}0$ **to** $n-k-1$ **do**
    $\psi_k^{(m)}(g_n){:=}(\psi_{k-1}^{(m+1)}(g_n) - \psi_{k-1}^{(m)}(g_n))/D_k^{(m)};$
  $D_{n-1}^{(0)}{:=}\psi_{n-2}^{(1)}(g_n) - \psi_{n-2}^{(0)}(g_n);\ \psi_{n-1}^{(0)}(s){:=}TN/D_{n-1}^{(0)};\ \psi_{n-1}^{(0)}(1){:=}TD/D_{n-1}^{(0)};$
  $T_0^{(n)}{:=}s_n;\ \psi_0^{(n)}(s){:=}s_n/g_1(n);\ \psi_0^{(n)}(1){:=}1/g_1(n);$
  for $k{:=}1$ **to** $n-1$ **do**
  begin
    $D_k^{(n-k)}{:=}\psi_{k-1}^{(n-k+1)}(g_{k+1}) - \psi_{k-1}^{(n-k)}(g_{k+1});$
    $\psi_k^{(n-k)}(s){:=}(\psi_{k-1}^{(n-k+1)}(s) - \psi_{k-1}^{(n-k)}(s))/D_k^{(n-k)};$
    $\psi_k^{(n-k)}(1){:=}(\psi_{k-1}^{(n-k+1)}(1) - \psi_{k-1}^{(n-k)}(1))/D_k^{(n-k)};$
    $T_k^{(n-k)}{:=}\psi_k^{(n-k)}(s)/\psi_k^{(n-k)}(1);$
    for $j{:=}k+2$ **to** n **do**
      $\psi_k^{(n-k)}(g_j){:=}(\psi_{k-1}^{(n-k+1)}(g_j) - \psi_{k-1}^{(n-k)}(g_j))/D_k^{(n-k)};$
  end
  $TN{:=}\psi_{n-1}^{(1)}(s) - \psi_{n-1}^{(0)}(s);\ TD{:=}\psi_{n-1}^{(1)}(1) - \psi_{n-1}^{(0)}(1);\ T_n^{(0)}{:=}TN/TD;$
  $\{$save $\psi_k^{(n-k)}(s),\ \psi_k^{(n-k)}(1),\ 0 \leqslant k \leqslant n-1,\ \psi_k^{(n-k)}(g_j),\ 0 \leqslant k \leqslant n-2,$
  $k+2 \leqslant j \leqslant n,\ TN,\ TD,$ discarding all others$\}$
  $\{$save all $T_k^{(n-k)}, 0 \leqslant k \leqslant n,\ D_k^{(l)},\ 1 \leqslant l+k \leqslant n,\ 1 \leqslant k \leqslant n-1\}$
end;

(The **for** statements of the form "**for** $k{:=}k_1$ **to** k_2 **do**" are not executed if $k_1 > k_2$.)

This algorithm will be called the efficient implementation for the Ford–Sidi algorithm. It is clear that this algorithm is mathematically equivalent to the E-algorithm and the Ford–Sidi algorithm.

Table 1
The numbers of arithmetic operation counts of three algorithms

Algorithm	Operation counts	$N=10$	$N=20$
The E-algorithm using (5) and (6)	$N^3+\frac{5}{2}N^2+\frac{3}{2}N$	1265	9030
The Ford–Sidi algorithm	$\frac{2}{3}N^3+4N^2+\frac{16}{3}N+2$	1122	7042
The present method	$\frac{2}{3}N^3+3N^2+\frac{10}{3}N$	1000	6600

The computation of $T_k^{(n-k)}, 0\leqslant n\leqslant N, 0\leqslant k\leqslant n$, by the efficient implementation for the Ford–Sidi algorithm, requires $\frac{1}{3}N^3 + N^2 + \frac{2}{3}N$ subtractions, and $\frac{1}{3}N^3 + 2N^2 + \frac{8}{3}N$ divisions, a total of $\frac{2}{3}N^3 + 3N^2 + \frac{10}{3}N$ arithmetic operations. Although operation counts of the present method and the Ford–Sidi algorithm are asymptotically equal, the present method is slightly more economical than the Ford–Sidi algorithm.

The number of arithmetic operation counts for the computation of $T_k^{(n-k)}$, $0\leqslant n\leqslant N$, $0\leqslant k\leqslant n$, by the E-algorithm using (5) and (6), the Ford–Sidi algorithm, and the present method are listed in Table 1.

Suppose that we accelerate the convergence of a usual sequence by a suitable method such as the Levin u transform in double precision, and that $T_N^{(0)}$ is an optimal extrapolated value. Then it is usually $10\leqslant N\leqslant 20$. (See, for example, [2,7].) Therefore, by Table 1, the present method is, in practice, 6–11% more economical than the Ford–Sidi algorithm.

Acknowledgements

The author would like to thank Prof. C. Brezinski for valuable comments and pointing out the Ref. [3]. The author would also like to thank Prof. A. Sidi for helpful comments, particularly remark (1) in subsection 2.2 is owed to him.

References

[1] C. Brezinski, A general extrapolation algorithm, Numer. Math. 35 (1980) 175–187.
[2] C. Brezinski, M. Redivo Zaglia, Extrapolation Methods, Theory and Practice, North-Holland, Amsterdam, 1991.
[3] C. Brezinski, M. Redivo Zaglia, A general extrapolation procedure revisited, Adv. Comput. Math. 2 (1994) 461–477.
[4] W.F. Ford, A. Sidi, An algorithm for a generalization of the Richardson extrapolation process, SIAM J. Numer. Anal. 24 (5) (1987) 1212–1232.
[5] T. Håvie, Generalized Neville type extrapolation schemes, BIT 19 (1979) 204–213.
[6] C. Schneider, Vereinfachte Rekursionen zur Richardson-Extrapolation in Spezialfällen, Numer. Math. 24 (1975) 177–184.
[7] D.A. Smith, W.F. Ford, Acceleration of linear and logarithmic sequence, SIAM J. Numer. Anal. 16 (2) (1979) 223–240.

ELSEVIER

Journal of Computational and Applied Mathematics 122 (2000) 231–250

JOURNAL OF COMPUTATIONAL AND APPLIED MATHEMATICS

www.elsevier.nl/locate/cam

Diophantine approximations using Padé approximations

M. Prévost

Laboratoire de Mathématiques Pures et Appliquées Joseph Liouville, Université du Littoral Côte d'Opale, Centre Universitaire de la Mi-Voix, Bâtiment H. Poincaré, 50 Rue F. Buisson B.P. 699, 62228 Calais Cédex, France

Received 9 June 1999; received in revised form 1 December 1999

Abstract

We show how Padé approximations are used to get Diophantine approximations of real or complex numbers, and so to prove the irrationality. We present two kinds of examples. First, we study two types of series for which Padé approximations provide exactly Diophantine approximations. Then, we show how Padé approximants to the asymptotic expansion of the remainder term of a value of a series also leads to Diophantine approximation.

1. Preliminary

Definition 1 (*Diophantine approximation*). Let x a real or complex number and $(p_n/q_n)_n$ a sequence of $\mathbb{Q}$ or $\mathbb{Q}(i)$.

If $\lim_{n\to\infty}|q_n x - p_n| = 0$ and $p_n/q_n \neq x$, $\forall n \in \mathbb{N}$, then the sequence $(p_n/q_n)_n$ is called a Diophantine approximation of x.

It is well known that Diophantine approximation of x proves the irrationality of x.

So, to construct Diophantine approximation of a number, a mean is to find rational approximation, for example with Padé approximation.

We first recall the theory of formal orthogonal polynomials and its connection with Padé approximation and ε-algorithm.

1.1. Padé approximants

Let h be a function whose Taylor expansion about $t=0$ is $\sum_{i=0}^{\infty} c_i t^i$. The Padé approximant $[m/n]_h$ to h is a rational fraction $N_m(t)/D_n(t)$ whose Taylor series at $t=0$ coincides with that of h up to

E-mail address: prevost@lmpa.univ-littoral.fr (M. Prévost)

0377-0427/00/$ - see front matter
PII: S0377-0427(00)00365-4

the maximal order, which is in general the sum of the degrees of numerator and denominator of the fraction, i.e,

$$\deg(N_m) \leqslant m, \qquad \deg(D_n) \leqslant n, \qquad D_n(t)h(t) - N_m(t) = \mathrm{O}(t^{m+n+1}), \quad t \to 0.$$

Note that the numerator N_m and the denominator D_n both depend on the index m and n.

The theory of Padé approximation is linked with the theory of orthogonal polynomials (see [10]): Let us define the linear functional c acting on the space $\mathscr{P}$ of polynomials as follows:

$$c : \mathscr{P} \to \mathbb{R} \quad (\text{or } \mathbb{C}),$$

$$x^i \to \langle c, x^i \rangle = c_i, \quad i = 0, 1, 2, \ldots \text{ and if } p \in \mathbb{Z},$$

$$c^{(p)} : \mathscr{P} \to \mathbb{R} \quad (\text{or } \mathbb{C}),$$

$$x^i \to \langle c^{(p)}, x^i \rangle := \langle c, x^{i+p} \rangle = c_{i+p}, \quad i = 0, 1, 2, \ldots \quad (c_i = 0, i < 0),$$

then the denominators of the Padé approximants $[m/n]$ satisfy the following orthogonality property:

$$\langle c^{(m-n+1)}, \; x^i \tilde{D}_n(x) \rangle = 0, \quad i = 0, 1, 2, \ldots, n-1,$$

where $\tilde{D}_n(x) = x^n D_n(x^{-1})$ is the reverse polynomial. Since the polynomials D_n involved in the expression of Padé approximants depend on the integers m and n, and since $\tilde{D}_n$ is orthogonal with respect to the shifted linear functional $c^{(m-n+1)}$, we denote

$$P_n^{(m-n+1)}(x) = \tilde{D}_n(x),$$

$$\tilde{Q}_n^{(m-n+1)}(x) = N_m(x).$$

If we set

$$R_{n-1}^{(m-n+1)}(t) := \left\langle c^{(m-n+1)}, \frac{P_n^{(m-n+1)}(x) - P_n^{(m-n+1)}(t)}{x - t} \right\rangle, \quad R_{n-1}^{(m-n+1)} \in \mathscr{P}_{n-1},$$

where $c^{(m-n+1)}$ acts on the letter x, then

$$N_m(t) = \left(\sum_{i=0}^{m-n} c_i t^i \right) \tilde{P}_n^{(m-n+1)}(t) + t^{m-n+1} \tilde{R}_{n-1}^{(m-n+1)}(t),$$

where $\tilde{R}_{n-1}^{(m-n+1)}(t) = t^{n-1} R_{n-1}^{(m-n+1)}(t^{-1})$, $\tilde{P}_n^{(m-n+1)}(t) = t^n P_n^{(m-n+1)}(t^{-1})$ and $\sum_{i=0}^{n-m} c_i t^i = 0, \; n < m$.

The sequence of polynomials $(P_k^{(n)})_k$, of degree k, exists if and only if $\forall n \in \mathbb{Z}$, the Hankel determinant

$$H_k^{(n)} := \begin{vmatrix} c_n & \cdots & c_{n+k-1} \\ \cdots & \cdots & \cdots \\ c_{n+k-1} & \cdots & c_{n+2k-2} \end{vmatrix} \neq 0,$$

where $c_n = 0$ if $n < 0$.

In that case, we shall say that the linear functional c is completely definite. For the noncompletely definite case, the interested reader is referred to Draux [15].

For extensive applications of Padé approximants to Physics, see Baker's monograph [5].

If c admits an integral representation by a nondecreasing function α, with bounded variation

$$c_i = \int_{\mathbb{R}} x^i \mathrm{d}\alpha(x),$$

then the theory of Gaussian quadrature shows that the polynomials P_n orthogonal with respect to c, have all their roots in the support of the function α and

$$h(t) - [m/n]_h(t) = \frac{t^{m-n+1}}{(\tilde{P}_n^{(m-n+1)}(t))^2} c^{(m-n+1)} \left(\frac{(\tilde{P}_n^{(m-n+1)}(x))^2}{1-xt} \right)$$

$$= \frac{t^{m-n+1}}{(\tilde{P}_n^{(m-n+1)}(t))^2} \int_{\mathbb{R}} x^{m-n+1} \frac{(\tilde{P}_n^{(m-n+1)}(x))^2}{1-xt} \mathrm{d}\alpha(x). \tag{1}$$

Note that if $c_0 = 0$ then $[n/n]_h(t) = t[n-1/n]_{h/t}(t)$ and if $c_0 = 0$ and $c_1 = 0$, then $[n/n]_h(t) = t^2[n-2/n]_{h/t^2}(t)$.

Consequence: If α is a nondecreasing function on $\mathbb{R}$, then

$$h(t) \neq [m/n]_f(t) \quad \forall t \in \mathbb{C} - \mathrm{supp}(\alpha).$$

1.2. Computation of Padé approximants with ε-algorithm

The values of Padé approximants at some point of parameter t, can be recursively computed with the ε-algorithm of Wynn. The rules are the following:

$$\varepsilon_{-1}^{(n)} = 0, \ \varepsilon_0^{(n)} = S_n, \quad n = 0, 1, \ldots,$$

$$\varepsilon_{k+1}^{(n)} = \varepsilon_{k-1}^{(n+1)} + \frac{1}{\varepsilon_k^{(n+1)} - \varepsilon_k^{(n)}}, \quad k, n = 0, 1, \ldots \quad \text{(rhombus rule)},$$

where $S_n = \sum_{k=0}^{n} c_k t^k$.

ε-values are placed in a double-entry array as following:

$$\begin{array}{ccccc}
\varepsilon_{-1}^{(0)} = 0 & & & & \\
 & \varepsilon_0^{(0)} = S_0 & & & \\
\varepsilon_{-1}^{(1)} = 0 & & \varepsilon_1^{(0)} & & \\
 & \varepsilon_0^{(1)} = S_1 & & \varepsilon_2^{(0)} & \\
\varepsilon_{-1}^{(2)} = 0 & & \varepsilon_1^{(1)} & & \varepsilon_3^{(0)} \\
 & \varepsilon_0^{(2)} = S_2 & & \varepsilon_2^{(1)} & \ddots \\
\varepsilon_{-1}^{(3)} = 0 & & \varepsilon_1^{(2)} & \vdots & \ddots \\
\vdots & \varepsilon_0^{(3)} = S_3 & \vdots & & \ddots
\end{array}$$

The connection between Padé approximant and ε-algorithm has been established by Shanks [26] and Wynn [35]:

Theorem 2. *If we apply ε-algorithm to the partial sums of the series $h(t)=\sum_{i=0}^{\infty} c_i t^i$, then*

$$\varepsilon_{2k}^{(n)} = [n+k/k]_h(t).$$

Many convergence results for ε-algorithm has been proved for series which are meromorphic functions in some complex domain, or which have an integral representation (Markov–Stieltjes function) (see [29,6,11] for a survey).

2. Diophantine approximation of sum of series with Padé approximation

Sometimes, Padé approximation is sufficient to prove irrationality of values of a series, as it can be seen in the following two results.

2.1. Irrationality of $\ln(1-r)$

We explain in the following theorem, how the old proof of irrationality of some logarithm number can be re-written in terms of ε-algorithm.

Theorem 3. *Let $r=a/b$, $a\in\mathbb{Z}$, $b\in\mathbb{N}$, $b\neq 0$, with* b.e.$(1-\sqrt{1-r})^2<1(\ln e=1)$ *Then ε-algorithm applied to the partial sums of $f(r):=-\ln(1-r)/r=\sum_{i=0}^{\infty} r^i/(i+1)$ satisfies that $\forall n\in\mathbb{N}$, $(\varepsilon_{2k}^{(n)})_k$ is a Diophantine approximation of $f(r)$.*

Proof. From the connection between Padé approximation, orthogonal polynomials and ε-algorithm, the following expression holds:

$$\varepsilon_{2k}^{(n)} = \sum_{i=0}^{n} \frac{r^i}{i+1} + r^{n+1}\frac{\tilde{R}_{k-1}^{(n+1)}(r)}{\tilde{P}_k^{(n+1)}(r)} = \frac{N_{n+k}(r)}{\tilde{P}_k^{(n+1)}(r)},$$

where

$$\tilde{P}_k^{(n+1)}(t) = t^k P_k^{(n+1)}(t^{-1}) = \sum_{i=0}^{k}\binom{k}{k-i}\binom{k+n+1}{i}(1-t)^i$$

is the reversed shifted Jacobi polynomial on [0,1], with parameters $\alpha=0$, $\beta=n+1$, and $\tilde{R}_{k-1}^{(n+1)}(t)=t^{k-1}R_{k-1}^{(n+1)}(t^{-1})$ with $R_{k-1}^{(n+1)}(t)=\langle c^{(n+1)}, \frac{P_k^{(n+1)}(x)-P_k^{(n+1)}(t)}{x-t}\rangle(\langle c^{(n+1)}, x^i\rangle:=1/(n+i+2))$ (c acts on the variable x).

Since $\tilde{P}_k^{(n+1)}(t)$ has only integer coefficients, $b^k\tilde{P}_k^{(n+1)}(a/b)\in\mathbb{Z}$.

The expression of $R_{k-1}^{(n+1)}(t)$ shows that $d_{n+k+1}b^k\tilde{R}_{k-1}^{(n+1)}(a/b)\in\mathbb{Z}$, where $d_{n+k+1}:=\mathrm{LCM}(1,2,\ldots,n+k+1)$ (LCM means lowest common multiple).

We prove now that the sequence $(\varepsilon_{2k}^{(n)})_k$ is a Diophantine approximation of $\ln(1-a/b)$.

The proof needs asymptotics for d_{n+k+1}, for $\tilde{P}_k^{(n+1)}(a/b)$ and for $(\varepsilon_{2k}^{(n)} - f(r))$ when k tends to infinity. $d_n = \mathrm{e}^{n(1+o(1))}$ follows from analytic number theory [1].

$\lim_k(\tilde{P}_k^{(n+1)}(x))^{1/k} = x(y+\sqrt{y^2-1})$, $x>1$, $y=2/x-1$, comes from asymptotic properties of Jacobi polynomials (see [30]), and $\lim_{k\to+\infty}(\varepsilon_{2k}^{(n)} - f(r))^{1/k} = (2/r-1-\sqrt{(2/r-1)^2-1})^2$ (error of Padé approximants to Markov–Stieltjes function).

So

$$\lim_{k\to+\infty}\sup\left|d_{n+k+1}b^k\tilde{P}_k^{(n+1)}(a/b)f(r) - d_{n+k+1}b^kN_{n+k}(a/b)\right|^{1/k}$$

$$\leqslant \lim_{k\to+\infty}\sup(d_{n+k+1})^{1/k}\limsup_k\left|b^k\tilde{P}_k^{(n+1)}(a/b)\right|^{1/k}\lim_{k\to+\infty}\sup\left|\varepsilon_{2k}^{(n)} + 1/r\ln(1-r)\right|^{1/k}$$

$$\leqslant \mathrm{e.b.r.}(2/r-1+\sqrt{(2/r-1)^2-1})(2/r-1-\sqrt{(2/r-1)^2-1})^2$$

$$= \mathrm{e.b.}(2/r-1-\sqrt{(2/r-1)^2-1}) = \mathrm{e.b.}(1-\sqrt{1-r})^2 < 1$$

by hypothesis, which proves that

$$\forall n\in\mathbb{N},\quad \lim_{k\to+\infty}(d_{n+k+1}b^k\tilde{P}_k^{(n+1)}(a/b)f(r) - d_{n+k+1}b^kN_{n+k}(a/b)) = 0.$$

Moreover,

$$\varepsilon_{2k}^{(n)} + 1/r\ln(1-r) = -\frac{r^{2k+n+1}}{(\tilde{P}_k^{(n+1)}(r))^2}\int_0^1 \frac{(P_k^{(n+1)}(x))^2}{1-xr}(1-x)^{n+1}\,\mathrm{d}x \neq 0.$$

So the sequence $(\varepsilon_{2k}^{(n)})_k$ is a Diophantine approximation of $\ln(1-a/b)$, if $\mathrm{b.e.}(1-\sqrt{1-a/b})^2 < 1$. □

2.2. *Irrationality of $\sum t^n/w_n$*

The same method as previously seen provides Diophantine approximations of $f(t) := \sum_{n=0}^{\infty} t^n/w_n$ when the sequence $(w_k)_k$ satisfies a second-order recurrence relation

$$w_{n+1} = sw_n - pw_{n-1},\quad n\in\mathbb{N}, \tag{2}$$

where w_0 and w_{-1} are given in $\mathbb{C}$ and s and p are some complex numbers.

We suppose that $w_n \neq 0$, $\forall n\in\mathbb{N}$ and that the two roots of the characteristic equation $z^2 - sz + p = 0$, α and β satisfy $|\alpha| > |\beta|$.

So w_n admits an expression in term of geometric sequences: $w_n = A\alpha^n + B\beta^n$, $n\in\mathbb{N}$.

The roots of the characteristic equation are assumed to be of distinct modulus ($|\alpha| > |\beta|$), so there exists an integer r such that $|\alpha/\beta|^r > |B/A|$.

Lemma 4 (see [25]). *If α, β, A, B are some complex numbers, and $|\alpha| > |\beta|$, then the function*

$$f(t) := \sum_{k=0}^{\infty}\frac{t^k}{A\alpha^k + B\beta^k}$$

admits another expansion

$$f(t)=\sum_{k=0}^{r-1}\frac{t^k}{A\alpha^k+B\beta^k}-\frac{t^r}{A\alpha^r}\sum_{k=0}^{\infty}\frac{[(-B/A)(\beta/\alpha)^{r-1}]^k}{t/\alpha-(\alpha/\beta)^k},$$

where $r\in\mathbb{N}$ *is chosen such that* $|\alpha|^r|A|>|\beta|^r|B|$.

With the notations of Section 1.1, the Padé approximant $[n+k-1/k]_f$ is

$$[n+k-1/k]_f(t)=\frac{\tilde{Q}_k^{(n)}(t)}{\tilde{P}_k^{(n)}(t)},$$

where $\tilde{P}_k^{(n)}(t)=t^kP_k^{(n)}(t^{-1})$.

In a previous papers by the author [24,25], it has been proved that for all $n\in\mathbb{Z}$, the sequence of Padé approximants $([n+k-1/k])_k$ to f converges on any compact set included in the domain of meromorphy of the function f, with the following error term:

$$\forall t\in\mathbb{C}\setminus\{\alpha(\alpha/\beta)^j, j\in\mathbb{N}\},\ \forall n\in\mathbb{N},\quad \limsup_k|f(t)-[n+k-1/k]_f(t)|^{1/k^2}\leqslant\frac{\beta}{\alpha}, \tag{3}$$

where α and β are the two solutions of $z^2-sz+p=0$, $|\alpha|>|\beta|$.

Theorem 5. *If* $\tilde{Q}_k^{(n)}(t)/\tilde{P}_k^{(n)}(t)$ *denotes the Padé approximant* $[n+k-1/k]_f$, *then*

$$\text{(a)}\quad \tilde{P}_k^{(n)}(t)=\sum_{i=0}^{k}\binom{k}{i}_q q^{i(i-1)/2}(-t/\alpha)^i\prod_{j=1}^{i}\frac{A+Bq^{n+k-j}}{A+Bq^{n+2k-j}},$$

where

$$q:=\beta/\alpha,\quad \binom{k}{i}_q:=\frac{(1-q^k)\dots(1-q^{k-i+1})}{(1-q)(1-q^2)\dots(1-q^i)},\quad 1\leqslant i\leqslant k\ (\textit{Gaussian binomial coefficient}),$$

$$\binom{k}{0}_q=1.$$

$$\text{(b)}\quad \left|\tilde{P}_k^{(n)}(t)-\prod_{j=0}^{k-1}(1-tq^j/\alpha)\right|\leqslant R|q|^k,\quad k\geqslant K_0$$

for some constant R *independent of* k *and* K_0 *is an integer depending on* A,B,q,n.

Moreover, if $s,p,w_{-1},w_0\in\mathbb{Z}(i)$, *for all common multiple* d_m of $\{w_0,w_1,\dots,w_m\}$

$$\text{(c)}\quad w_{n+k}\cdots w_{n+2k-1}\tilde{P}_k^{(n)}\in\mathbb{Z}(i)[t],\quad \forall n\in\mathbb{Z}/n+k-1\geqslant 0$$

and

(d) $d_{n+k-1}\, w_{n+k} \cdots w_{n+2k-1}\, \tilde{Q}_k^{(n)} \in \mathbb{Z}(i)[t], \quad \forall n \in \mathbb{Z} \,/ n+k-1 \geqslant 0.$

Proof. (a) is proved in [16] and (b) is proved in [25]. (c) and (d) comes from expression (a). □

The expression of w_n is

$$w_n = A\alpha^n + B\beta^n.$$

If A or B is equal to 0 then $f(t)$ is a rational function, so without loss of generality, we can assume that $AB \neq 0$.

The degrees of $\tilde{Q}_k^{(n)}$ and $\tilde{P}_k^{(n)}$ are, respectively, $k+n-1$ and k, so if we take $t \in \mathbb{Q}(i)$ with $vt \in \mathbb{Z}(i)$, the above theorem implies that the following sequence:

$$e_{k,n} := f(t) \times v^{k'}\, d_{n+k-1} w_{n+k} \cdots w_{n+2k-1} \tilde{P}_k^{(n)}(t) - v^{k'}\, d_{n+k-1} w_{n+k} \cdots w_{n+2k-1} \tilde{Q}_k^{(n)}(t),$$

where $k' = \max\{n+k-1, k\}$ is a Diophantine approximation to $f(t)$, if

(i) $\forall n \in \mathbb{Z}, \lim_{k\to\infty} e_{k,n} = 0$,
(ii) $[n+k-1/k]_f(t) \neq [n+k/k+1]_f(t)$.

For sake of simplicity, we only display the proof for the particular case $n = 0$.

We set

$$e_k := e_{k,0}, \qquad \tilde{Q}_k := \tilde{Q}_k^{(0)} \quad \text{and} \quad \tilde{P}_k := \tilde{P}_k^{(0)}.$$

From the asymptotics given in (3), we get

$$\limsup_k |e_k|^{1/k^2} \leqslant \limsup_k \left| f(t) - \frac{\tilde{Q}_k(t)}{\tilde{P}_k(t)} \right|^{1/k^2} \limsup_k \left| v^k d_{k-1} w_k \cdots w_{2k-1} \tilde{P}_k(t) \right|^{1/k^2} \tag{4}$$

$$\leqslant |p| \limsup |\rho_{k-1}|^{1/k^2}, \tag{5}$$

where $\rho_k := d_k / \prod_{i=0}^k w_i$.

We will get $\lim_{k\to\infty} e_k = 0$ if the following condition is satisfied:

$$\limsup_{k\to\infty} |\rho_{k-1}|^{1/k^2} < 1/|p|.$$

Moreover, from the Christoffel–Darboux identity between orthogonal polynomials, condition (ii) is satisfied since the difference

$$\tilde{Q}_{k+1}(t)\tilde{P}_k(t) - \tilde{P}_{k+1}(t)\tilde{Q}_k(t) = t^{2k} \frac{(-1)^k}{A+B} \prod_{i=1}^k AB\, p^{2i-2} (\alpha^i - \beta^i)^2 \frac{w_{i-1}^2}{w_{2i-1}^2 w_{2i} w_{2i-2}^2}$$

is different from 0.

The following theorem is now proved.

Theorem 6. *Let f be the meromorphic function defined by the following series:*

$$f(t) = \sum_{n=0}^{\infty} \frac{t^n}{w_n},$$

where $(w_n)_n$ *is a sequence of* $\mathbb{Z}(i)$ *satisfying a three-term recurrence relation*

$$w_{n+1} = s\, w_n - p\, w_{n-1}, \quad s, p \in \mathbb{Z}(i)$$

with the initial conditions: $w_{-1}, w_0 \in \mathbb{Z}(i)$. *If for each integer* m, *there exists a common multiple* d_m *for the numbers* $\{w_0, w_1, \ldots, w_m\}$ *such that* ρ_m *defined by*

$$\rho_m := \frac{d_m}{\prod_{i=0}^{m} w_i}$$

satisfies the condition

$$\limsup_m |\rho_m|^{1/m^2} < 1/|p|, \tag{6}$$

then for $t \in \mathbb{Q}(i)$, $t \neq \alpha(\alpha/\beta)^j$, $j = 0, 1, 2, \ldots$ *we have*

$$f(t) \notin \mathbb{Q}(i).$$

See [25] for application to Fibonacci and Lucas series. (If F_n and L_n are, respectively, Fibonacci and Lucas sequences, then $f(t) = \sum t^n/F_n$ and $g(t) = \sum t^n/L_n$ are not rational for all t rational, not a pole of the functions f or g, which is a generalization of [2].)

3. Diophantine approximation with Padé approximation to the asymptotic expansion of the remainder of the series

For sums of series f, Padé approximation to the function f does not always provide Diophantine approximation. Although the approximation error $|x - p_n/q_n|$ is very sharp, the value of the denominator q_n of the approximation may be too large such that $|q_n x - p_n|$ does not tend to zero when n tends to infinity.

Another way is the following.

Consider the series $f(t) = \sum_{i=0}^{\infty} c_i t^i = \sum_{i=0}^{n} c_i t^i + R_n(t)$. If, for some complex number t_0, we know the asymptotic expansion of $R_n(t_0)$ on the set $\{1/n^i, i = 1, 2, \ldots\}$, then it is possible to construct an approximation of $f(t_0)$, by adding to the partial sums $S_n(t_0) := \sum_{i=0}^{n} c_i t_0^i$, some Padé approximation to the remainder $R_n(t_0)$ for the variable n.

But it is not sure that we will get a Diophantine approximation for two reasons.

(1) the Padé approximation to $R_n(t_0)$ may not converge to $R_n(t_0)$,
(2) the denominator of the approximant computed at t_0, can converge to infinity more rapidly than the approximation error does converge to zero.

So, this method works only for few cases.

3.1. *Irrationality of* $\zeta(2), \zeta(3)$, $\ln(1+\lambda)$ *and* $\sum_n 1/(q^n + r)$

3.1.1. *Zeta function*

The Zeta function of Riemann is defined as

$$\zeta(s) = \sum_{n=1}^{\infty} \frac{1}{n^s}, \tag{7}$$

where the Dirichlet series on the right-hand side of (7) is convergent for $\mathrm{Re}(s) > 1$ and uniformly convergent in any finite region where $\mathrm{Re}(s) \geqslant 1 + \delta$, with $\delta > 0$. It defines an analytic function for $\mathrm{Re}(s) > 1$.

Riemann's formula

$$\zeta(s) = \frac{1}{\Gamma(s)} \int_0^\infty \frac{x^{s-1}}{e^x - 1}\,dx, \qquad \mathrm{Re}(s) > 1,$$

where

$$\Gamma(s) = \int_0^\infty y^{s-1} e^{-y}\,dy \text{ is the gamma function} \tag{8}$$

and

$$\zeta(s) = \frac{e^{-i\pi s}\Gamma(1-s)}{2i\pi} \int_{\mathscr{C}} \frac{z^{s-1}}{e^z - 1}\,dz \tag{9}$$

where $\mathscr{C}$ is some path in $\mathbb{C}$, provides the analytic continuation of $\zeta(s)$ over the whole s-plane.

If we write formula (7) as

$$\zeta(s) = \sum_{k=1}^{n} \frac{1}{k^s} + \sum_{k=1}^{\infty} \frac{1}{(n+k)^s}$$

and set $\Psi_s(x) := \Gamma(s) \sum_{k=1}^{\infty} (x/(1+kx))^s$ then

$$\zeta(s) = \sum_{k=1}^{n} \frac{1}{k^s} + \frac{1}{\Gamma(s)} \Psi_s(1/n). \tag{10}$$

The function $\sum_{k=1}^{\infty}(x/(1+kx))^s)$ is known as the generalized zeta-function $\zeta(s, 1+1/x)$ [32, Chapter XIII] and so we get another expression of $\Psi_s(x)$:

$$\Psi_s(x) = \int_0^\infty u^{s-1} \frac{e^{-u/x}}{e^u - 1}\,du, \quad x > 0,$$

whose asymptotic expansion is

$$\Psi_s(x) = \sum_{k=0}^{\infty} \frac{B_k}{k!} \Gamma(k+s-1) x^{k+s-1},$$

where B_k are the Bernoulli numbers.

Outline of the method: In (10), we replace the unknown value $\Psi_s(1/n)$ by some Padé-approximant to $\Psi_s(x)$, at the point $x = 1/n$. We get the following approximation:

$$\zeta(s) \approx \sum_{k=1}^{n} \frac{1}{k^s} + \frac{1}{\Gamma(s)} [p/q]_{\Psi_s}(x = 1/n). \tag{11}$$

We only consider the particular case $p = q$.

Case $\zeta(2)$: If $s=2$ then (10) becomes

$$\zeta(2)=\sum_{k=1}^{n}\frac{1}{k^2}+\Psi_2(1/n),$$

and its approximation (11):

$$\zeta(2)\approx\sum_{k=1}^{n}\frac{1}{k^2}+[p/p]_{\Psi_2}(x=1/n), \tag{12}$$

where

$$\Psi_2(x)=\sum_{k=0}^{\infty}B_kx^{k+1}=B_0x+B_1x^2+B_2x^3+\cdots \quad \text{(asymptotic expansion).} \tag{13}$$

The asymptotic expansion (13) is Borel-summable and its sum is

$$\Psi_2(x)=\int_0^\infty u\frac{\mathrm{e}^{-u/x}}{\mathrm{e}^u-1}\,\mathrm{d}u.$$

Computation of $[p/p]_{\Psi_2(x)/x}$: We apply Section 1.1, where function $f(x)=\Psi_2(x)/x$. The Padé approximants $[p/p]_f$ are linked with the orthogonal polynomial with respect to the sequence $B_0,B_1,B_2\ldots$.

As in Section 1, we define the linear functional B acting on the space of polynomials by

$$B:\mathscr{P}\to\mathbb{R}$$

$$x^i\to\langle B,x^i\rangle=B_i,\quad i=0,1,2,\ldots.$$

The orthogonal polynomials Ω_p satisfy

$$\langle B,x^i\Omega_p(x)\rangle=0,\quad i=0,1,\ldots,p-1. \tag{14}$$

These polynomials have been studied by Touchard ([31,9,28,29]) and generalized by Carlitz ([12,13]).

The following expressions

$$\Omega_p(x)=\sum_{2r\leqslant p}\binom{2x+p-2r}{p-2r}\binom{x}{r}^2$$

$$=(-1)^p\sum_{k=0}^{p}(-1)^k\binom{p}{k}\binom{p+k}{k}\binom{x+k}{k}=\sum_{k=0}^{p}\binom{p}{k}\binom{p+k}{k}\binom{x}{k} \tag{15}$$

hold (see [34,12]).

Note that the Ω_p's are orthogonal polynomials and thus satisfy a three-term recurrence relation.

The associated polynomials Λ_p of degree $p-1$ are defined as

$$\Lambda_p(t)=\left\langle B,\ \frac{\Omega_p(x)-\Omega_p(t)}{x-t}\right\rangle,$$

where B acts on x.

From expression (15) for Ω_p, we get the following formula for Λ_p:

$$\Lambda_p(t)=\sum_{k=0}^{p}\binom{p}{k}\binom{p+k}{k}\left\langle B,\frac{\binom{x}{k}-\binom{t}{k}}{x-t}\right\rangle.$$

The recurrence relation between the Bernoulli numbers B_i implies that

$$\left\langle B, \binom{x}{k} \right\rangle = \frac{(-1)^k}{k+1}.$$

Using the expression of the polynomial $(\binom{x}{k} - \binom{t}{k})/(x-t)$ on the Newton basis on $0, 1, \ldots, k-1$,

$$\frac{\binom{x}{k} - \binom{t}{k}}{x-t} = \binom{t}{k} \sum_{i=1}^{k} \frac{\binom{x}{i-1}}{i\binom{t}{i}},$$

we can write a compact formula for Λ_p:

$$\Lambda_p(t) = \sum_{k=1}^{p} \binom{p}{k}\binom{p+k}{k}\binom{t}{k} \sum_{i=1}^{k} \frac{(-1)^{i-1}}{i^2\binom{t}{i}} \in \mathscr{P}_{p-1}.$$

Approximation (12) for $\zeta(2)$ becomes

$$\zeta(2) \approx \sum_{k=1}^{n} \frac{1}{k^2} + t\frac{\tilde{\Lambda}_p(t)}{\tilde{\Omega}_p(t)}\bigg|_{t=1/n} = \sum_{k=1}^{n} \frac{1}{k^2} + \frac{\Lambda_p(n)}{\Omega_p(n)}.$$

Using partial decomposition of $1/\binom{n}{i}$ with respect to the variable n, it is easy to prove that

$$\frac{d_n}{i\binom{n}{i}} \in \mathbb{N}, \quad \forall i \in \{1, 2, \ldots, n\} \tag{16}$$

with $d_n := \mathrm{LCM}(1, 2, \ldots, n)$.

A consequence of the above result is

$$d_n^2 \Lambda_p(n) \in \mathbb{N}, \quad \forall p \in \mathbb{N}$$

and

$$d_n^2 \Omega_p(n)\zeta(2) - d_n^2(S_n \Omega_p(n) + \Lambda_p(n)) \tag{17}$$

is a Diophantine approximation of $\zeta(2)$, for all values of integer p, where S_n denotes the partial sums $S_n = \sum_{k=1}^{n} 1/k^2$. It remains to estimate the error for the Padé approximation:

$$\Psi_2(t) - [p/p]_{\Psi_2}(t) = \Psi_2(t) - [p-1/p]_{\Psi_{2/t}}(t).$$

Touchard found the integral representation for the linear functional B:

$$\langle B, x^k \rangle := B_k = -\mathrm{i}\frac{\pi}{2} \int_{\alpha - \mathrm{i}\infty}^{\alpha + \mathrm{i}\infty} x^k \frac{\mathrm{d}x}{\sin^2(\pi x)}, \quad -1 < \alpha < 0.$$

Thus, formula (1) becomes

$$t^{-1}\Psi_2(t)-[p-1/p]_{\Psi_{2/t}}(t)=-\mathrm{i}\frac{\pi}{2}\frac{t^{2p}}{\tilde{\Omega}_p^2(t)}\int_{\alpha-\mathrm{i}\infty}^{\alpha+\mathrm{i}\infty}\frac{\Omega_p^2(x)}{1-xt}\frac{\mathrm{d}x}{\sin^2(\pi x)},$$

and we obtain the error for the Padé approximant to Ψ_2:

$$\Psi_2(t)-[p/p]_{\Psi_2}(t)=-\mathrm{i}\frac{\pi}{2}\frac{t}{\Omega_p^2(t^{-1})}\int_{\alpha-\mathrm{i}\infty}^{\alpha+\mathrm{i}\infty}\frac{\Omega_p^2(x)}{1-xt}\frac{\mathrm{d}x}{\sin^2(\pi x)}$$

and the error for formula (17):

$$d_n^2\Omega_p(n)\zeta(2)-d_n^2(S_n\Omega_p(n)+\Lambda_p(n))=-d_n^2\mathrm{i}\frac{\pi}{2n}\frac{1}{\Omega_p(n)}\int_{\alpha-\mathrm{i}\infty}^{\alpha+\mathrm{i}\infty}\frac{\Omega_p^2(x)}{1-x/n}\frac{\mathrm{d}x}{\sin^2(\pi x)}. \tag{18}$$

If $p=n$, we get Apéry's numbers [4]:

$$b_n'=\Omega_n(n)=\sum_{k=0}^{n}\binom{n}{k}^2\binom{n+k}{k}$$

and

$$a_n'=S_n\Omega_n(n)+\Lambda_n(n)=\left(\sum_{k=1}^{n}\frac{1}{k^2}\right)b_n'+\sum_{k=1}^{n}\binom{n}{k}^2\binom{n+k}{k}\sum_{i=1}^{k}\frac{(-1)^{i-1}}{i^2\binom{n}{i}}.$$

The error in formula (18) becomes

$$d_n^2b_n'\zeta(2)-d_n^2a_n'=-d_n^2\mathrm{i}\frac{\pi}{2n}\frac{1}{b_n'}\int_{\alpha-\mathrm{i}\infty}^{\alpha+\mathrm{i}\infty}\frac{\Omega_n^2(x)}{1-x/n}\frac{\mathrm{d}x}{\sin^2\pi x} \tag{19}$$

In order to prove the irrationality of $\zeta(2)$, we have to show that the right-hand side of (19) tends to 0 when n tends to infinity, and is different from 0, for each integer n.

We have

$$\left|\int_{-1/2-\mathrm{i}\infty}^{-1/2+\mathrm{i}\infty}\frac{\Omega_n^2(x)}{1-x/n}\frac{\mathrm{d}x}{\sin^2\pi x}\right|\leqslant\left|\int_{-\infty}^{+\infty}\frac{\Omega_n^2(-\frac{1}{2}+\mathrm{i}u)}{1+1/2n}\frac{\mathrm{d}u}{\cosh^2\pi u}\right|\leqslant\frac{1}{1+1/2n}|\langle B,\Omega_n^2(x)\rangle|$$

since $\cosh^2\pi u$ is positive for $u\in\mathbb{R}$ and $\Omega_n^2(-\frac{1}{2}+iu)$ real positive for u real (Ω_n has all its roots on the line $-\frac{1}{2}+\mathrm{i}\mathbb{R}$, because $\Omega_n(-\frac{1}{2}+iu)$ is orthogonal with respect to the positive weight $1/\cosh^2\pi u$ on $\mathbb{R}$). The quantity $\langle B,\Omega_n^2(x)\rangle$ can be computed from the three term recurrence relation between the Ω_n's [31]:

$$\langle B,\Omega_n^2(x)\rangle=\frac{(-1)^n}{2n+1}.$$

The Diophantine approximation (19) satisfies

$$|d_n^2b_n'\zeta(2)-d_n^2a_n'|\leqslant d_n^2\frac{\pi}{(2n+1)^2}\times\frac{1}{b_n'}.$$

In [14], it is proved that $b'_n \sim A'((1+\sqrt{5})/2)^{5n}n^{-1}$ when $n \to \infty$, for some constant A'. From a result concerning $d_n = \mathrm{LCM}(1,2,\dots,n)$: $(d_n = \mathrm{e}^{(n(1+o(1))})$, we get

$$\lim_{n\to\infty} |d_n^2 b'_n \zeta(2) - d_n^2 a'_n| = 0, \tag{20}$$

where $d_n^2 b'_n$ and $d_n^2 a'_n$ are integers.

Relation (20) proves that $\zeta(2)$ is not rational.

Case $\zeta(3)$: If $s=3$ then equality (10) becomes

$$\zeta(3) = \sum_{k=1}^{n} \frac{1}{k^3} + \frac{1}{2}\Psi_3(1/n), \tag{21}$$

where

$$\Psi_3(x) = \int_0^\infty u^2 \frac{\mathrm{e}^{-u/x}}{\mathrm{e}^u - 1}\,\mathrm{d}u$$

whose asymptotic expansion is

$$\Psi_3(x) = \sum_{k=0}^{\infty} B_k(k+1)x^{k+2}.$$

Computation of $[p/p]_{\Psi_3(x)/x^2}$: Let us define the derivative of B by

$$\langle -B', x^k\rangle := \langle B, kx^{k-1}\rangle = kB_{k-1}, \quad k \geqslant 1,$$

$$\langle -B', 1\rangle := 0.$$

So, the functional B' admits an integral representation:

$$\langle B', x^k\rangle = \mathrm{i}\pi^2 \int_{\alpha-\mathrm{i}\infty}^{\alpha+\mathrm{i}\infty} x^k \frac{\cos(\pi x)}{\sin^3(\pi x)}\,\mathrm{d}x, \quad -1<\alpha<0.$$

Let $(\Pi_n)_n$ be the sequence of orthogonal polynomial with respect to the sequence

$$-B'_0{:=}0, \qquad -B'_1 = B_0, \qquad -B'_2 = 2B_1, \qquad -B'_3 = 3B_2, \dots .$$

The linear form B' is not definite and so the polynomials Π_n are not of exact degree n.

More precisely, Π_{2n} has degree $2n$ and $\Pi_{2n+1} = \Pi_{2n}$. For the general theory of orthogonal polynomials with respect to a nondefinite functional, the reader is referred to Draux [15]. If we take $\alpha = -\frac{1}{2}$, the weight $\cos\pi x/\sin^3(\pi x)\,\mathrm{d}x$ on the line $-\frac{1}{2}+\mathrm{i}\mathbb{R}$ becomes $\sinh \pi t/\cosh^3 \pi t\,\mathrm{d}t$ on $\mathbb{R}$, which is symmetrical around 0. So, $\Pi_{2n}(\mathrm{i}t-\frac{1}{2})$ only contains even power of t and we can write $\Pi_{2n}(\mathrm{i}t-\frac{1}{2}) = W_n(t^2)$, W_n of exact degree n. Thus W_n satisfies

$$\int_{\mathbb{R}} W_n(t^2)W_m(t^2)\frac{t\sinh\pi t}{\cosh^3\pi t}\,\mathrm{d}t = 0, \quad n \neq m.$$

The weight $t\sinh \pi t/\cosh^3 \pi t$ equals $(1/4\pi^3)|\Gamma(\frac{1}{2}+\mathrm{i}t)|^8|\Gamma(2\mathrm{i}t)|^2$ and has been studied by Wilson [33,3]:

$$n \geqslant 0, \quad \Pi_{2n}(y) = \sum_{k=0}^{n} \binom{n}{k}\binom{n+k}{k}\binom{y+k}{k}\binom{y}{k}. \tag{22}$$

Let Θ_{2n} the polynomial associated to Π_{2n}:

$$\Theta_{2n}(t) = \left\langle -B', \frac{\Pi_{2n}(x) - \Pi_{2n}(t)}{x-t} \right\rangle, \quad B' \text{ acts on } x.$$

For the computation of Θ_{2n}, we need to expand the polynomial

$$\frac{\binom{x+k}{k}\binom{x}{k} - \binom{t+k}{k}\binom{t}{k}}{x-t}.$$

On the Newton basis with the abscissa $\{0, 1, -1, \ldots, n, -n\}$

$$\frac{\binom{x+k}{k}\binom{x}{k} - \binom{t+k}{k}\binom{t}{k}}{x-t} = \sum_{i=1}^{2k} \frac{N_{2k}(t)}{N_i(t)} \frac{N_{i-1}(x)}{[(i+1)/2]},$$

where $N_0(x) := 1$, $N_1(x) = \binom{x}{1}$, $N_2(x) = \binom{x}{1}\binom{x+1}{1}, \ldots, N_{2i}(x) = \binom{x}{i}\binom{x+i}{i}$ $N_{2i+1}(x) = \binom{x}{i+1}\binom{x+i}{i}$.

By recurrence, the values $\langle -B', N_i(x)\rangle$ can be found in

$$i \in \mathbb{N}, \quad \langle -B', N_{2i}(x)\rangle = 0, \quad \langle -B', N_{2i+1}(x)\rangle = \frac{(-1)^i}{(i+1)^2}.$$

Using the linearity of B', we get the expression of Θ_{2n}:

$$\Theta_{2n}(t) = \sum_{k=0}^{n} \binom{n}{k}\binom{n+k}{k} \sum_{i=1}^{k} \frac{(-1)^{i+1}}{i^3} \frac{\binom{t+k}{k-i}\binom{t-i}{k-i}}{\binom{k}{i}^2} \in \mathscr{P}_{2n-2}. \tag{23}$$

Eq. (16) implies that

$$d_n^3 \Theta_{2n}(t) \in \mathbb{N}, \quad \forall t \in \mathbb{N}.$$

The link between Π_{2n}, Θ_{2n} and the Apéry's numbers a_n, b_n is given by taking $y = n$ in (22) and $t = n$ in (23):

$$\Pi_{2n}(n) = \sum_{k=0}^{n} \binom{n}{k}^2 \binom{n+k}{k}^2 = b_n,$$

$$\left(\sum_{k=1}^{n} \frac{1}{k^3}\right) \Pi_{2n}(n) + \frac{1}{2}\Theta_{2n}(n) = a_n.$$

Apéry was the first to prove irrationality of $\zeta(3)$. He only used recurrence relation between the a_n and b_n. We end the proof of irrationality of $\zeta(3)$ with the error term for the Padé approximation.

Let us recall equality (21),

$$\zeta(3) = \sum_{k=1}^{n} \frac{1}{k^3} + \frac{1}{2}\Psi_3\left(\frac{1}{n}\right)$$

in which we replace the unknown term $\Psi_3(1/n)$ by its Padé approximant $[2n/2n]_{\Psi_3}(x=1/n)$. It arises the following approximation for $\zeta(3)$:

$$\zeta(3) \approx \sum_{k=1}^{n} \frac{1}{k^3} + \frac{1}{2}\frac{\Theta_{2n}(n)}{\Pi_{2n}(n)}$$

and the expression

$$e_n = 2d_n^3 \Pi_{2n}(n)\zeta(3) - \left[\left(\sum_{k=1}^{n}\frac{1}{k^3}\right) 2\Pi_{2n}(n) + \Theta_{2n}(n)\right] d_n^3$$

will be a Diophantine approximation, if we prove that $\lim_n e_n = 0$ (since $\Pi_{2n}(n)$ and $d_n^3\Theta_{2n}(n)$ are integer).

Let us estimate the error e_n. The method is the same as for $\zeta(2)$:

$$\Psi_3(t) - [2n/2n]_{\Psi_3}(t) = \Psi_3(t) - t^2[2n-2/2n]_{\Psi_3/t^2}(t) = \Psi_3(t) - \frac{\Theta_{2n}(t^{-1})}{\Pi_{2n}(t^{-1})}.$$

The integral representation of B' gives

$$\Psi_3(t) - [2n/2n]_{\Psi_3}(t) = -\frac{t\pi^2 \mathrm{i}}{\Pi_{2n}^2(t^{-1})}\int_{\alpha-\mathrm{i}\infty}^{\alpha+\mathrm{i}\infty} \frac{\Pi_{2n}^2(x)}{1-xt}\frac{\cos \pi x}{\sin^3 \pi x}\,\mathrm{d}x.$$

The previous expression implies that the error $\Psi_3(t) - [2n/2n]_{\Psi_3}(t)$ is nonzero, and also that

$$|\Psi_3(t) - [2n/2n]_{\Psi_3}(t)| \leqslant \frac{\pi^2 t}{\Pi_{2n}^2(t^{-1})} \cdot \frac{1}{1+t/2} \cdot \int_{\mathbb{R}} W_n^2(u^2)\frac{u \sinh \pi u}{\cosh^3 \pi u}\,\mathrm{d}u, \quad t \in \mathbb{R}^+.$$

From the expression of the integral (see [33]) we get

$$|\Psi_3(1/n) - [2n/2n]_{\Psi_3}(1/n)| \leqslant \frac{4\pi^2}{(2n+1)^2 \Pi_{2n}^2(n)}.$$

The error term in the Padé approximation satisfies

$$\left|2\zeta(3) - 2\sum_{k=1}^{n}\frac{1}{k^3} - [2n/2n]_{\Psi_3}(1/n)\right| \leqslant \frac{4\pi^2}{(2n+1)^2 \Pi_{2n}^2(n)}$$

and the error term e_n satisfies

$$|e_n| = \left|2d_n^3 \Pi_{2n}(n)\zeta(3) - \left[2\left(\sum_{k=1}^{n}\frac{1}{k^3}\right)\Pi_{2n}(n) + \Theta_{2n}(n)\right] d_n^3\right| \leqslant \frac{8\pi^2}{(2n+1)^2}\frac{d_n^3}{\Pi_{2n}(n)}.$$

$\Pi_{2n}(n) = b_n$ implies that $\Pi_{2n}(n) = A(1+\sqrt{2})^{4n} n^{-3/2}$ [14], and so we get, since $d_n = \mathrm{e}^{n(1+\mathrm{o}(1))}$,

$$|2d_n^3 b_n \zeta(3) - 2d_n^3 a_n| \to 0, \quad n \to \infty, \tag{24}$$

where $2d_n^3 b_n$ and $2d_n^3 a_n$ are integers.

The above relation (24) shows that $\zeta(3)$ is irrational.

Of course, using the connection between Padé approximation and ε-algorithm, the Diophantine approximation of $\zeta(3)$ can be constructed by means of the following ε-array: $a_n/b_n = \sum_{k=1}^{n} 1/k^3 + \varepsilon_{4n}^{(0)}(T_m) = \varepsilon_{4n}^{(0)}(\sum_{k=1}^{n} 1/k^3 + T_m)$, where T_m is the partial sum of the asymptotic series (nonconvergent) $T_m = \frac{1}{2}\sum_{k=1}^{m} B_k(k+1)1/n^k$.

We get the following ε-arrays for $n = 1$,

$$\begin{bmatrix} 0 & & \\ 0 & 0 & \\ 1 & 1/2 & 2/5 = \varepsilon_4^{(0)} \\ 0 & 1/3 & \\ 1/2 & & \end{bmatrix}, \qquad 1 + \tfrac{1}{2} * \varepsilon_4^{(0)} = \tfrac{6}{5} = a_1/b_1 \quad \text{(Apery's numbers)},$$

and for $n = 2$,

$$\begin{bmatrix} 0 & & & & \\ 0 & 0 & & & \\ 1/4 & 1/6 & 2/13 & & \\ 1/8 & 3/20 & 2/13 & 2/13 & \\ 5/32 & 5/32 & 21/136 & 37/240 & 45/292 = \varepsilon_8^{(0)} \\ 5/32 & 5/32 & 2/13 & 53/344 & \\ 59/384 & 59/384 & 37/240 & & \\ 59/384 & 59/384 & & & \\ 79/512 & & & & \end{bmatrix}$$

(we have only displayed the odd columns), $1+1/2^3+1/2*\varepsilon_8^{(0)}=351/292=a_2/b_2$. ε-algorithm is a particular extrapolation algorithm as Padé approximation is particular case of Padé-type approximation. Generalization has been achieved by Brezinski and Hävie, the so-called E-algorithm. Diophantine approximation using E-algorithm and Padé-type approximation are under consideration.

3.1.2. *Irrationality of* $\ln(1+\lambda)$

In this part, we use the same method as in the preceding section:

$$\text{We set } \ln(1+\lambda) = \sum_{k=1}^{n} (-1)^{k+1}\frac{\lambda^k}{k} + \sum_{k=1}^{\infty} \frac{(-1)^{k+n+1}}{k+n}\lambda^{k+n}. \tag{25}$$

From the formula $1/(k+n) = \int_0^\infty e^{-(k+n)v}\,dv$, we get an integral representation for the remainder term in (25):

$$\sum_{k=1}^{\infty} (-1)^{k+n+1}\frac{\lambda^{k+n}}{k+n} = (-1)^n \int_0^\infty \lambda^{n+1}\frac{e^{-nv}}{e^v+\lambda}dv.$$

If we expand the function

$$\frac{1+\lambda}{e^v+\lambda}=\sum_{k=0}^{\infty} R_k(-\lambda)\frac{v^k}{k!},$$

where the $R_k(-\lambda)$'s are the Eulerian numbers [12], we get the following asymptotic expansion:

$$\sum_{k=1}^{\infty}(-1)^{k+n+1}\frac{\lambda^{k+n}}{k+n}=\frac{(-1)^n\lambda^{n+1}}{n(1+\lambda)}\left(\sum_{k=0}^{\infty} R_k(-\lambda)x^k\right)_{x=1/n}.$$

Let us set

$$\Phi_1(x)=\sum_{k=0}^{\infty} R_k(-\lambda)x^k.$$

Carlitz has studied the orthogonal polynomials with respect to $R_0(-\lambda)$, $R_1(-\lambda),\dots$.

If we define the linear functional R by

$$\langle R, x^k\rangle := R_k(-\lambda),$$

then the orthogonal polynomials P_n with respect to R, i.e.,

$$\langle R, x^k P_n(x)\rangle = 0, \quad k=0,1,\dots,n-1,$$

satisfy $P_n(x)=\sum_{k=0}^{n}(1+\lambda)^k\binom{n}{k}\binom{x}{k}$ [12].

The associated polynomials are

$$Q_n(t)=\sum_{k=0}^{n}(1+\lambda)^k\binom{n}{k}\left\langle R, \frac{\binom{x}{k}-\binom{t}{k}}{x-t}\right\rangle. \tag{26}$$

Carlitz proved that $\langle R, \binom{x}{k}\rangle=(-\lambda-1)^{-k}$ and thus, using (26),

$$Q_n(t)=\sum_{k=0}^{n}(1+\lambda)^k\binom{n}{k}\binom{t}{k}\sum_{i=1}^{k}\frac{1}{i\binom{t}{i}}\left(\frac{-1}{\lambda+1}\right)^{i-1}.$$

If we set $\lambda=p/q$, p and $q\in\mathbb{Z}$ and $t=n$, then

$$q^n d_n Q_n(n)\in\mathbb{Z}.$$

An integral representation for $R_k(-\lambda)$ is given by Carlitz:

$$R_k(-\lambda)=-\frac{1+\lambda}{2i\lambda}\int_{\alpha-i\infty}^{\alpha+i\infty} z^k\frac{\lambda^{-z}}{\sin\pi z}\,dz, \quad -1<\alpha<0, \tag{27}$$

and thus

$$\Phi_1(x)=-\frac{1+\lambda}{2i\lambda}\int_{\alpha-i\infty}^{\alpha+i\infty}\frac{1}{1-xz}\frac{\lambda^{-z}}{\sin\pi z}\,dz.$$

The orthogonal polynomial P_n satisfies [12]

$$\int_{\alpha-i\infty}^{\alpha+i\infty} P_n^2(z)\frac{\lambda^{-z}}{\sin\pi z}\,dz=\frac{+2i}{i+\lambda}(-\lambda)^{n+1},$$

and since $\mathrm{Re}(\lambda^{-z}\sin\pi z) > 0$ for $z \in -\frac{1}{2} + \mathrm{i}\mathbb{R}$, we obtain a majoration of the error for the Padé approximation to Φ_1:

$$x > 0, |\Phi_1(x) - [n-1/n]_{\Phi_1}(x)| \leqslant \frac{\lambda^n}{|1+x/2|}$$

and if $x = 1/n$, we get

$$\left|\Phi_1\left(\frac{1}{n}\right) - [n-1/n]_{\Phi_1}(1/n)\right| \leqslant \frac{|\lambda|^n}{1+1/2n}.$$

Let us replace in (25) the remainder term by its Padé approximant:

$$\ln(1+\lambda) \approx \sum_{k=1}^{n}(-1)^{k+1}\frac{\lambda^k}{k} + \frac{(-1)^n\lambda^{n+1}}{(1+\lambda)n}[n-1/n]_{\Phi_1}(1/n),$$

we obtain a Diophantine approximation for $\ln(1+p/q)$:

$$\left|\ln\left(1+\frac{p}{q}\right)d_nq^{2n}P_n(n) - d_nq^{2n}T_n(n)\right| \leqslant \frac{\lambda^{2n}d_nq^{2n}}{(n+2)P_n(n)}, \qquad (28)$$

where $T_n(n) = P_n(n)\sum_{k=1}^{n}(-1)^{k+1}p^k/kq^k + (-1)^{n+1}Q_n(n)q^n$.

From the expression of $P_n(x)$ we can conclude that

$$P_n(n) = \sum_{k=0}^{n}(1+\lambda)^k\binom{n}{k}^2 = \mathrm{Legendre}\left(n, \frac{2}{\lambda}+1\right)\lambda^n,$$

where Legendre (n,x) is the nth Legendre polynomial and thus

$$\frac{T_n(n)}{P_n(n)} = [n/n]_{\ln(1+x)}\ (x=1).$$

So, the classical proof for irrationality of $\ln(1+p/q)$ based on Padé approximants to the function $\ln(1+x)$ is recovered by formula (28).

Proof of irrationality of $\zeta(2)$ with alternated series: Another expression for $\zeta(2)$ is

$$\zeta(2) = 2\sum_{k=1}^{\infty}\frac{(-1)^{k-1}}{k^2}.$$

Let us write it as a sum

$$\zeta(2) = 2\sum_{k=1}^{n}\frac{(-1)^{k-1}}{k^2} + 2\sum_{k=1}^{\infty}\frac{(-1)^{k+n+1}}{(k+n)^2}.$$

Let Φ_2 be defined by $\Phi_2(x) = \sum_{k=0}^{\infty} R_k(-1)(k+1)x^k$. So

$$\zeta(2) = 2\sum_{k=1}^{n}\frac{(-1)^{k-1}}{k^2} + \frac{(-1)^n}{n^2}\Phi_2(1/n).$$

With the same method, we can prove that the Padé approximant $[2n/2n]_{\Phi_2}(x)$ computed at $x = 1/n$ leads to Apéry's numbers a'_n and b'_n and so proves the irrationality of $\zeta(2)$ with the integral representation for the sequence $(kR_{k-1}(-1))_k$:

$$kR_{k-1}(-1) = -\frac{\pi(1+\lambda)}{2i\lambda}\int_{\alpha-i\infty}^{\alpha+i\infty} z^k \frac{\cos \pi z}{\sin^2 \pi z}\,dz, \quad k \geqslant 1.$$

obtained with an integration by parts applied to (27).

3.1.3. Irrationality of $\sum 1/(q^n + r)$

In [7], Borwein proves the irrationality of $L(r) = \sum 1/(q^n - r)$, for q an integer greater than 2, and r a non zero rational (different from q^n, for any $n \geqslant 1$), by using similar method. It is as follows: Set

$$L_q(x) := \sum_{n=1}^{\infty} \frac{x}{q^n - x} = \sum_{n=1}^{\infty} \frac{x^n}{q^n - 1}, \quad |q| > 1.$$

Fix N a positive integer and write $L_q(r) = \sum_{n=1}^{N} r/(q^n - r) + L_q(r/q^N)$.

Then, it remains to replace $L_q(r/q^N)$ by its Padé approximant $[N/N]_{L_q}(r/q^N)$.

The convergence of $[N/N]_{L_q}$ to L_q is a consequence of the following formula:

$$\forall t \in \mathbb{C} \setminus \{q^j, j \in \mathbb{N}\}, \quad \forall n \in \mathbb{N}, \quad \limsup_N |L_q(t) - [N/N]_{L_q}(t)|^{1/3N^2} \leqslant 1/q.$$

p_n/q_n defined by $p_n/q_n := \sum_{n=1}^{N} r/(q^n - r) + [N/N]_{L_q}(r/q^N)$ leads to Diophantine approximation of $L_q(r)$ and so proves the irrationality of $L_q(r)$.

For further results concerning the function L_q, see [17–19].

Different authors used Padé or Padé Hermite approximants to get Diophantine approximation, see for example [8,20–23,27].

References

[1] K. Alladi, M.L. Robinson, Legendre polynomials and irrationality, J. Reine Angew. Math. 318 (1980) 137–155.

[2] R. André-Jeannin, Irrationalité de la somme des inverses de certaines suites récurrentes, C.R. Acad. Sci. Paris Sér. I 308 (1989) 539–541.

[3] G.E. Andrews, R. Askey, Classical orthogonal polynomials, in: C. Brezinski, A. Draux, A.P. Magnus, P. Maroni, A. Ronveaux (Eds.) Polynômes Orthogonaux et applications, Lecture notes in Mathematics, Vol. 1171, Springer, New York, 1985, pp. 36–62.

[4] R. Apéry, Irrationalité de $\zeta(2)$ et $\zeta(3)$, J. Arith. Luminy, Astérisque 61 (1979) 11–13.

[5] G.A. Baker Jr., Essentials of Padé approximants, Academic Press, New York, 1975.

[6] G.A. Baker Jr. P.R. Graves Morris, Padé approximants, Encyclopedia of Mathematics and its Applications, 2nd Edition, Cambridge University Press, Cambridge.

[7] P. Borwein, On the irrationality of $\sum 1/(q^n + r^n)$, J. Number Theory 37 (1991) 253–259.

[8] P. Borwein, On the irrationality of certain series, Math. Proc. Cambridge Philos Soc. 112 (1992) 141–146.

[9] F. Brafmann, On Touchard polynomials, Canad. J. Math. 9 (1957) 191–192.

[10] C. Brezinski, in: Padé-type approximation and general orthogonal polynomials, ISM, Vol. 50, Birkäuser Verlag, Basel, 1980.

[11] C. Brezinski, J. Van Iseghem, Padé Approximations, in: P.G. Ciarlet, J.L. Lions (Eds.), Handbook of Numerical Analysis, Vol. III, North Holland, Amsterdam, 1994.

[12] L. Carlitz, Some polynomials of Touchard connected with the Bernoulli numbers, Canad. J. Math. 9 (1957) 188–190.

[13] L. Carlitz, Bernouilli and Euler numbers and orthogonal polynomials, Duke Math. J. 26 (1959) 1–16.

[14] H. Cohen, Démonstration de l'irrationalité de $\zeta(3)$, d'après Apéry, séminaire de Théorie des nombres de Bordeaux, 5 octobre 1978.

[15] A. Draux, in: Polynômes orthogonaux formels — Applications, Lecture Notes in Mathematics, Vol. 974, Springer, Berlin, 1983.

[16] K.A. Driver, D.S. Lubinsky, Convergence of Padé approximants for a q-hypergeometric series Wynn's Power Series III, Aequationes Math. 45 (1993) 1–23.

[17] D. Duverney, Approximation Diophantienne et irrationalité de la somme de certaines séries de nombres rationnels", Thèse, Université de Lille I, 1993.

[18] D. Duverney, Approximants de Padé et U-dérivation, Bull. Soc. Math. France 122 (1994) 553–570.

[19] D. Duverney, Sur l'irrationalité de $\sum r^n/(q^n - r)$, C.R. Acad. Sci. Paris, Sér. I 320 (1995) 1–4.

[20] M. Hata, On the linear independance of the values of polylogarithmic functions, J. Math. Pures Appl. 69 (1990) 133–173.

[21] M. Huttner, Irrationalité de certaines intégrales hypergéométriques, J. Number Theory 26 (1987) 166–178.

[22] T. Matala Aho, On the irrationality measures for values of some q-hypergeometric series, Acta Univ. Oulu Ser. A Sci. Rerum Natur. 219 (1991) 1–112.

[23] E.M. Nikischin, On irrationality of the values of the functions $F(x,s)$, Math. USSR Sbornik 37 (1980) 381–388.

[24] M. Prévost, Rate of convergence of Padé approximants for a particular Wynn series, Appl. Numer. Math. 17 (1995) 461–469.

[25] M. Prévost, On the irrationality of $\sum t^n/(A\alpha^n + B\beta^n)$, J. Number Theory 73 (1998) 139–161.

[26] D. Shanks, Non linear transformations of divergent and slowly convergent series, J. Math. Phys. 34 (1955) 1–42.

[27] V.N. Sorokin, On the irrationality of the values of hypergeometric functions, Math. USSR Sb. 55 (1) (1986) 243–257.

[28] T.J. Stieltjes, Sur quelques intégrales définies et leur développement en fractions continues, Quart. J. Math. 24 (1880) 370–382; oeuvres, Vol. 2, Noordhoff, Groningen, 1918, pp. 378–394.

[29] T.J. Stieltjes, Recherches sur les fractions continues, Ann. Faculté Sci. Toulouse 8, (1894), J1-122; 9 (1895), A1-47; oeuvres, Vol. 2, pp. 398–566.

[30] G. Szegö, in: Orthogonal Polynomials, Amer. Math. Soc. Coll. Pub., Vol. XXIII, American Mathematical Society, Providence, RI, 1939.

[31] J. Touchard, Nombres exponentiels et nombres de Bernoulli, Canad. J. Math. 8 (1956) 305–320.

[32] G.N. Watson, E.T. Whittaker, A Course of Modern Analysis, Cambridge University Press, London, 1958.

[33] J. Wilson, Some hypergeometric orthogonal polynomials, SIAM J. Math. Anal. 11 (1980) 690–701.

[34] M. Wymann, L. Moser, On some polynomials of Touchard, Canad. J. Math. 8 (1956) 321–322.

[35] P. Wynn, l'ε-algorithmo e la tavola di Padé, Rend. Mat. Roma 20 (1961) 403.

ELSEVIER

Journal of Computational and Applied Mathematics 122 (2000) 251–273

JOURNAL OF COMPUTATIONAL AND APPLIED MATHEMATICS

www.elsevier.nl/locate/cam

The generalized Richardson extrapolation process GREP$^{(1)}$ and computation of derivatives of limits of sequences with applications to the $d^{(1)}$-transformation

Avram Sidi

Computer Science Department, Technion, Israel Institute of Technology, Haifa 32000, Israel

Received 3 May 1999; received in revised form 15 December 1999

Abstract

Let $\{S_m\}$ be an infinite sequence whose limit or antilimit S can be approximated very efficiently by applying a suitable extrapolation method E_0 to $\{S_m\}$. Assume that the S_m and hence also S are differentiable functions of some parameter ξ, $(\mathrm{d}/\mathrm{d}\xi)S$ being the limit or antilimit of $\{(\mathrm{d}/\mathrm{d}\xi)S_m\}$, and that we need to approximate $(\mathrm{d}/\mathrm{d}\xi)S$. A direct way of achieving this would be by applying again a suitable extrapolation method E_1 to the sequence $\{(\mathrm{d}/\mathrm{d}\xi)S_m\}$, and this approach has often been used efficiently in various problems of practical importance. Unfortunately, as has been observed at least in some important cases, when $(\mathrm{d}/\mathrm{d}\xi)S_m$ and S_m have essentially different asymptotic behaviors as $m \to \infty$, the approximations to $(\mathrm{d}/\mathrm{d}\xi)S$ produced by this approach, despite the fact that they are good, do not converge as quickly as those obtained for S, and this is puzzling. In a recent paper (A. Sidi, Extrapolation methods and derivatives of limits of sequences, *Math. Comp.*, 69 (2000) 305–323) we gave a rigorous mathematical explanation of this phenomenon for the cases in which E_0 is the Richardson extrapolation process and E_1 is a generalization of it, and we showed that the phenomenon has nothing to do with numerics. Following that we proposed a very effective procedure to overcome this problem that amounts to first applying the extrapolation method E_0 to $\{S_m\}$ and then differentiating the resulting approximations to S. As a practical means of implementing this procedure we also proposed the direct differentiation of the recursion relations of the extrapolation method E_0 used in approximating S. We additionally provided a thorough convergence and stability analysis in conjunction with the Richardson extrapolation process from which we deduced that the new procedure for $(\mathrm{d}/\mathrm{d}\xi)S$ has practically the same convergence properties as E_0 for S. Finally, we presented an application to the computation of integrals with algebraic/logarithmic endpoint singularities via the Romberg integration. In this paper we continue this research by treating Sidi's generalized Richardson extrapolation process GREP$^{(1)}$ in detail. We then apply the new procedure to various infinite series of logarithmic type (whether convergent or divergent) in conjunction with the $d^{(1)}$-transformation of Levin and Sidi. Both the theory and the numerical results of this paper too indicate that this approach is the preferred one for computing derivatives of limits of infinite sequences and series. © 2000 Elsevier Science B.V. All rights reserved.

MSC: 40A25; 41A60; 65B05; 65B10; 65D30

E-mail address: asidi@cs.technion.ac.il (A. Sidi)

0377-0427/00/$ - see front matter © 2000 Elsevier Science B.V. All rights reserved.
PII: S0377-0427(00)00362-9

1. Introduction and review of recent developments

Let $\{S_m\}$ be an infinite sequence whose limit or antilimit S can be approximated very efficiently by applying a suitable extrapolation method E_0 to $\{S_m\}$. Assume that the S_m and hence also S are differentiable functions of some parameter ξ, $(\mathrm{d}/\mathrm{d}\xi)S$ being the limit or antilimit of $\{(\mathrm{d}/\mathrm{d}\xi)S_m\}$, and that we need to approximate $(\mathrm{d}/\mathrm{d}\xi)S$. A direct way of achieving this would be by applying again a suitable extrapolation method E_1 to the sequence $\{(\mathrm{d}/\mathrm{d}\xi)S_m\}$, and this approach has often been used efficiently in various problems of practical importance. When S_m and $(\mathrm{d}/\mathrm{d}\xi)S_m$ have essentially different asymptotic behaviors as $m \to \infty$, the approximations to $(\mathrm{d}/\mathrm{d}\xi)S$ produced by applying E_1 to $\{(\mathrm{d}/\mathrm{d}\xi)S_m\}$ do not converge to $(\mathrm{d}/\mathrm{d}\xi)S$ as quickly as the approximations to S obtained by applying E_0 to $\{S_m\}$ even though they may be good. This is a curious and disturbing phenomenon that calls for an explanation and a befitting remedy, and both of these issues were addressed by the author in the recent paper [14] via the Richardson extrapolation. As far as is known to us [14] is the first work that handles this problem.

The procedure to cope with the problem above that was proposed in [14] amounts to *first applying the extrapolation method E_0 to $\{S_m\}$ and then differentiating the resulting approximations to S*. As far as practical implementation of this procedure is concerned, it was proposed in [14] to *actually differentiate the recursion relations satisfied by the method E_0*.

In the present work we continue this new line of research by extending the approach of [14] to GREP$^{(1)}$ that is the simplest case of the generalized Richardson extrapolation process GREP of Sidi [7]. Following this, we consider the application of the $d^{(1)}$-transformation, the simplest of the d-transformations of Levin and Sidi [6], to computing derivatives of sums of infinite series. Now GREP is a most powerful extrapolation procedure that can be applied to a very large class of sequences and the d-transformations are GREPs that can be applied successfully again to a very large class of infinite series. Indeed, it is known theoretically and has been observed numerically that GREP in general and the d-transformations in particular have scopes larger than most known extrapolation methods.

Before we go on to the main theme of this paper, we will give a short review of the motivation and results of [14]. This will also help establish some of the notation that we will use in the remainder of this work and set the stage for further developments. As we did in [14], here too we will keep the treatment general by recalling that infinite sequences are either directly related to or can be formally associated with a function $A(y)$, where y may be a continuous or discrete variable.

Let a function $A(y)$ be known and hence computable for $y \in (0,b]$ with some $b > 0$, the variable y being continuous or discrete. Assume, furthermore, that $A(y)$ has an asymptotic expansion of the form

$$A(y) \sim A + \sum_{k=1}^{\infty} \alpha_k y^{\sigma_k} \quad \text{as } y \to 0+, \tag{1.1}$$

where σ_k are known scalars satisfying

$$\sigma_k \neq 0,\ k = 1,2,\ldots; \quad \Re\sigma_1 < \Re\sigma_2 < \cdots; \quad \lim_{k\to\infty} \Re\sigma_k = +\infty, \tag{1.2}$$

and A and α_k, $k = 1,2,\ldots,$ are constants independent of y that are not necessarily known.

From (1.1) and (1.2) it is clear that $A=\lim_{y\to 0+}A(y)$ when this limit exists. When $\lim_{y\to 0+}A(y)$ does not exist, A is the antilimit of $A(y)$ for $y\to 0+$, and in this case $\Re\sigma_1\leqslant 0$ necessarily. In any case, A can be approximated very effectively by the Richardson extrapolation process that is defined via the linear systems of equations

$$A(y_l)=A_n^{(j)}+\sum_{k=1}^{n}\bar{\alpha}_k y_l^{\sigma_k},\quad j\leqslant l\leqslant j+n, \tag{1.3}$$

with the y_l picked as

$$y_l=y_0\omega^l,\quad l=0,1,\ldots,\quad \text{for some } y_0\in(0,b] \text{ and } \omega\in(0,1). \tag{1.4}$$

Here $A_n^{(j)}$ are the approximations to A and the $\bar{\alpha}_k$ are additional (auxiliary) unknowns. As is well known, $A_n^{(j)}$ can be computed very efficiently by the following algorithm due to Bulirsch and Stoer [2]:

$$A_0^{(j)}=A(y_j),\quad j=0,1,\ldots,$$

$$A_n^{(j)}=\frac{A_{n-1}^{(j+1)}-c_nA_{n-1}^{(j)}}{1-c_n},\quad j=0,1,\ldots,\quad n=1,2,\ldots, \tag{1.5}$$

where we have defined

$$c_n=\omega^{\sigma_n},\quad n=1,2,\ldots\ . \tag{1.6}$$

Let us now consider the situation in which $A(y)$ and hence A depend on some real or complex parameter ξ and are continuously differentiable in ξ for ξ in some set X of the real line or the complex plane, and we are interested in computing $(\mathrm{d}/\mathrm{d}\xi)A\equiv\dot{A}$. Let us assume in addition to the above that $(\mathrm{d}/\mathrm{d}\xi)A(y)\equiv\dot{A}(y)$ has an asymptotic expansion for $y\to 0+$ that is obtained by differentiating that in (1.1) term by term. (This assumption is satisfied at least in some cases of practical interest as can be shown rigorously.) Finally, let us assume that the α_k and σ_k, as well as $A(y)$ and A, depend on ξ and that they are continuously differentiable for $\xi\in X$. As a consequence of these assumptions we have

$$\dot{A}(y)\sim\dot{A}+\sum_{k=1}^{\infty}(\dot{\alpha}_k+\alpha_k\dot{\sigma}_k\log y)y^{\sigma_k}\quad \text{as } y\to 0+, \tag{1.7}$$

where $\dot{\alpha}_k\equiv(\mathrm{d}/\mathrm{d}\xi)\alpha_k$ and $\dot{\sigma}_k\equiv(\mathrm{d}/\mathrm{d}\xi)\sigma_k$. Obviously, $\dot{A}$ and the $\dot{\alpha}_k$ and $\dot{\sigma}_k$ are independent of y. As a result, the infinite sum on the right-hand side of (1.7) is simply of the form $\sum_{k=1}^{\infty}(\alpha_{k0}+\alpha_{k1}\log y)y^{\sigma_k}$ with α_{k0} and α_{k1} constants independent of y.

Note that when the σ_k do not depend on ξ, we have $\dot{\sigma}_k=0$ for all k, and, therefore, the asymptotic expansion in (1.7) becomes of exactly the same form as that given in (1.1). This means that we can apply the Richardson extrapolation process above directly to $\dot{A}(y)$ and obtain very good approximations to $\dot{A}$. This amounts to replacing $A(y_j)$ in (1.5) by $\dot{A}(y_j)$, keeping everything else the same. However, when the σ_k are functions of ξ, the asymptotic expansion in (1.7) is essentially different from that in (1.1). This is so since $y^{\sigma_k}\log y$ and y^{σ_k} behave entirely differently as $y\to 0+$. In this case the application of the Richardson extrapolation process directly to $\dot{A}(y)$ does not produce approximations to $\dot{A}$ that are of practical value.

The existence of an asymptotic expansion for $\dot{A}(y)$ of the form given in (1.7), however, suggests immediately that a generalized Richardson extrapolation process can be applied to produce approximations to $\dot{A}$ in an efficient manner. In keeping with the convention introduced by the author in [12], this extrapolation process is defined via the linear systems

$$B(y_l) = B_n^{(j)} + \sum_{k=1}^{\lfloor (n+1)/2 \rfloor} \bar{\alpha}_{k0} y_l^{\sigma_k} + \sum_{k=1}^{\lfloor n/2 \rfloor} \bar{\alpha}_{k1} y_l^{\sigma_k} \log y_l, \quad j \leqslant l \leqslant j+n, \tag{1.8}$$

where $B(y) \equiv \dot{A}(y)$, $B_n^{(j)}$ are the approximations to $B \equiv \dot{A}$, and $\bar{\alpha}_{k0}$ and $\bar{\alpha}_{k1}$ are additional (auxiliary) unknowns. (This amounts to "eliminating" from (1.7) the functions $y^{\sigma_1}, y^{\sigma_1} \log y, y^{\sigma_2}, y^{\sigma_2} \log y, \ldots,$ in this order.) With the y_l as in (1.4), the approximations $B_n^{(j)}$ can be computed very efficiently by the following algorithm developed in Sidi [12] and denoted the SGRom-algorithm there:

$$B_0^{(j)} = B(y_j), \quad j = 0, 1, \ldots,$$

$$B_n^{(j)} = \frac{B_{n-1}^{(j+1)} - \lambda_n B_{n-1}^{(j)}}{1 - \lambda_n}, \quad j = 0, 1, \ldots, \quad n = 1, 2, \ldots, \tag{1.9}$$

where we have now defined

$$\lambda_{2k-1} = \lambda_{2k} = c_k, \quad k = 1, 2, \ldots, \tag{1.10}$$

with the c_n as defined in (1.6).

Before going on, we would like to mention that the problem we have described above arises naturally in the numerical evaluation of integrals of the form $B = \int_0^1 (\log x) x^{\xi} g(x)\,\mathrm{d}x$, where $\Re \xi > -1$ and $g \in C^{\infty}[0,1]$. It is easy to see that $B = (\mathrm{d}/\mathrm{d}\xi)A$, where $A = \int_0^1 x^{\xi} g(x)\,\mathrm{d}x$. Furthermore, the trapezoidal rule approximation $B(h)$ to B with stepsize h has an Euler–Maclaurin (E–M) expansion that is obtained by differentiating with respect to ξ the E–M expansion of the trapezoidal rule approximation $A(h)$ to A. With this knowledge available, B can be approximated by applying a generalized Richardson extrapolation process to $B(h)$. Traditionally, this approach has been adopted in multidimensional integration of singular functions as well. For a detailed discussion see [3,9].

If we arrange the $A_n^{(j)}$ and $B_n^{(j)}$ in two-dimensional arrays of the form

$$\begin{array}{lllll} Q_0^{(0)} & & & & \\ Q_0^{(1)} & Q_1^{(0)} & & & \\ Q_0^{(2)} & Q_1^{(1)} & Q_2^{(0)} & & \\ Q_0^{(3)} & Q_1^{(2)} & Q_2^{(1)} & Q_3^{(0)} & \\ \vdots & \vdots & \vdots & \vdots & \ddots \end{array} \tag{1.11}$$

then the diagonal sequences $\{Q_n^{(j)}\}_{n=0}^{\infty}$ with fixed j have much better convergence properties than the column sequences $\{Q_n^{(j)}\}_{j=0}^{\infty}$ with fixed n. In particular, the following convergence results are known:

1. The column sequences satisfy

$$A_n^{(j)} - A = \mathrm{O}(|c_{n+1}|^j) \quad \text{as } j \to \infty,$$
$$B_{2m+s}^{(j)} - B = \mathrm{O}(j^{1-s}|c_{m+1}|^j) \quad \text{as } j \to \infty,\ s = 0,1. \tag{1.12}$$

2. Under the additional condition that

$$\Re\sigma_{k+1} - \Re\sigma_k \geqslant d > 0,\ k = 1,2,\ldots, \quad \text{for some fixed } d \tag{1.13}$$

and assuming that α_k, $\dot{\alpha}_k$, and $\alpha_k\dot{\sigma}_k$ grow with k at most like $\exp(\beta k^\eta)$ for some $\beta \geqslant 0$ and $\eta < 2$, the diagonal sequences satisfy, for all practical purposes,

$$A_n^{(j)} - A \doteq \mathrm{O}\left(\prod_{i=1}^{n} |c_i|\right) \quad \text{as } n \to \infty,$$
$$B_n^{(j)} - B \doteq \mathrm{O}\left(\prod_{i=1}^{n} |\lambda_i|\right) \quad \text{as } n \to \infty. \tag{1.14}$$

The results pertaining to $A_n^{(j)}$ in (1.12) and (1.14), with real σ_k, are due to Bulirsch and Stoer [2]. The case of complex σ_k is contained in [12], and so are the results on $B_n^{(j)}$. Actually, [12] gives a complete treatment of the general case in which

$$A(y) \sim A + \sum_{k=1}^{\infty}\left[\sum_{i=0}^{q_k} \alpha_{ki}(\log y)^i\right] y^{\sigma_k} \quad \text{as } y \to 0+, \tag{1.15}$$

where q_k are known arbitrary nonnegative integers, and α_{ki} are constants independent of y, and the σ_k satisfy the condition

$$\sigma_k \neq 0,\ k = 1,2,\ldots, \quad \Re\sigma_1 \leqslant \Re\sigma_2 \leqslant \cdots, \quad \lim_{k\to\infty} \Re\sigma_k = +\infty \tag{1.16}$$

that is much weaker than that in (1.2). Thus, the asymptotic expansions in (1.1) and (1.7) are special cases of that in (1.15) with $q_k = 0$, $k = 1,2,\ldots$, and $q_k = 1$, $k = 1,2,\ldots$, respectively.

Comparison of the diagonal sequences $\{A_n^{(j)}\}_{n=0}^{\infty}$ and $\{B_n^{(j)}\}_{n=0}^{\infty}$ (with j fixed) with the help of (1.14) reveals that the latter has inferior convergence properties, even though the computational costs of $A_n^{(j)}$ and $B_n^{(j)}$ are almost identical. (They involve the computation of $A(y_l)$, $j \leqslant l \leqslant j+n$, and $B(y_l)$, $j \leqslant l \leqslant j+n$, respectively). As a matter of fact, from (1.6), (1.10), and (1.13) it follows that the bound on $|A_{2m}^{(j)} - A|$ is smaller than that of $|B_{2m}^{(j)} - B|$ by a factor of $\mathrm{O}(\prod_{i=1}^{m} |c_{m+i}/c_i|) = \mathrm{O}(\omega^{dm^2})$ as $m \to \infty$. This theoretical observation is also supported by numerical experiments. Judging from (1.14) again, we see that, when $\Re\sigma_{k+1} - \Re\sigma_k = d$ for all k in (1.13), $B_{\lfloor\sqrt{2}\,n\rfloor}^{(j)}$ will have an accuracy comparable to that of $A_n^{(j)}$. This, however, increases the cost of the extrapolation substantially, as the cost of computing $A(y_l)$ and $B(y_l)$ increases drastically with increasing l in most cases of interest. This quantitative discussion makes it clear that the inferiority of $B_n^{(j)}$ relative to $A_n^{(j)}$ is actually mathematical and has nothing to do with numerics.

From what we have so far it is easy to identify the Richardson extrapolation of (1.3) as method E_0 and the generalized Richardson extrapolation of (1.8) as method E_1. We now turn to the new procedure "$(\mathrm{d}/\mathrm{d}\xi)E_0$".

Let us now approximate $\dot{A}$ by $(\mathrm{d}/\mathrm{d}\xi)A_n^{(j)} = \dot{A}_n^{(j)}$. This can be achieved computationally by differentiating the recursion relation in (1.5), the result being the following recursive algorithm:

$$A_0^{(j)} = A(y_j) \quad \text{and} \quad \dot{A}_0^{(j)} = \dot{A}(y_j), \; j = 0, 1, \ldots,$$

$$A_n^{(j)} = \frac{A_{n-1}^{(j+1)} - c_n A_{n-1}^{(j)}}{1 - c_n} \quad \text{and}$$

$$\dot{A}_n^{(j)} = \frac{\dot{A}_{n-1}^{(j+1)} - c_n \dot{A}_{n-1}^{(j)}}{1 - c_n} + \frac{\dot{c}_n}{1 - c_n}(A_n^{(j)} - A_{n-1}^{(j)}), \quad j = 0, 1, \ldots, \quad n = 1, 2, \ldots . \tag{1.17}$$

Here $\dot{c}_n \equiv (\mathrm{d}/\mathrm{d}\xi)c_n$, $n = 1, 2, \ldots$. This shows that we need two tables of the form given in (1.11), one for $A_n^{(j)}$ and another for $\dot{A}_n^{(j)}$. We also see that the computation of the $\dot{A}_n^{(j)}$ involves both $\dot{A}(y)$ and $A(y)$.

The column sequences $\{\dot{A}_n^{(j)}\}_{j=0}^{\infty}$ converge to $\dot{A}$ almost in the same way the corresponding sequences $\{A_n^{(j)}\}_{j=0}^{\infty}$ converge to A, cf. (1.12). We have

$$\dot{A}_n^{(j)} - \dot{A} = \mathrm{O}(j|c_{n+1}|^j) \quad \text{as } j \to \infty. \tag{1.18}$$

The diagonal sequences $\{\dot{A}_n^{(j)}\}_{n=0}^{\infty}$ converge to $\dot{A}$ also practically the same way the corresponding $\{A_n^{(j)}\}_{n=0}^{\infty}$ converge to A, subject to the mild conditions that $\sum_{i=1}^{\infty} |\dot{c}_i| < \infty$ and $\sum_{i=1}^{n} |\dot{c}_i/c_i| = \mathrm{O}(n^a)$ as $n \to \infty$ for some $a \geqslant 0$, in addition to (1.13). We have for all practical purposes, cf. (1.14),

$$\dot{A}_n^{(j)} - \dot{A} \doteq \mathrm{O}\left(\prod_{i=1}^{n} |c_i|\right) \quad \text{as } n \to \infty. \tag{1.19}$$

The stability properties of the column and diagonal sequences of the $\dot{A}_n^{(j)}$ are likewise analyzed in [14] and are shown to be very similar to those of the $A_n^{(j)}$. We refer the reader to [14] for details.

This completes our review of the motivation and results of [14]. In the next section we present the extension of the procedure of [14] to GREP$^{(1)}$. We derive the recursive algorithm for computing the approximations and for assessing their numerical stability. In Section 3 we discuss the stability and convergence properties of the new procedure subject to a set of appropriate sufficient conditions that are met in many cases of interest. The main results of this section are Theorem 3.3 on stability and Theorem 3.4 on convergence and both are optimal asymptotically. In Section 4 we show how the method and theory of Sections 2 and 3 apply to the summation of some infinite series of logarithmic type via the $d^{(1)}$-transformation. Finally, in Section 5 we give two numerical examples that illustrate the theory and show the superiority of the new approach to derivatives of limits over the direct one. In the first example we apply the new approach to the computation of the derivative of the Riemann zeta function. In the second example we compute $(\mathrm{d}/\mathrm{d}\xi)F(\xi, \frac{1}{2}; \frac{3}{2}; 1)$, where $F(a, b; c; z)$ is the Gauss hypergeometric function. This example shows clearly that our approach is very effective for computing derivatives of special functions such as the hypergeometric functions with respect to their parameters.

2. GREP$^{(1)}$ and its derivative

2.1. General preliminaries on GREP$^{(1)}$

As GREP$^{(1)}$ applies to functions $A(y)$ that are in the class $\mathrm{F}^{(1)}$, we start by describing $\mathrm{F}^{(1)}$.

Definition 2.1. We shall say that a function $A(y)$, defined for $0 < y \leqslant b$, for some $b > 0$, where y can be a discrete or continuous variable, belongs to the set $\mathrm{F}^{(1)}$, if there exist functions $\phi(y)$ and $\beta(y)$ and a constant A, such that

$$A(y) = A + \phi(y)\beta(y), \tag{2.1}$$

where $\beta(x)$, as a function of the continuous variable x and for some $\eta \leqslant b$, is continuous for $0 \leqslant x \leqslant \eta$, and, for some constant $r > 0$, has a Poincaré-type asymptotic expansion of the form

$$\beta(x) \sim \sum_{i=0}^{\infty} \beta_i x^{ir} \quad \text{as } x \to 0+. \tag{2.2}$$

If, in addition, the function $B(t) \equiv \beta(t^{1/r})$, as a function of the continuous variable t, is infinitely differentiable for $0 \leqslant t \leqslant \eta^r$, we shall say that $A(y)$ belongs to the set $\mathrm{F}^{(1)}_{\infty}$. Note that $\mathrm{F}^{(1)}_{\infty} \subset \mathrm{F}^{(1)}$.

Remark. $A = \lim_{y\to 0+} A(y)$ whenever this limit exists. If $\lim_{y\to 0+} A(y)$ does not exist, then A is said to be the antilimit of $A(y)$. In this case $\lim_{y\to 0+} \phi(y)$ does not exist as is obvious from (2.1) and (2.2).

It is assumed that the functions $A(y)$ and $\phi(y)$ are computable for $0 < y \leqslant b$ (keeping in mind that y may be discrete or continuous depending on the situation) and that the constant r is known. The constants A and β_i are not assumed to be known. The problem is to find (or approximate) A whether it is the limit or the antilimit of $A(y)$ as $y \to 0+$, and GREP$^{(1)}$, the extrapolation procedure that corresponds to $\mathrm{F}^{(1)}$, is designed to tackle precisely this problem.

Definition 2.2. Let $A(y) \in \mathrm{F}^{(1)}$, with $\phi(y)$, $\beta(y)$, A, and r being exactly as in Definition 2.1. Pick $y_l \in (0, b]$, $l = 0, 1, 2, \ldots$, such that $y_0 > y_1 > y_2 > \cdots$, and $\lim_{l\to\infty} y_l = 0$. Then $A_n^{(j)}$, the approximation to A, and the parameters $\bar{\beta}_i$, $i = 0, 1, \ldots, n-1$, are defined to be the solution of the system of $n + 1$ linear equations

$$A_n^{(j)} = A(y_l) + \phi(y_l) \sum_{i=0}^{n-1} \bar{\beta}_i y_l^{ir}, \quad j \leqslant l \leqslant j + n, \tag{2.3}$$

provided the matrix of this system is nonsingular. It is this process that generates the approximations $A_n^{(j)}$ that we call GREP$^{(1)}$.

www.ingramcontent.com/pod-product-compliance
Lightning Source LLC
Chambersburg PA
CBHW061357210326
41598CB00035B/6012
9780444505972